Graduate Texts in Physics

Graduate Texts in Physics publishes core learning/teaching material for graduate- and advanced-level undergraduate courses on topics of current and emerging fields within physics, both pure and applied. These textbooks serve students at the MS- or PhD-level and their instructors as comprehensive sources of principles, definitions, derivations, experiments and applications (as relevant) for their mastery and teaching, respectively. International in scope and relevance, the textbooks correspond to course syllabi sufficiently to serve as required reading. Their didactic style, comprehensiveness and coverage of fundamental material also make them suitable as introductions or references for scientists entering, or requiring timely knowledge of, a research field.

Krastan B. Blagoev · Herbert Levine
Editors

Physics of Molecular and Cellular Processes

 Springer

Editors
Krastan B. Blagoev
Johns Hopkins University
Baltimore, MD, USA

Herbert Levine
Northeastern University
Boston, MA, USA

ISSN 1868-4513　　　　　　　ISSN 1868-4521　(electronic)
Graduate Texts in Physics
ISBN 978-3-030-98608-7　　　ISBN 978-3-030-98606-3　(eBook)
https://doi.org/10.1007/978-3-030-98606-3

This Springer imprint is published by the registered company Springer Nature Switzerland AG
The registered company address is: Gewerbestrasse 11, 6330 Cham, Switzerland

Preface

A number of years ago, the two of us (together with the late Kamal Shukla) penned an editorial entitled "We need theoretical physics approaches to study living systems". This editorial, published in Physical Biology, posited that the mind-boggling complexity of living systems necessitates an approach, which focuses on using advanced theory to connect simple yet informative models to real-world predictions. Alternative strategies such as trying, presumably these days with the help of machine learning, to generalize directly from ever multiplying quantities of data or by constructing computational simulacra of complete biological systems would falter under their own weights. Instead, tractable theoretical models would be the key to obtaining a real understanding.

Given this belief, the question arises as to how then can we educate the next generation of theorists to meet this challenge? It is critical to note that the goal is not to just train people to use off-the-shelf tools. The best analogy here is with microscopy; all biologists must learn to effectively use advanced microscopy, but it takes much more fundamental training to be the physicists who invent these tools. Similarly, we certainly hope that everyone interested in living systems will be able to run simple simulations and appreciate their outputs, and plug-in parameters into predetermined predictive models. But to be the creator of new concepts and techniques requires much more fundamental training.

This book is hopefully part of an answer to the above question. We have collected chapters from a number of leading theorists who have successfully shown how to apply the idea of theoretical physics approaches to living systems. These chapters will guide students through a number of topics at the forefront of the field. Since the successful application of theoretical physics to living systems is most advanced at the molecular scale, there is some emphasis on this specific area. However, there are also detailed chapters on cellular scale issues such as networks and active motility. Of course, these topics do not cover the entire spectrum of activities; we did not attempt to write an encyclopedia. Nonetheless, mastering the material here will not only provide detailed familiarity with the specific subject matter but also should impart an overall feeling as to how to go about formulating and utilizing theoretical methods to tackle all manner of additional areas of interest.

Our main intended audience is centered around physical science graduate students or perhaps advanced undergraduates who are curious to see how material covered in their studies is being used in the study of living matter. This material includes the basics of chemical kinetics, polymer physics, and especially equilibrium and non-equilibrium statistical mechanics. Thus, we expect some familiarity with the basic principles of these subjects. We do not assume any sophisticated biology background, and any missing information can be easily obtained from any number of excellent introductory biology textbooks. And of course, aiming for students does not preclude more advanced science professionals; the newness of this field means that that the number of learners is much greater than the number of experts.

As opposed to being a true textbook, this book is organized as a series of chapters written by working scientists. We are grateful to all the chapter authors for their willingness to spend the time and effort to contribute to this undertaking. This choice leads to an unavoidable divergence of styles in the various chapters. As opposed to being a bug, we would like to think of this as a feature; readers will be exposed to a diversity of opinions as to how to go about conducting research devoted to living systems. This will be essential to impart the aforementioned feeling. Our book will be most successful if it motivates others to find new ways to apply theory, not merely if it just teaches others how to follow in our already-trodden paths.

Preparing a book requires the assistance of dedicated professionals devoted to the nuts and bolts of typesetting, editing, figure copyrights, etc. in short all those things that go beyond the scientific writing itself. We are grateful to Louise Miller of Northeastern University for spearheading this effort and creating a final product out of raw inputs. We are also grateful to the editorial staff at Springer-Verlag for encouraging us to make it through this project and then bringing it to final fruition. And of course, we are grateful for all the discussions we have had over the years with colleagues and especially students about the topics covered in this book.

To quote Rabbi Hanina in the Talmud (Taanit 7a:)

הרבה למדתי מרבותי ומחברי יותר ומתלמידי יותר מכולם

"I have learned much from my teachers and even more from my colleagues, but from my students most of all".

Baltimore, MD   Krastan B. Blagoev
Brookline, MA   Herbert Levine
April 2021

Introduction

This book is about the use of ideas from theoretical physics in service of the understanding of living systems. These ideas arise from disparate sub-disciplines such as statistical physics, nonlinear dynamics, and continuum mechanics including fluid and elasticity. Some background in these subjects is assumed, at the level of a graduate student in physical science.

To help the reader get started, we collect here a set of useful definitions and concepts. First, an *equilibrium system* is one in which the probability of the degrees of freedom being observed to take on certain values is given by the *Boltzmann distribution*

$$P(\{x_i\}) \sim e^{-E(\{x_i\})/kT} \tag{1}$$

where E is the energy and T the temperature. This expression leads immediately to the probability distribution of a composite object X constructed from the aforementioned microscopic variables

$$P(X) \sim \sum_{\{x_i\}} e^{-E(\{x_i\})/kT} \, \delta(X - f(\{x_i\})) \equiv e^{-F(X)/kT} \tag{2}$$

where F is the free energy. Systems in equilibrium have time-independent probabilities given by their free energies, even as the microscopic variables thrash about. Much of the work in fields such as protein folding involve finding the relevant *free energy landscape* for suitably chosen X.

When an equilibrium system is slightly perturbed, it relaxes back to equilibrium via the action of restoring forces. In general, these forces are proportional to the derivatives of the free energy with respect to the variables X_j, thereby driving the system back to free energy minima. This gives rise to the basic response equation

$$\frac{dX_j}{dt} = -\sum_k \mu_{jk} \frac{\partial F}{\partial X_k} \tag{3}$$

where μ is a positive-definite symmetric matrix of dissipation coefficients. This equation needs to be augmented by a noise term so as to recover the full equilibrium distribution. Thus, the fundamental *Langevin equation* takes the form

$$\frac{dX_j}{dt} = -\sum_k \mu_{jk}\frac{\partial F}{\partial X_k} + \eta_j(t) \tag{4}$$

where η is white-noise of the form $< \eta)_i(t)\eta_j(t') >= 2\mu_{ij}kT$. The relationship between the fluctuation amplitude and the dissipation coefficients is known as the Einstein relation or more generally as the *fluctuation-dissipation theorem*.

There is a well-known connection between the Langevin equation for a single stochastic trajectory of the dynamical variables and the Fokker-Planck equation for the probability distribution of those variables. The latter takes the form of the partial differential equation

$$\frac{\partial P}{\partial t} = \frac{\partial}{\partial X_i}\left(\mu_{ij}\frac{\partial F}{\partial X_j}P\right) + kT\mu_{ij}\frac{\partial^2 P}{\partial X_i\partial X_j} \tag{5}$$

where the Einstein relationship has been used to relate the first term on the right (the drift term) to the second diffusion term. This relationship guarantees that the steady-state solution of this equation takes the form of the equilibrium distribution $P_{eq} \sim \exp(-F/kT)$.

There are some systems where additional terms need to be added beyond the dissipative piece. These are called *reactive terms*. A good example of this type of term arises in the *Navier-Stokes* equation for incompressible fluids

$$\rho\left(\frac{\partial \mathbf{v}}{\partial t} + (\mathbf{v}\cdot\nabla)\mathbf{v}\right) = -\nabla p + \mu\nabla^2\mathbf{v} \tag{6}$$

where the noise term has been dropped. The second term on the left-hand side does not give any dissipation associated with relaxation to equilibrium; that is the role of the viscosity term. In fact this nonlinear term appears already in the Euler equations, which can be derived from a Hamiltonian and hence conserves energy.

So far, we have used a framework appropriate to continuum degrees of freedom, for example the coordinates of the atoms comprising a biomolecule. Often, the free energy has deep minima at certain configurations and the system in question is almost always found at one of these minima; in the previous case, these minima might be a folded state and an unfolded ensemble. If this is the case, we can replace the Fokker-Planck dynamics by a set of transitions between discrete states. This *master equation* governing the dynamics then becomes the *continuous time Markov process*

$$\frac{dp_i}{dt} = \sum_j (T_{ij}p_j - T_{ji}p_i) \tag{7}$$

where $p_i(t)$ is the probability of finding the system in state i at time t and T_{ij} are transition rates. The fact that this system arose from an equilibrium system has a critical consequence. The easiest way to understand this consequence is to use a simple estimate of the transition probability as being proportional to the relative chance of finding the system at the transition state, which is the configuration of highest free energy (denoted as $\hat{F}_{ij}$) on the path leading from state i to state j or back again. This estimate gives

$$T_{ij} \sim e^{(\hat{F}_{ij}-F_i)/kT} \quad ; \quad T_{ji} \sim e^{(\hat{F}_{ij}-F_j)/kT} \tag{8}$$

This then says that the steady-state equilibrium solution $p_i^{eq} \sim e^{-F_i/kT}$ automatically satisfies the *principle of detailed balance* for each pair of states i and j,

$$T_{ij} p_j^{eq} = T_{ji} p_i^{eq} \tag{9}$$

(no summation). Physically, this means that there is no net probability flux between any pairs of states. A related finding is that there cannot be any flux around a closed path in state space. Around such a loop, the product of transition rates going one way around the loop must equal the product going the other way. This is because the ratio of the rates going in opposite directions in one link is, according to Eq. 7, related directly to the free energy difference of the states. By the time, we go all the way around, the ratio of the rate products is just related to the free energy difference between starting and ending points, which is obviously zero for a closed loop.

Most applications of physical science to living systems need to go beyond the aforementioned equilibrium assumption. Even for biomolecules, many processes require coupling the desired dynamics to an external reaction such as ATP hydrolysis, which is assumed to be far from equilibrium; we will see this explicitly for the case of molecular motors. This coupling neatly avoids the restriction regarding loops, because a loop in state space for the biomolecule conformational degrees of freedom is actually not a loop, because one additional ATP molecule has been hydrolyzed. More generally, as we go up in scale and start considering cellular or tissue-level dynamics, exchange of energy with the environment obviates any possibility of using equilibrium physics. Cells, tissues, and organisms operate as quintessential *open systems*. Leaving the realm of free energy-based thinking, the emphasis is now based on explicitly solving dynamical equations upon which there are no particular constraints other than those due to symmetry considerations. Often, one can make do with deterministic equations, with clear exceptions arising in situations where the number of "individuals" involved in a process are small; one then has to invoke number fluctuations, referred to as *demographic noise*; this will be particularly relevant in our chapter on evolutionary processes.

Solving equations of motions for interacting deterministic degrees of freedom brings us to the field of nonlinear dynamics. There are very few general methods for solving such equations, but there are some important ideas that help organize what eventually boils down to numerical approaches. First, the temporal behavior of solutions can be classified as steady-state, oscillatory, and chaotic. If the system in question is spatially-extended, one can have homogeneous solutions or patterns of various kinds. A key concept is that of *bifurcation*, which is a point in parameter space where solutions either switch stability, are born in pairs, or annihilate each other. These can be classified by exactly what is occurring in the neighborhood of that point. For example, a *supercritical Hopf bifurcation* occurs when a steady state solution goes unstable and past the bifurcation point is replaced by a stable limit cycle. A nice example of a bifurcation of proposed relevance to biology is connected to the *Turing instability*; here a homogeneous solution of a coupled set of reaction-diffusion equation becomes unstable to a solution, which is still constant in time but now varies periodically in space. This is an example of *non-equilibrium pattern formation* and has been suggested as being responsible for phenomena such as the designs on the surface of tropical fish. A final example is a combination of the previous two, namely, an instability to a periodic state in both space and time, namely a wave. The sloshing back and forth of the MIN protein involved in determining the location at which E. Coli divides is believed to occur due to such an instability.

The last topic we wish to briefly introduce is that of active media. Recall that we already introduced the Navier-Stokes equation for providing a *continuum mechanics* approach to the collective dynamics of molecules in a liquid. What happens, though, if the objects whose collective dynamics we are studying are birds or fish, or on a smaller scale, bacteria or amoebae? A normal flowing fluid has dissipation and it will return to its equilibrium state unless we keep applying a perturbation such as an external pressure gradient; this is the role of the viscosity term in the equation and would be the source of a negative linear term in **v** on the right side of the equation if the motion was occurring on top of a frictional surface or through a porous medium. But living systems have their own internal sources of energy and can keep moving by compensating for the energy dissipation. Such system is an example of an *active medium*, a topic that has become very popular in the physics community. One of the approaches to this subject replaces the negative linear damping term in a continuum treatment by a positive linear term that then must be balanced by the addition of higher-order damping. Another possibility is adding a velocity-independent driving force on the right-hand side; the direction of the force (called the polarization) has to be determined by a second coupled equation for this motility force. Of course, one could just study these systems by the analog of molecular dynamics, utilizing experimentally measured rules for how individual objects decide on the direction of their motility force. The beginning of the field of active media can arguably be traced to a seminal paper, which studied a collection of these *self-propelled particles* in which each such particle tries to move in the average direction of motion of its near neighbors.

With these introductory ideas in hand, we hope the readers will make their ways through the chapters of this book. We do warn the readers that we make no pretense of completeness. The field of the physics of living systems is incredibly broad and it would take many such volumes to begin to touch on all of its aspects. But, the chapters do reveal the workings of the field and provide a good set of examples as to how to make progress, balancing the reality of biological complexity with the desire for understandability and quantitation enabled by physical science. So, go to it!

Contents

Contributors

Albert Réka Pennsylvania State University, State College, PA, USA

Bishop Terrance T. Southern Illinois University, Carbondale, IL, USA

Blagoev Krastan B. Johns Hopkins University, Baltimore, MD, USA

Deng Youyuan Rice University, Houston, TX, USA

Gan Xiao Pennsylvania State University, State College, PA, USA

Kessler David A. Department of Physics, Bar-Ilan University, Ramat-Gan, Israel

Kolomeisky Anatoly B. Rice University, Houston, TX, USA

Krapf Diego Georgetown University, Washington, DC, USA

Levine Herbert Northeastern University, Boston, MA, USA

Mugnai Mauro L. The University of Texas at Austin, Austin, TX, USA

Onuchic José N. Rice University, Houston, TX, USA

Ovryn Ben New York Institute of Technology, NewYork, NY, USA

Takaki Ryota Max Planck Institute, Dresden, Germany

Thirumalai D. The University of Texas at Austin, Austin, TX, USA

Wang Qian University of Science and Technology of China, Hefei, Anhui, China

Whitford Paul Charles Northeastern University, Boston, MA, USA

Chapter 1
Nonequilibrium Physics of Molecules and Cells

Krastan B. Blagoev

1.1 Thermodynamics

A thermodynamic system consists of particles and/or fields that are bound in space. The rest is the environment. The system can be open in which case the system can exchange matter and radiation with the environment as well as heat and work can be done on the system. A closed system can exchange heat and work can be done on it, but no matter or radiation is exchanged with the environment. An isolated system cannot exchange anything with the environment and work cannot be done on it. A fundamental concept in thermodynamics is the state of a system and its thermodynamic variables. Thermodynamic variables are macroscopic variables like number of particles, pressure, volume, and temperature. A state of a thermodynamic system is specified by the values of these variables. A system is said to be in equilibrium if all thermodynamic variables are constant in time and all macroscopic flows are zero. If the thermodynamic variables are constant, but there are macroscopic constant flaws, the system is in a steady state. If the thermodynamic variables or flaws vary in time the system is nonequilibrium The zeroth law of thermodynamics states that two systems in thermodynamic equilibrium with a third system are in thermodynamic equilibrium with each other. Thus, we can introduce a small system called thermometer that will equilibrate fast with big systems and measure the temperature of other systems by putting the thermometer in touch with the measured system. Historically, the equilibration of the temperature of two objects in contact was characterized by the heat that flows from the warmer to the colder body. The internal energy of a system U can be changed by heating the system, i.e., adding heat Q, by performing mechanical work W on the system or by adding molecules to the system. A small change in the internal energy therefore can be written as

K. B. Blagoev (✉)
Johns Hopkins University, Baltimore, MD, USA
e-mail: krastan@jhu.edu

© Springer Nature Switzerland AG 2022
K. B. Blagoev and H. Levine (eds.), *Physics of Molecular and Cellular Processes*,
Graduate Texts in Physics, https://doi.org/10.1007/978-3-030-98606-3_1

$$dU = \delta Q + \delta W + \mu dN. \tag{1.1}$$

Here, δQ and δW are not exact differentials, but small quantities. An alternative definition used in physics is to take the system to perform work on the environment. In this case, the sign in front of the work term is negative, leading to a decrease in the internal energy when the system does work, where μdN is the chemical work and μ is the chemical potential. The first law of thermodynamics states that energy is conserved. We can write the work done by the system in terms of the changes in the mechanical properties of the system: volume, length, surface area, etc. We will use the volume with corresponding force, the pressure, and the surface area whose generalized force is the surface tension. The work is then

$$\delta W = -PdV + \sigma dA. \tag{1.2}$$

The sign in front of the pressure term is negative, since when we apply pressure on the system the volume contracts so $dV < 0$, but the internal energy increases. When we stretch a membrane the stretching, it causes an increase in the internal energy of the membrane and the surface area is expanded, $dA > 0$. Here, the surface tension, σ, is the force per unit length or, more generally, the energy per unit surface area. Next, we need to quantify the heat term. To define the thermodynamic temperature, Thomson and Clausius considered a reverse Carnot-cycle (cooling machine) operating between two reservoirs with an infinitesimal difference in temperatures. The Clausius theorem states that heat cannot be transferred from a colder reservoir to a hotter one without work being performed on the system. The work required to do this δW divided by the transferred heat δQ is proportional to the temperature difference, i.e.,

$$\delta W = \frac{\delta Q}{T} dT, \tag{1.3}$$

which is the definition of the thermodynamic temperature T introduced by Thomson. For the reversible Carnot cycle as well as for any reversible cycle,

$$\oint \frac{\delta Q}{T} = 0 \tag{1.4}$$

and the function under the integral is an exact differential. The heat then is defined through

$$dS = \frac{\delta Q}{T}, \tag{1.5}$$

where S is the entropy. For an irreversible cycle,

$$\oint \frac{\delta Q}{T} < 0 \tag{1.6}$$

and the function under the integral is not an exact differential. Suppose we have an irreversible process and we would like to calculate the entropy difference between the final and the initial state. This can be done by constructing a reversible process between the two states and computing the change of entropy along the reversible path. This will be the difference between the entropies also for the irreversible path since the entropy is an exact differential and the change of entropy depends only on the initial and final state. The value of the integral along a reversible path is maximum and thus will be larger than the value along an irreversible path. Thus,

$$dS \geq \frac{\delta Q}{T},$$ (1.7)

where equality holds for reversible process. If a system is isolated, i.e., no heat exchange can take place, $\delta Q = 0$ and therefore

$$\int dS = \Delta S \geq 0.$$ (1.8)

The second law of thermodynamics states that for any thermodynamic process the entropy change of an isolated system is greater than or equal to zero and is zero for reversible processes. The equilibrium state is a state with maximum entropy and this will be used to describe equilibrium distributions in statistical physics in the next sections. Because heat is associated with unstructured energy, intuitively one can expect the entropy to have something to do with disorder and this will be quantified later. The entropy is additive for independent systems, i.e., the entropy of n independent systems is equal to the sum of the entropies of the individual systems. Now, we can write (1.1) as

$$TdS \geq \delta Q = dU + PdV - \sigma dA - \mu dN.$$ (1.9)

For reversible processes,

$$dU = TdS - PdV + \sigma dA + \mu dN$$ (1.10)

and

$$T = \frac{\partial U}{\partial S}$$ (1.11)

$$P = -\frac{\partial U}{\partial V}$$ (1.12)

$$\sigma = \frac{\partial U}{\partial A}$$ (1.13)

$$\mu = \frac{\partial U}{\partial N}.$$ (1.14)

When taking the partial derivatives, we keep all other extensive parameters fixed. Like the entropy, the internal energy is also additive. This means that

$$U(xS, xV, xA, xN) = xU(S, V, A, N) \tag{1.15}$$

for arbitrary x. To obtain an expression for U also known as the Euler equation, we differentiate with respect to x

$$\frac{d}{dx}(xU) = U = \frac{\partial U}{\partial(xS)}\frac{d(xS)}{dx} + \frac{\partial U}{\partial(xV)}\frac{d(xV)}{dx} + \frac{\partial U}{\partial(xA)}\frac{d(xA)}{dx} + \frac{\partial U}{\partial(xN)}\frac{d(xN)}{dx}. \tag{1.16}$$

Taking $x = 1$ and using the expressions for the derivatives, we obtain Euler's equation

$$U = TS - PV + \sigma A + \mu N. \tag{1.17}$$

This equation can be used, e.g., to find the entropy S. The internal energy is convenient to use when we keep constant S, V, A, and N. If the process that we are interested in is at constant T, V, A, and N, the convenient thermodynamic potential is the Helmholtz free energy

$$F = U - TS. \tag{1.18}$$

Therefore, from (1.9), we have

$$dF \leq -SdT - PdV + \sigma dA + \mu dN. \tag{1.19}$$

At constant T, P, σ, and N, the Gibbs free energy is defined as

$$G = U - TS + PV - \sigma A = \mu N, \tag{1.20}$$

$$dG \leq -SdT + VdP - Ad\sigma + \mu dN. \tag{1.21}$$

At constant S, P or σ, and N, the enthalpy is defined as

$$H = U + PV - \sigma A, \tag{1.22}$$

$$dH \leq TdS + VdP - Ad\sigma + \mu dN. \tag{1.23}$$

Similarly, when the system is open and T, V, A, and μ are constant, the convenient thermodynamic potential is the grand potential

$$\Omega = U - TS - \mu N, \tag{1.24}$$

$$d\Omega \leq -SdT - PdV + \sigma dA - Nd\mu. \tag{1.25}$$

Important quantities are the heat capacities/specific heats and the compressibilities. The heat capacity is the heat required to raise the temperature of a substance by a unit of temperature and the specific heat is the heat needed to raise the temperature of a specific amount of such substance by that unit of energy. The heat capacity is

$$C = \lim_{\Delta T \to 0} \frac{\delta Q}{\Delta T}. \tag{1.26}$$

From the first law of thermodynamics: at constant volume, the

$$C_V = \left(\frac{\partial U}{\partial T}\right)_V. \tag{1.27}$$

Similarly, at constant pressure,

$$C_P = \left(\frac{\partial H}{\partial T}\right)_P. \tag{1.28}$$

The compressibility is

$$\kappa_T = -\frac{1}{V}\left(\frac{\partial V}{\partial P}\right)_T. \tag{1.29}$$

1.1.1 Phase Transitions

Physical substances exist in different forms at different conditions, e.g., gas, liquid, solid, or plasma state. Which state is realized, depends on the values of the intensive thermodynamic parameters, e.g., temperature, pressure, etc., of the system. The transition from one state to another as the thermodynamic parameters are varied is called phase transition. Phase transitions are classified as continuous or discontinuous. For continuous phase transitions, the derivative of the Gibbs free energy is a continuous function of the intensive thermodynamic variables like pressure, surface tension, and temperature, while for discontinuous or first-order phase transitions, the derivatives are discontinuous functions. For a thermodynamic state to be stable, it has to be a minimum of the free energy. At the transition line, the two phases coexist and, at equilibrium at this line, the chemical potentials and temperatures of these phases are equal. In some pure systems, there are points where three phases can coexist. An important concept is the critical point. This is a point at the end of a coexistence curve. Landau and Lifshitz [1] argued that critical point is impossible when the symmetries of the two phases are not related like in the liquid–solid phase transition. Recently, however, molecular dynamic simulations of monatomic systems have challenged the universality of this argument [2] opening the possibility that inside cells the recently observed nucleation processes might possess critical points. To quantify some of these notions, consider the Gibbs free energy

$$G = U - TS - PV = \mu N. \tag{1.30}$$

At constant T, P, and σ, $G = \mu dN$. It follows that at the coexistence curve, the two phases have the same Gibbs free energy and the chemical potentials are equal. For first-order phase transitions, the derivatives

$$V = -\frac{\partial G}{\partial P}, \; S = -\frac{\partial G}{\partial T}, \; etc., \tag{1.31}$$

are discontinuous, while at second- and higher order phase transitions they are continuous, but higher order derivatives are discontinuous. Recently, phase separation of cellular components has been observed inside cells and it has been argued that a liquid–liquid phase transition is responsible for the observed phenomena. Liquid–liquid phase transitions have been studied in water. When pure water is cooled at $T = 0°C$, the free energy of water and ice becomes equal. Below $0°C$, ice has lower energy, however, for pure water, one can cool water tens of degrees below the phase transition temperature, although this supercooled water is in an unstable state. How does the phase transformation proceed as the transition point is reached and passed? Similar phenomena exist in supersaturated solutions. A supersaturated solution of a substance A, e.g., protein in a substance (solvent) B, e.g., water is a solution in which the dissolved material is more than the solvent can dissolve. In general, when the critical saturation is reached, crystallization will occur through a process called nucleation. This process is usually the way first-order phase transitions occur. When impurities are present nucleation occurs at a higher temperature than in pure substances. We will come back to nucleation after the introduction of the statistical description of many-body systems. When a system is in contact with a heat reservoir at temperature T, the system is characterized by the Helmholtz free energy $F = U - TS$. Phase transitions in such systems occur due to the tendency of the system to minimize the internal energy (order) and to maximize its entropy (disorder) with the constraint that the free energy is a minimum.

1.2 Foundations of Statistical Physics

1.2.1 Liouville Theorem for Hamiltonian Systems

This section provides background material from classical equilibrium statistical mechanics. First, we consider an isolated many-body system. The state of the system is given by the particle coordinates and momenta $\mathbf{q}_i$ and $\mathbf{p}_i$, which span phase space and for N particles in 3 dimensions has $6N$ dimensions. Assuming that the Hamiltonian of the system $H(p, q)$ is known, (here we use the short notations where p stands for $\mathbf{p}_1, \ldots, \mathbf{p}_N$ and q stands for $\mathbf{q}_1, \ldots, \mathbf{q}_N$), one can write the Hamilton equations of motion

$$q' = \frac{\partial H}{\partial p} \tag{1.32}$$

$$p' = -\frac{\partial H}{\partial q}, \tag{1.33}$$

and we have assumed that the Hamiltonian does not explicitly depend on time and prime is the time derivative. Such systems are called conservative and the Hamiltonian is a constant of motion equal to the energy of the system

$$H(p, q) = E. \tag{1.34}$$

We will also assume that the forces in the system are velocity independent. For a large system of particles, these equations might be solvable using molecular dynamic simulations. This approach will be discussed later. Another approach is to use statistical description of the system and its dynamics consistent with the Hamilton equations. For this, first, we introduce a phase space density:

$$\rho(p, q, t) Dp Dq, \tag{1.35}$$

which is the probability of finding the system in the vicinity of the point (p, q) in the $6N$-dimensional phase space at time t. Here, $Dp = d^{3N}p$, $Dq = d^{3N}q$ are the measures in phase space and in the vicinity, respectively, of the point $p_i^a \in [p_i^a, p_i^a + dp_i^a]$ and $q_i^a \in [q_i^a, q_i^a + dq_i^a]$ for every $i = 1, \ldots, N$ and $a = 1, 2, 3$. The index i is the particle index, $a = 1, 2, 3$ is the coordinate of the particle in the physical 3D space. The normalization of ρ follows from the requirement that the probability of finding the system anywhere in phase space at any time is one:

$$\int \rho(p, q, t) Dp Dq = 1, \tag{1.36}$$

where the integral is over the entire phase space. The probability of finding the system in a subspace or region in phase space is an integral of ρ over that subspace or region. The ensemble average at time t of a physical observable, which is a function of (p, q), over this probability density is given by

$$<s(t)> = \frac{\int s(p, q)\rho(p, q, t) Dp Dq}{\rho(p, q, t) Dp Dq}. \tag{1.37}$$

For a stationary ensemble,

$$\frac{\partial \rho}{\partial t} = 0 \tag{1.38}$$

for fixed (p, q) and therefore

$$\frac{\partial <s(t)>}{\partial t} = 0. \tag{1.39}$$

The probability is conserved and, therefore, if we focus on the flow of probability in a closed region of phase space with volume V, the change of the probability of finding the system inside this volume at time t is

$$P(V) = \int_V \rho(p, q, t) Dp Dq \tag{1.40}$$

and must be related to the total flow of probability across the boundary of V, ∂V:

$$\frac{d P(V)}{dt} = -\int_{\partial V} \rho\left(\mathbf{R}(t), t\right) \frac{d\mathbf{R}(t)}{dt} \cdot d\sigma, \tag{1.41}$$

where $d\sigma$ is a surface element and we have introduced the $6N$-dimensional vector:

$$\mathbf{R}(t) = (p_1^1, p_1^2, p_1^3, p_2^1, \ldots, q_{3N}^3). \tag{1.42}$$

On the other hand, the change of probability from our definition of ρ is

$$\frac{d P(V)}{dt} = \frac{\partial}{\partial t} \int_V \rho\left(\mathbf{R}(t), t\right) dR = \int_V \frac{\partial \rho\left(\mathbf{R}(t), t\right)}{\partial t} dR. \tag{1.43}$$

Using Gauss's theorem, the surface integral can be transformed into a volume integral, i.e.,

$$\int_{\partial V} \rho\left(\mathbf{R}(t), t\right) \frac{d\mathbf{R}(t)}{dt} \cdot d\sigma = \int_V \nabla \cdot \left[\rho\left(\mathbf{R}(t), t\right) \frac{d\mathbf{R}(t)}{dt}\right] dR. \tag{1.44}$$

From the last two equalities and the fact that V is time-independent, the equation satisfied by ρ is

$$\frac{\partial \rho\left(\mathbf{R}(t), t\right)}{\partial t} = -\nabla \cdot \left[\rho\left(\mathbf{R}(t), t\right) \frac{d\mathbf{R}(t)}{dt}\right]. \tag{1.45}$$

An important result following from Hamilton's equations is that the time evolution of the probability density of a mechanical system in phase space behaves like an incompressible fluid. To see this, first, consider how a volume element changes in time:

$$d\mathbf{R}(t) = J\left(\mathbf{R}(t), \mathbf{R}(0)\right) d\mathbf{R}(0), \tag{1.46}$$

where J is the Jacobian of the transformation and which is the determinant of the Jacobian matrix, $\hat{J}$ with elements

$$\hat{J}_{\alpha\beta} = \frac{\partial R_\alpha(t)}{\partial R_\beta(0)}, \tag{1.47}$$

$\alpha, \beta = 1, \ldots, 6N$. Then the Jacobian can be written as

$$J(\mathbf{R}(t), \mathbf{R}(0))d\mathbf{R}(0) = det[\hat{J}] = e^{Tr\ln\hat{J}}. \tag{1.48}$$

The last equality follows from the property that if λ is the eigenvalue of a matrix $\hat{M}$, then e^λ is the eigenvalue of the matrix $e^{\hat{M}}$, which can be seen from the identities

$$det[\hat{M}] = \prod \lambda_i \tag{1.49}$$

and

$$e^{Tr\ln\hat{M}} = e^{\sum \ln\lambda_i} = e^{\ln\prod\lambda_i} = \prod \lambda_i. \tag{1.50}$$

Another useful relation is

$$tr\hat{M} = \ln(det[e^{\hat{M}}]). \tag{1.51}$$

The rate of change of the Jacobian is

$$\frac{dJ}{dt} = \frac{d}{dt}det[\hat{J}] = Tr\left(\hat{J}^{-1}\frac{d\hat{J}}{dt}\right)e^{Tr\ln\hat{J}}, \tag{1.52}$$

and because

$$Tr[\hat{A}\hat{B}] = Tr[A_{ij}B_{jk}] = A_{ij}B_{ji}, \tag{1.53}$$

it follows that

$$\frac{dJ}{dt} = J\hat{J}^{-1}_{\alpha\beta}\frac{d\hat{J}_{\beta\alpha}}{dt}, \tag{1.54}$$

where we use the convention that we sum over repeated indexes, i.e.,

$$\hat{A}_{ij}\hat{B}_{jk} = \sum_j^{6N} \hat{A}_{ij}\hat{B}_{jk} \tag{1.55}$$

and

$$\hat{A}_{ij}\hat{B}_{ji} = \sum_i^{6N}\sum_j^{6N} \hat{A}_{ij}\hat{B}_{jk}. \tag{1.56}$$

The matrix elements of $\hat{J}^{-1}$ are

$$\hat{J}^{-1}_{\alpha\beta} = \frac{\partial R_\alpha(0)}{\partial R_\beta(t)} \tag{1.57}$$

and

$$\frac{d\hat{J}_{\alpha\beta}}{dt} = \frac{\partial V_\alpha(t)}{\partial R_\beta(0)}, \tag{1.58}$$

where the velocity is $V_\alpha(t) = \frac{dR_\alpha(t)}{dt}$. Therefore, we obtain

$$
\begin{aligned}
\frac{dJ}{dt} &= J \frac{\partial R_\alpha(0)}{\partial R_\beta(t)} \frac{\partial V_\beta(t)}{\partial R_\alpha(0)} = J \frac{\partial R_\alpha(0)}{\partial R_\beta(t)} \frac{\partial V_\beta(t)}{\partial R_\gamma(t)} \frac{\partial R_\gamma(t)}{\partial R_\alpha(0)} \\
&= J \left(\frac{\partial R_\alpha(0)}{\partial R_\beta(t)} \frac{\partial R_\gamma(t)}{\partial R_\alpha(0)} \right) \frac{\partial V_\beta(t)}{\partial R_\gamma(t)}.
\end{aligned}
\tag{1.59}
$$

Summing with respect to the index α, the expression in the parenthesis, one obtains

$$
\frac{\partial R_\alpha(0)}{\partial R_\beta(t)} \frac{\partial R_\gamma(t)}{\partial R_\alpha(0)} = \hat{J}_{\alpha\beta}^{-1} \hat{J}_{\gamma\alpha} = \hat{J}_{\gamma\alpha} \hat{J}_{\alpha\beta}^{-1} = \delta_{\gamma\beta}.
\tag{1.60}
$$

Therefore, the rate of change of the Jacobian is

$$
\frac{dJ}{dt} = J \delta_{\gamma\beta} \frac{\partial V_\beta(t)}{\partial R_\gamma(t)} = J \nabla \cdot \mathbf{R}'(t) = J \partial^\alpha R_\alpha'(t),
\tag{1.61}
$$

where $\partial^\alpha = \frac{\partial}{\partial R_\alpha}$. For Hamiltonian systems,

$$
\partial^\alpha R_\alpha'(t) = \sum_{i=1}^{3N} \left(\frac{\partial}{\partial q_i} \frac{\partial q_i}{\partial t} + \frac{\partial}{\partial p_i} \frac{\partial p_i}{\partial t} \right) = \sum_{i=1}^{3N} \left(\frac{\partial}{\partial q_i} \frac{\partial H}{\partial p} - \frac{\partial}{\partial p_i} \frac{\partial H}{\partial q_i} \right) = 0,
\tag{1.62}
$$

where we used Hamilton's equations and the commutativity of the p_i and q_i derivatives. Therefore that the right-hand side is zero, i.e., the phase space is incompressible and thus

$$
\frac{dJ}{dt} = 0.
\tag{1.63}
$$

The initial condition for the Jacobian is $J(R(0), R(0)) = 1$ and, therefore for arbitrary times, $J = 1$, i.e., the phase space volume is conserved. The equation for ρ was

$$
\frac{\partial \rho(\mathbf{R}(t), t)}{\partial t} = -\nabla \cdot \left[\rho(\mathbf{R}(t), t) \frac{d\mathbf{R}(t)}{dt} \right] = -(div \rho(\mathbf{R}(t)) \cdot \mathbf{R}'(t) - \rho(\mathbf{R}(t)) \partial^\alpha R_\alpha'(t).
\tag{1.64}
$$

Since the second term from (1.63) is zero, the equation for ρ at a fixed point in phase space $\mathbf{R}$ is

$$
\frac{\partial \rho(\mathbf{R}(t), t)}{\partial t} = -\mathbf{R}'(t) \cdot (div \rho(\mathbf{R}(t))).
\tag{1.65}
$$

The total time derivative of the probability density is

$$
\frac{d\rho}{dt} = \frac{\partial \rho}{\partial t} + \mathbf{R}'(t) \cdot (div \rho(\mathbf{R}(t)))
\tag{1.66}
$$

and therefore

$$\frac{d\rho}{dt} = 0. \tag{1.67}$$

This is the Liouville theorem stating that the probability density is constant following a phase space point along its trajectory.

Instead of following the time trajectory of the mechanical system, we can consider a very large number of identical systems at different time points. A system is ergodic if the time average of a physical observable is equal to the ensemble average, i.e.,

$$\lim_{t \to \infty} \frac{1}{t} \int_0^t A\left(p\left(\tau\right), q\left(\tau\right)\right) d\tau = \frac{\int \delta \left(H((p,q) - E\right) A(p,q)DpDq}{\int \delta \left(H\left(p,q\right) - E\right) DpDq}, \tag{1.68}$$

where the integral in the right-hand side is over all phase space and the Dirac δ-function guarantees that the system is integrated only on the constant energy hypersurface, S, in phase space determined by the constant energy. Ergodic systems visit every allowed phase space point arbitrarily close in the infinite time limit (a system described by a phase space trajectory cannot fill phase space, since a $1D$ line that does not intersect itself cannot fully fill higher dimensional spaces). This statement is often used as definition of ergodic system and then (1.68) used as a consequence. Here, we take the operational definition, i.e., the equality (1.68) as the definition of ergodic system. Systems with other conservation laws besides the total energy are not ergodic. For example, non-interacting particles in a box with reflecting walls is not ergodic, because the momentum of each particle is conserved. Because the probability density ρ does not change as the system moves through phase space, it means that everywhere in phase space ρ is the same

$$\rho(p, q, t) = constant. \tag{1.69}$$

To normalize ρ we define the constant to be equal to $vol[S]$, the volume of the hyper-surface S, i.e.,

$$\rho(p, q, t) = \frac{1}{vol[S]} \tag{1.70}$$

for $(p, q) \in S$ and $\rho = 0$, otherwise. In such an ensemble of systems, what is the probability of finding the system at a phase space point? The answer is given by noticing that the probability density is constant and thus the probability of finding the system at any phase space point is constant. This probability ensemble is called the microcanonical ensemble and (p, q) a microstate of the system. The microcanonical ensemble is built upon the so-called postulate of equal a priori probabilities: For an isolated macroscopic system in equilibrium, all microscopic states corresponding to the same set of macroscopic observables are equally probable. Equation (1.65) can be cast using the Liouville operator. Noting that

$$-\mathbf{R}'(t) \cdot (div\rho(\mathbf{R}(t), t)) = -\left(\frac{dp_i}{dt}\frac{\partial\rho}{\partial p_i} + \frac{dq_i}{dt}\frac{\partial\rho}{\partial q_i}\right) = \left(\frac{\partial H}{\partial p_i}\frac{\partial\rho}{\partial q_i} - \frac{\partial H}{\partial q_i}\frac{\partial\rho}{\partial p_i}\right) =$$
(1.71)

$= -\{\rho, H\}$, where we have introduced the Poisson bracket. Introducing the Liouville operator $\hat{L}\rho = -i\{\rho, H\}$ the Liouville (1.65) is

$$\frac{\partial\rho}{\partial t} = -i\hat{L}\rho.$$
(1.72)

For a time-independent Hamiltonian, the solution of this equation is

$$\rho(p, q, t) = e^{-it\hat{L}}\rho(p.q, 0).$$
(1.73)

Because the Liouville operator is Hermitian, it has real eigenvalues. Therefore, the probability density will oscillate and not decay in phase space. Following Boltzmann, we denote by W the number of microstates of a physical system in a small energy range δE around the energy E. Boltzmann defined the entropy of an isolated system in equilibrium as

$$S = k \, log \, W,$$
(1.74)

where $k = 1.38064852x10^{-23} J/K = 8.6173303 \times 10^{-5}$ eV/K is the Boltzmann constant. The entropy must be maximum at thermodynamic equilibrium and also be additive. Gibbs introduced the following expression for the entropy in terms of the probability distribution:

$$S = -k \int \rho(p, q) \, ln \, \rho(p, q) Dp Dq.$$
(1.75)

For the microcanonical distribution given by $\rho = 1/W$, the Gibbs expression reduces to the Boltzmann formula for S, which equals

$$-k \int \rho(p, q) \, ln \, \rho(p, q) Dp Dq = k \int Dp Dq \frac{1}{W} ln \, W = k \int Dp Dq \frac{d \, ln \, W}{dW} = k \, ln \, W.$$
(1.76)

So far, we have considered an insulated system in which the energy and the particle number are constant. What happens in dissipative systems for which the energy is not conserved? Living systems are such systems and they are driven by energy sources or they lose energy due to the viscosity of the environment. For example, interstrand DNA excitations are rapidly damped due to the viscosity of water. For such systems, the Liouville theorem is not satisfied and phase space either shrinks or expands. However, phase space will shrink only down to the volume spanned by the thermal fluctuations. Thus, in the case of DNA fluctuations, the phase space volume remains constant when the molecule is not driven.

Physical systems are often subject to constraints. There are two types of constraints: holonomic and nonholonomic. Holonomic constraints restrict the motion of the system to a subspace in configuration space, i.e., space of particle positions. Nonholonomic constraints are all other constraints, e.g., constraints on the velocities. Liouville theorem is true for holonomic constraints, but for nonholonomic constraints, the system is not Hamiltonian and in general does not obey the theorem [3, 4]. Another question is if non-Hamiltonian dynamical systems obey some kind of generalized Liouville theorem. The derivation of Liouville's theorem relied on Hamilton's equations (1.33). While at a fundamental level, if one considers all degrees of freedom, any physical system is Hamiltonian, many dynamical systems and especially living organisms are described by few degrees of freedom and thus are not described by Hamiltonian dynamics. The question is if one can generalize Liouville's theorem for such systems.

1.2.2 Stability of Nonlinear Dynamical Systems

In this section, we present stability analysis of the solutions of a system of two nonlinear differential equations. We consider first-order differential equations, because a system of higher order equations can be reduced to a larger system of first order differential equations by introducing new functions, e.g., the second-order differential equation

$$\frac{d^2 f}{dt^2} = F(t) \qquad (1.77)$$

can be reduced to two first-order differential equations by introducing the new function $g(x) = \frac{df}{dx}$

$$\frac{dg}{dt} = F(t) \qquad (1.78)$$

$$\frac{df}{dt} = G(t). \qquad (1.79)$$

To illustrate how to perform the stability analysis, we consider the system of equations

$$\frac{dx}{dt} = F(x(t), y(t)) \qquad (1.80)$$

$$\frac{dy}{dt} = G(x(t), y(t)). \qquad (1.81)$$

If F and G are continuously differentiable with respect to x and y in some domain, then there is a unique solution to the system of first-order ordinary differential equations. A weaker condition is that F and G are Lipschitz continuous, but for our discussion, we assume that they are continuously differentiable. Fixed points are solutions describing steady states, i.e., the derivatives of x and y and therefore the functions F and G vanish. Without loss of generality, we can assume that such point is at the origin of the x, y coordinate system, i.e., $\mathbf{r}_0 = (0, 0)$. A steady-state solution of the dynamical system $\mathbf{R}$ is stable if for any $\epsilon > 0$ exist $\delta > 0$ such that every solution that is initially within distance δ from $\mathbf{R}$, i.e., $|\mathbf{R}(t) - \mathbf{r}(t_0)| < \delta$ remains within distance ϵ after t_0, i.e., $|\mathbf{R} - \mathbf{r}(t)| < \epsilon$ for $t > t_0$. It is asymptotically stable if starting close to the stable solution $\mathbf{R}$ it gets arbitrarily close to $\mathbf{R}$ as $t \to \infty$. To study the behavior of the system in the neighborhood of the origin, we expand to first order F and G around that point

$$\frac{dx}{dt} = \frac{\partial F(x, y)}{\partial x}(0, 0)x + \frac{\partial F(x, y)}{\partial y}(0, 0)y \tag{1.82}$$

$$\frac{dy}{dt} = \frac{\partial G(x, y)}{\partial x}(0, 0)x + \frac{\partial G(x, y)}{\partial y}(0, 0)y. \tag{1.83}$$

In matrix notations, this system is

$$\frac{d\mathbf{r}}{dt} = H\mathbf{r}, \tag{1.84}$$

where the matrix H consists of the four derivatives above evaluated at the origin. Diagonalizing the matrix H, we find two eigenvalues λ_i:

$$\lambda_{1,2} = \frac{1}{2}\left(\frac{\partial F}{\partial x} + \frac{\partial G}{\partial y} \pm \left(\left(\frac{\partial F}{\partial x} + \frac{\partial G}{\partial y}\right)^2 - 4\left(\frac{\partial F}{\partial x}\frac{\partial G}{\partial y} - \frac{\partial F}{\partial y}\frac{\partial G}{\partial x}\right)\right)^{1/2}\right). \tag{1.85}$$

The linearized equations are diagonal in the eigenvector basis, $\mathbf{p}$ and $\mathbf{q}$, of H. Let the matrix elements of the matrix H be h_{ij} with $i, j = 1, 2$. The eigenvalues can also be expressed using the trace and the determinant of H. If T is the trace and D the determinant is D, then the eigenvalues are

$$\lambda_1, 2 = \frac{T}{2} \pm \sqrt{\frac{T}{4} - D}. \tag{1.86}$$

The most general solution of the linearized equation when the two eigenvalues are different is

$$\mathbf{r}(t) = c_1\mathbf{r}_1 e^{\lambda_1 t} + c_2\mathbf{r}_2 e^{\lambda_2 t} \tag{1.87}$$

with c_i constants. Since the matrix H is real, it has either two real eigenvalues or two complex conjugate complex eigenvalues. (That this is not the case for a general matrix and one can find a complex matrix that has one real and one complex eigenvalues). Depending on the eigenvalues, the linearized system has different behavior at the origin or any other point around which we linearize the system.

1. $\lambda_1, \lambda_2 < 0$: The point is stable and also asymptotically stable.
2. $\lambda_1 = 0, \lambda_2 < 0$: The point is stable, but not asymptotically stable.
3. $\lambda_1 < 0$ or $\lambda_2 > 0$: The point is unstable.

4. $\lambda_i = \mu_i + iv_i$, $\mu_i < 0$: The point is asymptotically stable.
5. $\lambda_i = \mu_i + iv_i$, $\mu_i = 0$: No conclusions.

The case when there is one degenerate real eigenvalue, i.e., a real double root of the characteristic equation the system cannot be diagonalized, but can be represented in its normal Jordan form. Consider

$$\frac{dx}{dt} = \lambda x(t) + y(t) \tag{1.88}$$

$$\frac{dy}{dt} = y(t). \tag{1.89}$$

The solution is

$$x(t) = (c_1 + c_2 t)e^{\lambda t} \tag{1.90}$$

$$x(t) = c_2 e^{\lambda t}. \tag{1.91}$$

(Check it). If the eigenvalue $\lambda \geq 0$ the point is unstable. If $\lambda < 0$ then $te^{-|\lambda|t}$ first increases, then goes to a maximum value, which can be made arbitrarily close to zero by choosing small c_i coefficients.

1.2.3 Phase Space Dynamics of Dynamical Systems

In this section, we return to Liouville's theorem and ask if for a dynamical system described by

$$\frac{d\mathbf{R}}{dt} = \mathbf{F}(\mathbf{R}). \tag{1.92}$$

In general, for a non-Hamiltonian system, (1.62) is invalid, i.e.,

$$\partial^\alpha R'_\alpha(t) = \partial^\alpha F_\alpha = \kappa \neq 0. \tag{1.93}$$

From (1.61), we have for the Jacobian

$$\frac{1}{J}\frac{dJ}{dt} = \frac{d\ln J}{dt} = \kappa, \tag{1.94}$$

where κ is the phase space compressibility. For incompressible flow, i.e., Hamiltonian flow, the compressibility is zero. This differential equation can be solved by separating the variables and the solution is

$$J(\mathbf{R}(t), \mathbf{R}(0)) = e^{\int_0^t \kappa(\mathbf{R}(t))d\tau}, \tag{1.95}$$

because $J(\mathbf{R}(0), \mathbf{R}(0)) = 1$. Because κ is a total derivative of the logarithm of the Jacobian, it can be written as a the time derivative of a function, e.g.,

$$\kappa(\mathbf{R}) = W'(\mathbf{R}) \tag{1.96}$$

and therefore from (1.93)

$$d\mathbf{R}(t) = e^{W(\mathbf{R}(t)) - W(\mathbf{R}(0))} d\mathbf{R}(0). \tag{1.97}$$

Thus, we can write

$$e^{-W(\mathbf{R}(t))} d\mathbf{R}(t) = e^{-W(\mathbf{R}(0))} d\mathbf{R}(0), \tag{1.98}$$

and it seems that we have found invariant measure on phase space. This however is not true in general, because the function κ can be singular or the integral of κ might not exist. This is the case for repelling or attracting periodic orbits or fixed points. When invariant measure exists one can also define a metric g on phase space and the invariant measure can be written as

$$\sqrt{g(\mathbf{R}(0))} d\mathbf{R}(t) = \sqrt{g(\mathbf{R}(0))} d\mathbf{R}(0), \tag{1.99}$$

where

$$\sqrt{g} = e^{-W}. \tag{1.100}$$

It turns out that Liouville's theorem is again satisfied and for a detailed discussion of this topic see [5].

1.2.4 Canonical Ensemble

The microcanonical ensemble was introduced to describe isolated systems. However, most physical, and therefore biological, systems are in contact with their environment. Systems that can exchange energy by work or heat with the environment, but do not exchange particles or molecules are described by the canonical ensemble in which case T, V, A, and N are constant. Systems that can also exchange matter with their environment are described by the grand canonical ensemble.

Closed systems can exchange heat with their environment. Consider a very large system and a much smaller subsystem that can exchange heat but not matter with the rest of the system. The subsystem has many fewer degrees of freedom compared to the total system. This system is described thermodynamically by the Helmholtz free energy $F = U - ST$. At equilibrium, we ask what is the probability distribution of the small subsystem. From here, the subsystem will be called the system and the rest of the system the environment. At equilibrium, while the energy of the system will fluctuate, the average energy will be constant

$$E = const = < E > = \int H(p,q)\rho(p,q)DpDq. \qquad (1.101)$$

In addition, the probability of finding the system anywhere in its phase space must be normalized, i.e.,

$$\int \rho(p,q)DpDq = 1. \qquad (1.102)$$

We can use the maximum entropy principle to obtain the distribution function. To incorporate the above two constraints in the variational principle, we introduce two Lagrange multipliers, λ_1 and λ_2. Thus, the variational problem is to maximize

$$S = -k \int \rho(p,q) \, ln \, \rho(p,q)DpDq \qquad (1.103)$$

subject to the constraints above. The variational principle is

$$\frac{\delta}{\delta\rho} \int (-k\rho(p,q) \, ln \, \rho(p,q) + \lambda_1\rho(p,q) + \lambda_2 H(p,q)\rho(p,q)) \, DpDq = 0. \qquad (1.104)$$

The resulting equation is

$$\lambda_1 + \lambda_2 H(p,q) - (kln \, \rho(p,q) + k) = 0. \qquad (1.105)$$

Therefore, we obtain for the distribution function of a closed system at equilibrium

$$\rho(p,q) = e^{\frac{\lambda_1}{k} + \frac{\lambda_2}{k}H(p,q)-1}. \qquad (1.106)$$

From the normalization of ρ, we get

$$\int e^{\frac{\lambda_1}{k} + \frac{\lambda_2}{k}H(p,q)-1} DpDq = 1 \qquad (1.107)$$

and therefore

$$\int e^{\frac{\lambda_2}{k}H(p,q)} DpDq = e^{-\lambda_1+1}. \qquad (1.108)$$

The constant

$$Z = e^{-\frac{\lambda_1}{k}+1} = \int e^{\frac{\lambda_2}{k}H(p,q)} DpDq \qquad (1.109)$$

is called the partition function of the canonical distribution. Next, we use the second constraint

$$< E >= \int H(p,q)\rho(p,q)DpDq = \int H(p,q)e^{\frac{\lambda_1}{k}+\frac{\lambda_2}{k}H(p,q)-1}DpDq =$$
(1.110)

$$e^{\frac{\lambda_1}{k}-1}\int H(p,q)e^{\frac{\lambda_2}{k}H(p,q)}DpDq = \frac{\int H(p,q)e^{\frac{\lambda_2}{k}H(p,q)}DpDq}{Z}.$$
(1.111)

If we multiply the equation, which resulted from the variation by ρ and integrate, we obtain

$$\int \rho(p,q) \times [\lambda_1 + \lambda_2 H(p,q) - (kln\,\rho(p,q) + k)] = 0$$
(1.112)

and

$$\lambda_1 + \lambda_2 < E > -S - k = -klnZ + \lambda_2 < E > +S = 0.$$
(1.113)

Since $F = U - ST$, it follows that

$$\lambda_2 = -\frac{1}{T}$$
(1.114)

$$-klnZ = F$$
(1.115)

$$< E > = U.$$
(1.116)

Therefore, the distribution for equilibrium closed system called the canonical distribution is

$$\rho(p,q) = \frac{e^{-\frac{H(p,q)}{kT}}}{Z(T,V)}$$
(1.117)

with the partition function

$$Z = \int e^{-\frac{H(p,q)}{kT}} DpDq.$$
(1.118)

Given the free energy $F = -kTZ(T,V)$, one can calculate the entropy from the equilibrium expression for F

$$dF = -SdT - PdV + \sigma dA + \mu dN,$$
(1.119)

$$S = -\left(\frac{\partial F}{\partial T}\right)_{V,N} = \left(\frac{\partial [kTlnZ(T,V)]}{\partial T}\right)_{V,N}.$$
(1.120)

Similarly, the pressure, the surface tension, and the chemical potential are

$$P = -\left(\frac{\partial F}{\partial V}\right)_{T,A,N} \tag{1.121}$$

$$\sigma = \left(\frac{\partial F}{\partial A}\right)_{T,V,N} \tag{1.122}$$

$$\mu = \left(\frac{\partial F}{\partial N}\right)_{T,V,A}. \tag{1.123}$$

In a closed system, the temperature is constant fixed by the external environment, but the internal energy fluctuates. To estimate the energy fluctuations as a function of the number of molecules at equilibrium, we note that

$$U = <E> = \frac{1}{Z}\int H(p,q)e^{-\beta H(p,q)}DpDq = -\frac{1}{Z}\frac{\partial}{\partial \beta}\int e^{-\beta H(p,q)}DpDq,$$
$$\tag{1.124}$$

which equals $= -\frac{\partial \ln Z}{\partial \beta}$, and

$$<E^2> = \frac{1}{Z}\frac{\partial^2 Z}{\partial \beta^2} = \left(\frac{\partial \ln Z}{\partial \beta}\right)^2 + \frac{\partial^2 \ln Z}{\partial \beta^2}. \tag{1.125}$$

Therefore, the fluctuations are equal to

$$<(E- <E>)^2> = <E^2> - <E>^2 = \frac{\partial^2 \ln Z}{\partial \beta^2} = -\frac{\partial U}{\partial \beta} = kT^2C_V, \tag{1.126}$$

where C_V is the constant volume specific heat, which is proportional to the number of molecules in the system. The energy in the system is also proportional to the number of particles and therefore

$$\delta E = \sqrt{<E^2> - <E>^2} \propto \frac{1}{N^{1/2}}. \tag{1.127}$$

1.2.4.1 Grand Canonical Ensemble

For open systems, i.e., systems that can exchange heat and matter with the environment the grand potential $\Omega(T, V, A, \mu)$ (where T, V, A, and μ are kept constant) is the appropriate thermodynamic potential. To derive the equilibrium probability distribution function, we require that the average energy and particle number be constant. Thus, in addition to the normalization and fixed energy constraints in the derivation of the canonical distribution, we need a third Lagrange multiplier to account for the fixed average number of particles

$$<N> = const = \int N\rho(p,q)DpDq. \tag{1.128}$$

Using the same calculations as in the derivation of the canonical distribution, we obtain for the distribution function of the grand canonical ensemble, describing open systems

$$\rho(p, q) = \frac{e^{-\beta(H(p,q)-\mu N)}}{Z(T, V)} \tag{1.129}$$

with the partition function

$$Z = \int e^{-\beta(H(p,q)-\mu N)} Dp Dq, \tag{1.130}$$

where $\beta = kT$. In this case, the grand potential is related to the grand canonical partition function

$$\Omega(T, V, A, \mu) = -kT \ln Z. \tag{1.131}$$

1.2.5   Correlation and Response Functions

When doing experiments on a many-body system, one often perturbs the system and looks for the response of the system to this disturbance. When the perturbation is small, the system can return to its previous state, or if it is at an unstable fixed point, it can move to a different state. If a system is at equilibrium or a steady state, a small perturbation can lead to a small change in the systems parameters and thus a linear approximation may be sufficient to describe the response. To characterize the response of a system to an external perturbation at a given point and time, we would like to know the change of the measured parameter at every other point and subsequent time. For example, if we perturb the density at a space point $\mathbf{r}$ at time t

$$c(\mathbf{r}, t) = \sum_{i=1}^{N} \delta(\mathbf{r} - \mathbf{r}_i(t)), \tag{1.132}$$

one would like to know the density at all other points $\mathbf{r}'$ and time t'. Here, N is the number of particles or molecules, $\mathbf{r}_i(t)$ is the position of the ith particle at time t, $\mathbf{r}$ is a position vector, and c is the density. An important quantity is the density–density correlation function

$$C(\mathbf{r}, \mathbf{r}'; t, t') = \frac{1}{\rho} \left\langle \sum_{i=1}^{N} \sum_{j=1}^{N} \delta(\mathbf{r} - \mathbf{r}_i(t)) \delta(\mathbf{r}' - \mathbf{r}_j(t')) \right\rangle = \frac{1}{\rho} \langle c(\mathbf{r}, t) c(\mathbf{r}', t') \rangle, \tag{1.133}$$

where the average is taken over the probability distribution and $\rho = \frac{N}{V}$. In neutron scattering experiments and in diffusion MRI experiments, the Fourier transform of this function is measured. In X-ray scattering experiments, the Fourier transform of the equal-time $(t = t')$ correlation function is measured. For steady states, the correlation function depends only on the time difference and we can set $t' = 0$. The density–density correlation function is closely related to the van Hove correlation function. Consider the correlation function:

$$G(\mathbf{r}, \mathbf{r}', t) = \frac{1}{\rho} \left\langle \sum_{i=1}^{N} \sum_{j=1}^{N} \delta \left(\mathbf{r}' + \mathbf{r} - \mathbf{r}_i(t)\right) \delta \left(\mathbf{r}' - \mathbf{r}_j(0)\right) \right\rangle$$
$$= \frac{1}{\rho} \langle c(\mathbf{r}' + \mathbf{r}, t) c(\mathbf{r}', 0) \rangle. \tag{1.134}$$

If we integrate with respect to $\mathbf{r}'$, the δ-function will lead to setting $\mathbf{r}' = \mathbf{r}_j(0)$ and we obtain the van Hove correlation function

$$G(\mathbf{r}, t) = \frac{1}{\rho} \left\langle \int d^3 r' \sum_{i=1}^{N} \sum_{j=1}^{N} \delta \left(\mathbf{r}' + \mathbf{r} - \mathbf{r}_i(t)\right) \delta \left(\mathbf{r}' - \mathbf{r}_j(0)\right) \right\rangle = \tag{1.135}$$

$$\frac{1}{N} \left\langle \int d^3 r' \sum_{i=1}^{N} \sum_{j=1}^{N} \delta \left(\mathbf{r} - \mathbf{r}_i(t) + \mathbf{r}_j(0)\right) \right\rangle = \frac{1}{N} \langle c(\mathbf{r}, t) c(\mathbf{0}', 0) \rangle. \tag{1.136}$$

We will often work in Fourier space, i.e., instead of $\mathbf{r}$ and t, we will use the wave vector $\mathbf{q}$ or $\mathbf{k}$ and the angular frequency ω. The Fourier transform is a linear operator defined here as

$$f(\mathbf{k}, t) = \int_V \phi(\mathbf{r}, t) e^{-i\mathbf{k} \cdot \mathbf{r}} d^3 r. \tag{1.137}$$

The Fourier transform in time is defined as

$$f(\mathbf{r}, \omega) = \frac{1}{2\pi} \int_{-\infty}^{\infty} \phi(\mathbf{r}, t) e^{i\omega t} dt. \tag{1.138}$$

Therefore,

$$f(\mathbf{k}, \omega) = \frac{1}{2\pi} \int \phi(\mathbf{r}, t) e^{i(\omega t - \mathbf{k} \cdot \mathbf{r})} d^3 r dt. \tag{1.139}$$

The angular frequency $\omega = 2\pi f$, where f is the linear frequency measured in Hz. The wave vector $\mathbf{k} = \frac{2\pi \hat{n}}{\lambda}$, where $\hat{n}$ is the unit vector perpendicular to the surfaces of constant phase and λ is the wavelength. In a homogeneous lossless medium, $\mathbf{k}$ is in the direction of wave propagation, which is defined as the direction of energy propagation. This is not always the case for complex materials. With the above definitions of the Fourier transform, the inverse Fourier transform is

$$\phi(\mathbf{r}, t) = \frac{1}{8\pi^3} \int \phi(\mathbf{q}, \omega) e^{-i(\omega t - \mathbf{k} \cdot \mathbf{r})} d^3 k d\omega. \tag{1.140}$$

The Fourier transform of the Dirac delta function is

$$\delta(x - a) = \frac{1}{2\pi} \int_{-\infty}^{\infty} e^{iq(x-a)} dq, \tag{1.141}$$

where $\delta(x) = \infty$ if $x = 0$ and 0 otherwise and

$$\int_{-\infty}^{\infty} \delta(x)dx = 1. \tag{1.142}$$

A useful property of the Dirac δ-function is

$$\int_{-\infty}^{\infty} f(x)\delta(x - a)dx = f(a). \tag{1.143}$$

Now we can define the correlation functions in Fourier space. The Fourier transform of the concentration (or density) is

$$c(\mathbf{k}, t) = \sum_{i=1}^{N} \int_V \delta(\mathbf{r} - \mathbf{r}_i(t)) e^{-i\mathbf{k}\cdot\mathbf{r}} d^3r = \sum_{i=1}^{N} e^{i\mathbf{k}\cdot\mathbf{r}_i(t)}. \tag{1.144}$$

Consider more general correlation function between two space–time dynamic variables $a(\mathbf{r}, t)$ and $b(\mathbf{r}, t)$. Their correlation function is

$$C_{ab}(\mathbf{r}, \mathbf{r}'; t, t') = \langle a(\mathbf{r}, t)b(\mathbf{r}', t')\rangle. \tag{1.145}$$

The correlations function between the Fourier transforms is

$$C_{ab}(\mathbf{k}, \mathbf{k}'; t, t') = \langle a(\mathbf{k}, t)b^*(\mathbf{k}', t')\rangle = \langle a(\mathbf{k}, t)b(-\mathbf{k}', t')\rangle. \tag{1.146}$$

For stationary states, the correlation functions do not depend on the origin of time, but only on the time difference $t - t'$. For homogeneous systems, the correlation function depends only on the different $\mathbf{r} - \mathbf{r}'$. Therefore, translational invariance implies that only equal wave vectors correlations are nonzero, i.e., $\mathbf{k} = \mathbf{k}'$. If the system is also isotropic, the correlation functions only depend on the magnitudes $|\mathbf{r}| = r$ and $|\mathbf{k}| = k$. Returning back to the van Hove correlation function, we see that it is the time-dependent density–density correlation function. It is convenient to write it as the sum of a "self" and distinct term as $G = G_s + G_d$. From

$$G(\mathbf{r}, t) = \frac{1}{N}\left\langle \int d^3r \sum_{i=1}^{N}\sum_{j=1}^{N} \delta\left(\mathbf{r}' + \mathbf{r} - \mathbf{r}_i(t)\right)\delta\left(\mathbf{r} - \mathbf{r}_j(0)\right)\right\rangle$$

$$= \frac{1}{N}\left\langle \sum_{i=1}^{N}\sum_{j=1}^{N} \delta(\mathbf{r} + (\mathbf{r}_i(0) - \mathbf{r}_j(t)))\right\rangle, \tag{1.147}$$

and separating the terms for which $i = j$, we obtain

$$G(\mathbf{r}, t) = \frac{1}{N} \left\langle \sum_{i=1}^{N} \delta(\mathbf{r} + (\mathbf{r}_i(0) - \mathbf{r}_i(t))) \right\rangle + \frac{1}{N} \left\langle \sum_{i=1}^{N} \sum_{j \neq i}^{N} \delta(\mathbf{r} + (\mathbf{r}_i(0) - \mathbf{r}_j(t))) \right\rangle.$$

$$(1.148)$$

At $t = 0$, $G_s(\mathbf{r}, 0) = \delta(\mathbf{r})$ and $G_d(\mathbf{r}, 0) = \rho g(\mathbf{r})$, where $g(\mathbf{r})$ is the radial distribution function (or also called pair correlation function). The van Hove function is the probability density of finding a particle i in the neighborhood of $\mathbf{r}$ at time t given that particle j was at the origin at time 0. In Fourier space, the intermediate scattering function related to the scattering cross section in inelastic scattering experiments is the Fourier transform of the van Hove density–density correlation function:

$$F(\mathbf{k}, t) = \int d^3r \, G(\mathbf{r}, t) e^{-i\mathbf{k} \cdot \mathbf{r}} = \frac{1}{N} \langle c(\mathbf{k})(t) c(-\mathbf{k}, 0) \rangle. \tag{1.149}$$

The Fourier transform of the intermediate scattering function is the dynamic structure factor

$$S(\mathbf{k}, \omega) = \frac{1}{2\pi} \int dt \, F(\mathbf{k}, t) e^{i\omega t} = \frac{1}{2\pi} \int dt d^3r \, G(\mathbf{r}, t) e^{i(\omega t - \mathbf{k} \cdot \mathbf{r})}. \tag{1.150}$$

The differential cross section in coherent inelastic scattering experiment is related to the dynamic structure factor. If a is the scattering length and E_i and E_f are the initial and final energies, respectively, the differential cross section is

$$\frac{d^2\sigma}{d\omega d\Omega} = a^2 \sqrt{\frac{E_f}{E_i}} S(\mathbf{k}, \omega). \tag{1.151}$$

The integral of the dynamic structure factor over all frequencies is the static structure factor

$$\int_{-\infty}^{\infty} d\omega S(\mathbf{k}, \omega) = S(\mathbf{k}). \tag{1.152}$$

1.2.6 Linear Response Theory for Hamiltonian Systems

In the 1930s, Onsager hypothesized that if an equilibrium system is weakly perturbed the relaxation to equilibrium after the perturbation is removed is equivalent to the way equilibrium fluctuations relax. Consider a perturbation to the Hamiltonian which couples to the system's degrees of freedom. It can be magnetic field for magnetic systems, density perturbation, electrical field for polarizable systems, etc. The perturbation is turned on in the distant past and turned off at $t = 0$. The perturbed Hamiltonian is written as

$$H' = H + \Delta H, \tag{1.153}$$

where ΔH is small and $f(t)$ is the perturbation of the energy. For spatially varying perturbations, let the perturbation field coupled to a variable $A(\mathbf{r})$ be $f(\mathbf{r}, t)$. Then the macroscopic change in energy of the system is

$$\Delta H = -\int_V d^3r \, A(\mathbf{r}) f(\mathbf{r}, t). \tag{1.154}$$

The full linear response of a system to a perturbation can be represented as a superposition of the linear response of the system to a sum of plane waves and therefore it is enough to work with a single plane wave

$$f(\mathbf{r}, t) = \frac{1}{V} f(\mathbf{k}) e^{i(\mathbf{k} \cdot \mathbf{r} - \omega t)}. \tag{1.155}$$

Inserting this expression in the expression for ΔH leads to the energy perturbation.

$$\Delta H = -A(-\mathbf{k}) f(\mathbf{k}) e^{-i\omega t}. \tag{1.156}$$

For simplicity, first, we consider spatially independent perturbation and therefore the energy perturbation is

$$\Delta H(t) = -Af(t) = -Af_0 e^{-i\omega t}. \tag{1.157}$$

We will consider the relaxation of a macroscopic variable $B = B_0 + \Delta B$ that is perturbed by Af and we assume ΔB to be small. We want to understand how it relaxes to its equilibrium value. The phase space distribution function evolves according to

$$\frac{\partial \rho}{\partial t} = \{H, \rho\}. \tag{1.158}$$

At $t = -\infty$, the distribution function is ρ. At $t = -\infty$. ΔH is adiabatically turned on and now the new distribution is ρ'. The evolution of ρ' is described by

$$\frac{\partial \rho'}{\partial t} = \{H, \rho'\} + \{\Delta H, \rho'\}. \tag{1.159}$$

Consider the linear approximation $\rho' = \rho + \Delta\rho$. Then

$$\frac{\partial \rho}{\partial t} + \frac{\partial \Delta\rho'}{\partial t} = \{H, \rho\} + \{\Delta H, \rho\} + \{H, \Delta\rho\} + \{\Delta H, \Delta\rho\}. \tag{1.160}$$

The first terms on both sides of the equation cancel and in the linear approximation the last term on the right side of the equation is close to zero. Therefore, we obtain

$$\frac{\partial \Delta \rho'}{\partial t} = \{\Delta H, \rho\} + \{H, \Delta \rho\} = \{H, \Delta \rho\} - f(t)\{A, \rho\} = i\hat{L} - f(t)\{A, \rho\},$$

(1.161)

The solution of the homogeneous equation is

$$\Delta \rho(p, q, t) = e^{i(t-t')\hat{L}} \Delta \rho(p, q, t').$$

(1.162)

The solution of the inhomogeneous equation (the solution is found in the same way we found the solution for ρ) is

$$\Delta \rho(p, q, t) = -\int_{-\infty}^{t} dt' \, e^{i(t-t')\hat{L}}\{A(p, q), \rho(p, q)\} f(t').$$

(1.163)

The equation of motion for B is given by the Poisson equation

$$\frac{\partial B}{\partial t} = \frac{\partial B}{\partial p}\frac{\partial p}{\partial t} + \frac{\partial B}{\partial p}\frac{\partial p}{\partial t} = -\frac{\partial B}{\partial p}\frac{\partial H}{\partial q} + \frac{\partial B}{\partial p}\frac{\partial H}{\partial q} = \frac{\partial H}{\partial q}\frac{\partial B}{\partial p} - \frac{\partial H}{\partial q}\frac{\partial B}{\partial p} = \{B, H\}.$$

(1.164)

We can assume that the average of B at equilibrium is zero. The average of ΔB is given by

$$\langle \Delta B \rangle = \int Dp\, Dq\, \Delta \rho(p.q) B(p, q)$$

$$= -\int Dp\, Dq \int_{-\infty}^{t} dt' \, f(t') B(p, q) e^{i(t-t')\hat{L}}\{A(p, q), \rho(p, q)\}.$$

Taking into account that

$$\{A(p, q), \rho(p, q)\} = \sum_{i=1}^{3N}\left(\frac{\partial \rho}{\partial p}\frac{\partial A}{\partial q} - \frac{\partial \rho}{\partial q}\frac{\partial A}{\partial p}\right)$$

$$= -\beta \sum_{i=1}^{3N}\left(\frac{\partial H}{\partial p}\frac{\partial A}{\partial q} - \frac{\partial H}{\partial q}\frac{\partial A}{\partial p}\right)\rho = -\beta(i\hat{L}A)\rho = -\beta A'\rho$$

(1.165)

(where we used that in the canonical ensemble $\rho \propto e^{-\beta H}$ and A' is the time derivative) and the fact that the Liouville operator is Hermitian with respect to the inner product in Hilbert space of dynamical variables, i.e.,

$$(B, \hat{L}A') = (A, \hat{L}B)^* = (\hat{L}B, A),$$

(1.166)

we obtain

$$\langle \Delta B \rangle = -\int Dp\, Dq \int_{-\infty}^{t} dt' \, f(t') B(p, q) e^{i(t-t')\hat{L}}\{A(p, q), \rho(p, q)\} = \quad (1.167)$$

$$\beta \int Dp\, Dq \int_{-\infty}^{t} dt' \, f(t')\rho(p, q) A' e^{i(t-t')\hat{L}} B.$$

(1.168)

In the last transformation, we used the fact that the scalar product between two dynamical variables in the Hilbert space of dynamical variables is defined as:

$$(B, A) = \int Dp\,Dq\,B^*A, \tag{1.169}$$

where $*$ is complex conjugation. Note that the dynamical variables as functions of coordinates and momenta are real functions. Introducing the function

$$\phi_{BA}(t) = \int Dp\,Dq\,\rho(p, q)A'e^{i(t-t')\hat{L}}. \tag{1.170}$$

One can write the response of the system's state variable B to the perturbation f as

$$\langle\Delta B\rangle(t) = \int_{-\infty}^{t} dt'\, \phi_{BA}(t-t')f(t'). \tag{1.171}$$

The aftereffect function ϕ is defined as

$$\phi_{BA}(t) = \beta\langle B(t)A'(t)\rangle = -\beta\langle B'(t)A(t)\rangle. \tag{1.172}$$

Here, the average is over the unperturbed system since B evolves according to the unperturbed $\hat{L}$.

Next, we can generalize these results to spatially varying perturbations. Straightforward generalization for $\delta B(\mathbf{r}, t)$ is

$$\langle B(\mathbf{r}, t)\rangle = \int_{-\infty}^{t} dt' \int d^3r'\phi_{BA}(\mathbf{r}-\mathbf{r}', t-t')f(\mathbf{r}', t'). \tag{1.173}$$

The unperturbed system is spatially uniform and, in this case, we can work in Fourier space

$$\langle B(\mathbf{k}, t)\rangle = \int_{-\infty}^{t} dt'\phi_{BA}(\mathbf{k}', t-t')f(\mathbf{k}, t'), \tag{1.174}$$

where

$$\phi(\mathbf{k}, t) = -\frac{\beta}{V}\langle B'(\mathbf{k}, t)A(-\mathbf{k})\rangle. \tag{1.175}$$

Here, we used that for translationally invariant systems in Fourier space nonzero correlations exist only between equal wave vectors, i.e.,

$$C_{ab}(\mathbf{k}, \mathbf{k}'; t) = \langle a(\mathbf{k}, t)a(-\mathbf{k}', t)\rangle\delta(\mathbf{k}-\mathbf{k}'). \tag{1.176}$$

For simplicity, we consider isotropic liquid. Using a single plane wave perturbation, one has

$$\langle B(\mathbf{k}, t) \rangle = \int_{-\infty}^{t} \phi_{BA}(k, t - t') f(\mathbf{k}) e^{-i(\omega + i\epsilon)t'} dt' \tag{1.177}$$

$$= f(\mathbf{k}) e^{-i(\omega + i\epsilon)t} \int_{-\infty}^{t} \phi_{BA}(k, t - t') e^{i(\omega + i\epsilon)(t - t')} dt' \tag{1.178}$$

$$= f(\mathbf{k}) e^{-i(\omega + i\epsilon)t} \int_{0}^{\infty} \phi_{BA}(k, \tau) e^{i(\omega + i\epsilon)\tau} d\tau. \tag{1.179}$$

Taking the limit $\epsilon \to 0^+$, we obtain

$$\langle B(\mathbf{k}, t) \rangle = f(\mathbf{k}) e^{-i\omega t} \lim_{\epsilon \to 0^+} \int_{0}^{\infty} \phi_{BA}(k, \tau) e^{i(\omega + i\epsilon)\tau} d\tau = \chi_{BA}(k, \omega) f(\mathbf{k}) e^{-i\omega t}. \tag{1.180}$$

The complex dynamic response function also known as the dynamic susceptibility is defined as

$$\chi_{BA}(k, \omega) = Re\chi_{BA}(k, \omega) + iIm\chi_{BA}(k, \omega) = \lim_{\epsilon \to 0^+} \int_{0}^{\infty} \phi_{BA}(k, \tau) e^{i(\omega + i\epsilon)\tau} d\tau. \tag{1.181}$$

Using the expression for ϕ_{BA}

$$\phi(\mathbf{k}, t) = \frac{\beta}{V} \langle B'(\mathbf{k}, t) A(-\mathbf{k}) \rangle, \tag{1.182}$$

we have

$$\chi_{BA}(k, \omega) = \lim_{\epsilon \to 0^+} \int_{0}^{\infty} \frac{\beta}{V} \langle B'(\mathbf{k}, t) A(-\mathbf{k}) \rangle e^{i(\omega + i\epsilon)\tau} d\tau \tag{1.183}$$

$$= \lim_{\epsilon \to 0^+} \int_{0}^{\infty} \frac{\beta}{V} e^{i(\omega + i\epsilon)\tau} d\langle B(\mathbf{k}, t) A(-\mathbf{k}) \rangle \tag{1.184}$$

$$= \frac{\beta}{V} \lim_{\epsilon \to 0^+} e^{i(\omega + i\epsilon)\tau} \langle B(\mathbf{k}, t) A(-\mathbf{k}) \rangle \Big|_{0}^{\infty}$$

$$- \frac{\beta}{V} \lim_{\epsilon \to 0^+} \int_{0}^{\infty} i(\omega + i\epsilon) \langle B(\mathbf{k}, t) A(-\mathbf{k}) \rangle e^{i(\omega + i\epsilon)\tau} d\tau \tag{1.185}$$

$$= \frac{\beta}{V} [\langle B(\mathbf{k}, 0) A(-\mathbf{q}) \rangle - (i\omega - \epsilon) \langle B(\mathbf{k}, \omega + i\epsilon) A(-\mathbf{k}) \rangle] \tag{1.186}$$

$$= \frac{\beta}{V} [C_{BA}(k, 0) - i(\omega + i\epsilon) \tilde{C}_{BA}(k, \omega + i\epsilon)], \tag{1.187}$$

which describes the linear response of the system in terms of correlation functions. Here,

$$\tilde{C}_{BA}(k, z) = \int_{0}^{\infty} C_{AB}(t) e^{izt} dt \tag{1.188}$$

is the Laplace transform of the correlation function and $z = \omega + i\epsilon$.

1.2.7 Fluctuation–Dissipation Theorem

The fluctuation–dissipation theorem is a mathematical expression that quantifies Onsager's insight that at equilibrium small perturbations to the systems by external probes relax to equilibrium in the same way as the spontaneous fluctuations in the system. Here, we follow the excellent presentation by Hansen [6]. To formulate the theorem, first, consider dynamical variables $A(p(t), q(t))$, $B(p(t), q(t))$, $C(p(t), q(t))$, etc. These variables might depend on all or some of the canonical coordinates and momenta. The time evolution of any such variable is generated by the Liouville evolution operator:

$$\frac{dA(p(t), q(t))}{dt} = \{A, H\} = e^{i\hat{L}t} A(p(0), q(0)), \tag{1.189}$$

because $\frac{\partial A}{\partial t} = 0$, which follows from the fact that A is not explicit function of time. Recall that the Poisson bracket is defined as

$$\{A, B\} = \sum_{i=1}^{N} \left(\frac{\partial A}{\partial q_i} \frac{\partial B}{\partial p_i} - \frac{\partial A}{\partial p_i} \frac{\partial B}{\partial q_i} \right). \tag{1.190}$$

For the distribution function, we have

$$\frac{\partial \rho}{\partial t} = -i\hat{L}\rho \tag{1.191}$$

with

$$-i\hat{L}\rho = \{H, \rho\}. \tag{1.192}$$

Therefore, the probability density evolves according to

$$\rho(p, q, t) = e^{-i\hat{L}t} \rho(p, q, 0). \tag{1.193}$$

As in quantum mechanics, there are two ways of describing the time evolution of dynamical variables (Schroedinger and Heisenberg picture). In the first description, the probability density evolves in time and its evolution is given by (1.193). In this case, the time evolution of the average A is

$$\langle A(t) \rangle = \int dp\, dq\, A(p(0), q(0)) \rho(p, q, t). \tag{1.194}$$

In analogy with the Heisenberg picture, the dynamical variables evolve according to (1.189) and the average evolves according to

$$\langle A(t) \rangle = \int dp\, dq\, A(p(t), q(t)) \rho(p, q, 0). \tag{1.195}$$

Using the convention $t' > t''$, the time correlation function between two dynamical variables is

$$C_{AB}(t', t'') = \langle A(p(t'), q(t')) B(p(t''), q(t'')) \rangle \tag{1.196}$$

$$= \lim_{t \to \infty} \frac{1}{t} \int_0^t A(p(t' + t), q(t' + t)) B(p(t'' + t), q(t'' + t)) dt \tag{1.197}$$

$$= \int \rho(p, q, 0) B^*(p(0), q(0)) e^{-i(t' - t'')\hat{L}} A(p(0), q(0)) dp dq. \tag{1.198}$$

Here, we use again the canonical distribution. This correlation function is time translation-invariant because the equilibrium distribution ρ does not depend on time. Therefore, we can write it as $C_{AB}(\tau + t) B(\tau)$. It is common to chose $\tau = 0$ and therefore we have

$$C_{AB}(t) = \langle A(t) B^*(0) \rangle. \tag{1.199}$$

Since the correlation function is stationary with respect to τ, i.e., the same for any choice of τ, it follows that

$$\frac{d\langle A(\tau + t) B(\tau) \rangle}{dt} = \langle A'(\tau + t) B^*(\tau) \rangle + \langle A(\tau + t)(B^*)'(\tau) \rangle = 0, \tag{1.200}$$

and therefore $\langle A'(t) B^*(0) \rangle = -\langle A(t)(B^*)'(0) \rangle$ leading to $\langle A'(t) A^*(0) \rangle = 0$. The static correlation is defined as

$$\lim_{t \to 0} C_{AB}(t) = \langle AB^* \rangle \tag{1.201}$$

and, for very large times, the two variables are uncorrelated, i.e.,

$$\lim_{t \to \infty} C_{AB}(t) = \langle A(\infty) \rangle \langle B^*(0) \rangle. \tag{1.202}$$

It is convenient to work with the connected correlation function, which describes the correlations of the fluctuations

$$C_{AB}^c(t) = \langle (A(t) - \langle A \rangle)(B^*(0) - \langle B^* \rangle) \rangle = \langle A(t) B^*(0) \rangle_c \tag{1.203}$$

and therefore

$$\lim_{t \to \infty} C_{AB}^c(t) = 0. \tag{1.204}$$

The connected autocorrelation function describes the decay of fluctuations towards equilibrium and is maximum at $t = 0$ and then decays to zero.

To obtain the fluctuation–dissipation theorem, we need to understand the analytical properties of response functions restricted by causality. As before, consider a perturbation by an external force f_i that couples to a state variable A_i. The corresponding Hamiltonian is $H' = H - \sum A_i f_i$, where H is the unperturbed Hamiltonian. Let $a_i = A_i - A_i^{eq}$ represents the fluctuations around A^{eq}, the equilibrium value of A. It can be shown that from microscopic reversibility follows that

$$\langle a_i(t)a_j(0)\rangle = \langle a_i(0)a_j(t)\rangle. \tag{1.205}$$

The same relation also exists for the variable a_i at r two different points in space:

$$\langle a(\mathbf{r}, t)a(\mathbf{r}', 0)\rangle = \langle a(\mathbf{r}, 0)a(\mathbf{r}', t)\rangle. \tag{1.206}$$

These relations are known as the Onsager reciprocal relations.

When the deviations from equilibrium a of a state variable A is proportional to the applied force f, one can write the general expression:

$$\langle a(t)\rangle = \int_{-\infty}^{\infty} \chi(t - t')f(t')dt'. \tag{1.207}$$

The real function $k(t - t')$ is the response function. Causality requires that $k(t - t') = 0$ for $t - t' < 0$, because the response cannot occur before the force is applied. Introducing the Fourier transforms:

$$\langle a(t)\rangle = \int_{-\infty}^{\infty} \langle \tilde{a}(\omega')\rangle e^{-i\omega' t} d\omega' \tag{1.208}$$

$$\chi(t - t') = \int_{-\infty}^{\infty} \tilde{\chi}(\bar{\omega}) e^{-i\bar{\omega}t} d\bar{\omega} \tag{1.209}$$

$$f(t') = \int_{-\infty}^{\infty} \tilde{f}(\omega) e^{-i\omega t} d\omega, \tag{1.210}$$

one obtains:

$$\int_{-\infty}^{\infty} \langle \tilde{a}(\omega')\rangle e^{-i\omega' t} d\omega' = \int_{-\infty}^{\infty} dt' \left(\int_{-\infty}^{\infty} \tilde{\chi}(\bar{\omega}) e^{-i\bar{\omega}t} d\bar{\omega} \int_{-\infty}^{\infty} \tilde{f}(\omega) e^{-i\omega t} d\omega \right) = \tag{1.211}$$

$$\int_{-\infty}^{\infty} \int_{-\infty}^{\infty} \left(\int_{-\infty}^{\infty} dt' \left(e^{-i(\omega - \bar{\omega})t'} \right) \tilde{\chi}(\bar{\omega}) \tilde{f}(\omega) e^{-i\omega t} \right) d\omega d\bar{\omega} = \tag{1.212}$$

$$\int_{-\infty}^{\infty} \int_{-\infty}^{\infty} \delta(\omega - \bar{\omega}) \tilde{\chi}(\bar{\omega}) \tilde{f}(\omega) e^{-i\omega t} d\omega d\bar{\omega} = \int_{-\infty}^{\infty} \tilde{\chi}(\omega) \tilde{f}(\omega) e^{-i\omega t} d\omega, \tag{1.213}$$

where we used the definition of the δ-function

$$\delta(\omega) = \int_{-\infty}^{\infty} e^{-i\omega t} dt \tag{1.214}$$

and the property of the δ-function

$$\int_{-\infty}^{\infty} F(\bar{\omega})\delta(\omega - \bar{\omega})d\bar{\omega} = F(\omega). \qquad (1.215)$$

Comparing the first and the last expressions, we have the simple relation between the Fourier transforms

$$\langle a(\omega) \rangle = \tilde{\chi}(\omega)\tilde{f}(\omega). \qquad (1.216)$$

The inverse response function is a complex function

$$\tilde{\chi}(\omega) = \frac{1}{2\pi}\int \chi(t)e^{i\omega t}dt. \qquad (1.217)$$

Causality leads to restrictions on the possible types of functions that can be response functions. To see which functions can be response functions, we analytically continue $\chi(\omega)$ to the complex plane $z = \omega + i\epsilon$ with $\epsilon > 0$. Because the response is finite for finite perturbative force, the function

$$\chi(z) = \int_0^{\infty} \chi(t)e^{izt}dt \qquad (1.218)$$

does not have singularities in the upper half of the complex plane, but not necessarily in the lower half plane. We can use Cauchy theorem if we introduce a new function

$$f(z) = \frac{\chi(z)}{z - x}, \qquad (1.219)$$

where x is real. This function has a pole on the real axis at $\omega = x$. Integrating $f(z)$ along the contour shown in the figure, we have

$$\oint_C \frac{\chi(z)}{z - x} = 0, \qquad (1.220)$$

because there are no poles inside the area inside the contour C. As ϵ goes to infinity, the function $\chi(z)$ goes to zero, because $e^{izt} = e^{i\omega t}e^{-\epsilon t}$ and $\epsilon > 0$. Now, we can write the integral along the contour C into several parts:

$$\oint_C \frac{\chi(z)}{z - x}dz = \int_{-\infty}^{x-r} \frac{\chi(\omega)}{\omega - x} + \int_{x+r}^{\infty} \frac{\chi(\omega)}{\omega - x} + \int_{\pi}^{0} ire^{i\phi}\frac{\chi(x + re^{i\phi})}{x + re^{i\phi} - x}d\phi = 0. \qquad (1.221)$$

The last term is obtained as follows. Let $z = x + re^{i\phi}$. Then $dz = ire^{i\phi}d\phi$, because the small radius $r = const$. Introducing the principle value integral:

$$P\int_{-\infty}^{\infty} \frac{\chi(\omega)}{\omega - x}d\omega = \lim_{r \to 0}\left(\int_{-\infty}^{x-r} \frac{\chi(\omega)}{\omega - x} + \int_{x+r}^{\infty} \frac{\chi(\omega)}{\omega - x}\right). \qquad (1.222)$$

Using that in the limit $r \to 0$

$$\int_{\pi}^{0} i r e^{i\phi} \frac{\chi(x + r e^{i\phi})}{x + r e^{i\phi} - x} d\phi = -i\pi \chi(x), \qquad (1.223)$$

it follows that

$$\chi(x) = \frac{1}{i\pi} P \int \frac{\chi(\omega)}{\omega - x} d\omega. \qquad (1.224)$$

The susceptibility is a complex function $\chi(\omega) = \chi'(\omega) + i\chi''(\omega)$

$$\chi'(\omega) + i\chi''(\omega) = \frac{1}{i\pi} P \int \frac{\chi'(\omega)}{\omega - x} d\omega + \frac{1}{\pi} P \int \frac{\chi''(\omega)}{\omega - x} d\omega. \qquad (1.225)$$

Therefore,

$$\chi'(\omega) = \frac{1}{\pi} P \int \frac{\chi''(\omega)}{\omega - x} d\omega \qquad (1.226)$$

and

$$\chi''(\omega) = -\frac{1}{\pi} P \int \frac{\chi'(\omega)}{\omega - x} d\omega. \qquad (1.227)$$

These are the Kramers–Kronig relations and they allow us to find the real part of the susceptibility when we know the imaginary part. We will also need the Sokhotski–Plemelj formula. Let $f(x)$ be a function that is nonsingular and smooth around a point on the real axis x_0. Then

$$\lim_{\epsilon \to 0} \int_{-\infty}^{\infty} \frac{f(x) dx}{x - x_0 \pm i\epsilon} = P \int_{-\infty}^{\infty} \frac{f(x) dx}{x - x_0} \mp i\pi f(x_0). \qquad (1.228)$$

The Sokhotski–Plemelj formula is sometimes written as

$$\lim_{\epsilon \to 0} \frac{1}{x - x_0 \pm i\epsilon} = P \frac{1}{x - x_0} \mp i\pi \delta(x - x_0), \qquad (1.229)$$

but it only makes sense when it is integrated with a smooth nonsingular function. Going back to the connected correlation function

$$C_{AB}^{c}(t) = \langle AB \rangle_c = \langle (A(t) - \langle A \rangle)(B - \langle B \rangle) \rangle = \langle AB \rangle - \langle A \rangle \langle B \rangle, \qquad (1.230)$$

we can define the Fourier transform, which describes the power spectrum

$$\tilde{C}_{AB}^{c}(\omega) = \frac{1}{2\pi} \int_{-\infty}^{\infty} C_{AB}^{c}(t) e^{i\omega t} dt \qquad (1.231)$$

and its Laplace transform

Fig. 1.1 Integration contour

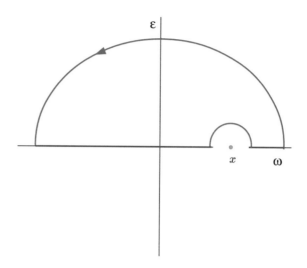

$$\bar{C}_{AB}^{c}(z) = \int_{0}^{\infty} C_{AB}^{c}(t)e^{izt}\,dt, \tag{1.232}$$

where z is complex. Since the connected correlation function is bounded, the Laplace transform is analytic in the upper half of the complex plane. The Laplace transform of the connected correlation function is also the Hilbert transform of its Fourier transform, i.e.,

$$\bar{C}_{AB}^{c}(z) = \int_{-\infty}^{\infty} \frac{C_{AB}^{c}(\omega)}{z - \omega}\,d\omega. \tag{1.233}$$

We also have

$$\lim_{\epsilon \to 0} Re\bar{C}_{AA}(\omega + i\epsilon) = \lim_{\epsilon \to 0} Re\left(i \int_{-\infty}^{\infty} \frac{\tilde{C}_{AA}^{c}(\omega')}{\omega + i\epsilon - \omega'}\,d\omega'\right) = \pi\tilde{C}_{AA}^{c}(\omega) \tag{1.234}$$

because

$$P \int_{-\infty}^{\infty} \frac{\tilde{C}_{AA}^{c}(\omega')}{\omega - \omega'}\,d\omega' \tag{1.235}$$

is real and when multiplied by i becomes imaginary (Fig. 1.1).

Now recall that when the system is perturbed the average of a state variable B evolves according to

$$\langle B(\mathbf{r}, t)\rangle = \int_{-\infty}^{t} dt' \int d^{3}r' \phi_{BA}(\mathbf{r} - \mathbf{r}', t - t')f(\mathbf{r}', t'). \tag{1.236}$$

When the unperturbed system is spatially uniform the Fourier transform is

$$\langle B(\mathbf{k}, t) \rangle = \int_{-\infty}^{t} dt' \phi_{BA}(\mathbf{k}', t - t') f(\mathbf{k}, t') \tag{1.237}$$

with

$$\phi(\mathbf{k}, t) = -\frac{\beta}{V} \langle B'(\mathbf{k}, t) A(-\mathbf{k}) \rangle. \tag{1.238}$$

For a single plane wave perturbation, we had

$$\langle B(\mathbf{k}, t) \rangle = f(\mathbf{k}) e^{-i\omega t} \lim_{\epsilon \to 0^+} \int_0^\infty \phi_{BA}(k, \tau) e^{i(\omega + i\epsilon)\tau} d\tau = \chi_{BA}(k, \omega) f(\mathbf{k}) e^{-i\omega t}, \tag{1.239}$$

where

$$\chi_{BA}(k, \omega) = Re\chi_{BA}(k, \omega) + i Im\chi_{BA}(k, \omega) = \lim_{\epsilon \to 0^+} \int_0^\infty \phi_{BA}(k, \tau) e^{i(\omega + i\epsilon)\tau} d\tau \tag{1.240}$$

is the dynamic response function. Using the expression for ϕ_{BA}

$$\phi(\mathbf{k}, t) = \frac{\beta}{V} \langle B'(\mathbf{k}, t) A(-\mathbf{k}) \rangle, \tag{1.241}$$

we obtained

$$\chi_{BA}(k, \omega) = \frac{\beta}{V} [C_{BA}(k, 0) - i(\omega + i\epsilon) \tilde{C}_{BA}(k, \omega + i\epsilon)] \tag{1.242}$$

Now using (1.234), we obtain

$$C_{AA}(\mathbf{k}, \omega) = \frac{VkT}{\pi \omega} Im\chi_{AA}(\mathbf{k}, \omega), \tag{1.243}$$

which is the fluctuation–dissipation theorem. If we apply the above theory to a perturbation of the Hamiltonian

$$\Delta H = \frac{1}{V} c_{-\mathbf{k}} \phi_{\mathbf{k}} e^{-i\omega t} \tag{1.244}$$

of the concentration $c_{\mathbf{k}}$, the fluctuation–dissipation theorem is

$$S(\mathbf{k}, \omega) = -\frac{kT}{\pi \omega} Im\chi_{AA}(\mathbf{k}, \omega), \tag{1.245}$$

where the change in sign is due to the change of the sign of the perturbation ΔH (the density response is defined in terms of a potential and not a field) and $S(\mathbf{k}, \omega)$ is the dynamic structure function.

1.2.8 Diffusion

Consider a liquid and molecules dissolved in the liquid. Since concentration gradients will cause currents (linear response), Fick's first law states

$$\mathbf{J}(\mathbf{r}, t) = -D\nabla c(\mathbf{r}, t), \tag{1.246}$$

where D is a constant called diffusion coefficient and $\nabla c(\mathbf{r}, t) = (\frac{\partial c}{\partial x}, \frac{\partial c}{\partial y}, \frac{\partial c}{\partial z})$ is the gradient partial differential operator. When the molecules are not created or destroyed, the continuity equation is

$$\frac{\partial c(\mathbf{r}, t)}{\partial t} = -\nabla \cdot \mathbf{J}(\mathbf{r}, t). \tag{1.247}$$

Combining the two equations, we obtain the diffusion equation

$$\frac{\partial c(\mathbf{r}, t)}{\partial t} = D\nabla^2 c(\mathbf{r}, t), \tag{1.248}$$

where the Laplacian of c is $\nabla^2 c = \frac{\partial^2 c}{\partial x^2} + \frac{\partial^2 c}{\partial y^2} + \frac{\partial^2 c}{\partial z^2}$. In some cases, the diffusion coefficient depends on the position in the system, i.e., $D(\mathbf{r})$ and then the diffusion equation is

$$\frac{\partial c(\mathbf{r}, t)}{\partial t} = \nabla D(\mathbf{r}) \cdot \nabla c(\mathbf{r}, t). \tag{1.249}$$

The diffusion equation is a partial differential equation and, to solve it, one needs to specify the initial density $c(\mathbf{r}, 0)$ in space and the boundary conditions. For homogeneous isotropic infinite media, or for very large system and short times, the solution is

$$c(\mathbf{r}, t) = \frac{c_0}{(4\pi Dt)^{3/2}} e^{-\frac{r^2}{4Dt}}, \tag{1.250}$$

where the initial concentration is $c(\mathbf{r}, 0) - c_0 \delta(\mathbf{r})$. The mean square displacement is

$$\langle \Delta r^2(t) \rangle = 6Dt. \tag{1.251}$$

In 1D, the right-hand side of this expression is $2Dt$ and, in 2D, it is $4Dt$. We see that as time progresses the concentration goes to zero. In finite systems, the solution is different. For example, in a cubic container in the large time limit, the concentration becomes constant everywhere in the container. The diffusion coefficient is a response function or also called transport coefficient and can be related to the velocity–velocity autocorrelation function. Consider the mean square displacement

$$\langle \Delta r^2(t) \rangle = \langle (\mathbf{r}(t) - \mathbf{r}(0))^2 \rangle. \tag{1.252}$$

The displacement is

$$\mathbf{r}(t) - \mathbf{r}(0) = \int_0^t v(t')dt',$$
(1.253)

from which follows that

$$\langle \Delta r^2(t) \rangle = \int_0^t \int_0^t \langle \mathbf{v}(t') \cdot \mathbf{v}(t'') \rangle dt' dt''$$
(1.254)

and the mean square displacement is related to the velocity autocorrelation function. At equilibrium or steady state, the origin of time can be chosen arbitrary and thus we can use $\langle \mathbf{v}(t' - t'')\mathbf{v}(0) \rangle$ as we did before. Changing the time variable of integration $\tau = t' - t''$ and $dt'' = -d\tau$, we obtain

$$\langle \Delta r^2(t) \rangle = \int_0^t dt' \int_{t'-t}^{t'} \langle \mathbf{v}(\tau) \cdot \mathbf{v}(0) \rangle d\tau.$$
(1.255)

Changing the order of integration and using that $\langle \mathbf{v}(\tau) \cdot \mathbf{v}(0) \rangle = \langle \mathbf{v}(-\tau) \cdot \mathbf{v}(0) \rangle$, we obtain

$$\langle \Delta r^2(t) \rangle = 2 \int_0^t (t - t') \langle \mathbf{v}(t') \cdot \mathbf{v}(0) \rangle dt',$$
(1.256)

The normalized velocity autocorrelation function is

$$\alpha(t) = \frac{\langle \mathbf{v}(t) \cdot \mathbf{v}(0) \rangle}{\langle \mathbf{v}(0) \cdot \mathbf{v}(0) \rangle}.$$
(1.257)

In terms of $\alpha(t)$, the mean square displacement is

$$\langle \Delta r^2(t) \rangle = 6v_0 \int_0^t (t - t')\alpha(t')dt',$$
(1.258)

where the thermal velocity is $v_0^2 = 3\langle \mathbf{v}(0) \cdot \mathbf{v}(0) \rangle$. The diffusion coefficient can be defined as

$$D = \lim_{t \to \infty} \frac{\langle \Delta r^2(t) \rangle}{6t}$$
(1.259)

and therefore one obtains one of the Green–Kubo relations

$$D = v_0^2 \int_0^\infty \alpha(t)dt.$$
(1.260)

1.2.8.1 Einstein Relation

In 1905, Einstein derived a relation between the diffusion coefficient and the mobility, which is an example of the fluctuation–dissipation theorem. The fluctuations are the fluctuations of the particle position described by the diffusion coefficient and the mean square displacement and the response is the response of the particle to an external field described by the mobility. To derive the Einstein relation, first, consider a small particle in a fluid in an external field, electrical or gravitational, and assume that the field is in the opposite direction of the y-axis. The particle will drift in the direction of the field and, as it starts to accelerate, it will experience more collisions in the front than in the back and this will slow it down. Thus, the particle will drift with some average, constant velocity called the drift velocity. When the response is linear, the drift velocity is proportional to the field, which is true for small velocities and fields. The coefficient of proportionality is the mobility μ and is defined as

$$\langle \mathbf{v} \rangle = \mu \mathbf{F}_{ext}. \tag{1.261}$$

In the liquid, besides this force due to the external field, there is also a random force whose origin are the collisions between the liquid molecules and the particle and the force related to the difference in the number of collisions in the front and the back of a moving particle. This force can be written as a sum of two parts: a drag force due to the friction between the particle and the molecules of the fluid and a random force set by the fast collisions. The Newton equation of motion is therefore

$$\mathbf{F} = -\beta \mathbf{v}(t) + \mathbf{F}_{ext} + \boldsymbol{\zeta}(t) = m \frac{d\mathbf{v}(t)}{dt}. \tag{1.262}$$

This is the Langevin equation for this system. The ensemble average of the random component of the force is zero, but the autocorrelation function is not, i.e.,

$$\langle \boldsymbol{\zeta}(t) \rangle = 0 \tag{1.263}$$
$$\langle \boldsymbol{\zeta}(t)\boldsymbol{\zeta}(t') \rangle \neq 0. \tag{1.264}$$

In the absence of the external field, the equation for the average velocity is

$$\frac{d\langle \mathbf{v}(t) \rangle}{dt} = -\frac{\beta}{m} \langle \mathbf{v}(t) \rangle \tag{1.265}$$

and the solution is

$$\langle \mathbf{v}(t) \rangle = \mathbf{v}(0)e^{-\frac{t}{\tau}}. \tag{1.266}$$

The relaxation time $\tau = m/\beta$ for a spherical particle obeying the Stokes equation is equal to $m/(6\pi \eta a)$, where η is the viscosity and a is the particle radius (for a self-diffusion $\beta = 4\pi \eta a$). The relaxation time is the characteristic time for the initial conditions to die out every time the particle starts to move. Consider a fluid in a container and let us calculate the flux across a circle with area S due to an external field that sets a concentration gradient $c(y)$ in a direction perpendicular to the area of the circle (see the figure). The particles will drift in the direction of the gradient from higher concentrations towards lower and in time Δt the number of particles that will cross S is equal to the number of particles in the cylinder C with base S and hight $\langle v_y \rangle \Delta t$, i.e. $N = cS \langle v_y \rangle \Delta t$. The flux is the number of particles crossing a surface per unit area per unit time and, therefore, it is

$$\mathbf{J} = \langle v_y \rangle c(y) \hat{y}. \tag{1.267}$$

The potential energy is related to the force acting on the particles by

$$f_y = -grad\ U \tag{1.268}$$

and therefore

$$\langle v_y \rangle = \mu \mathbf{f} = -\mu \frac{\partial U}{\partial y} \tag{1.269}$$

$$\mathbf{J} = -\mu c(y) \frac{\partial U}{\partial y}. \tag{1.270}$$

The reduction in molecules at the top of the cylinder will create a diffusive flux in the opposite direction, i.e.,

$$\mathbf{J}_{diff} = -D \frac{\partial c(y)}{\partial y} \hat{y}. \tag{1.271}$$

At equilibrium, the magnitude of the two fluxes must be equal, i.e.,

$$D \frac{\partial c(y)}{\partial y} = -\mu c(y) \frac{\partial U}{\partial y}. \tag{1.272}$$

At equilibrium the probability of finding a particle at a position y is proportional to the Boltzmann factor, i.e.,

$$c(y) = c(0) e^{-\frac{U(y)}{kT}}. \tag{1.273}$$

The equation then leads to

$$\frac{D}{kT} c(y) \frac{\partial U(y)}{\partial y} = \mu\ c(y) \frac{\partial U(y)}{\partial y} \tag{1.274}$$

and therefore the Einstein relation is (Fig. 1.2)

$$D = \mu kT. \tag{1.275}$$

Fig. 1.2 Fluxes

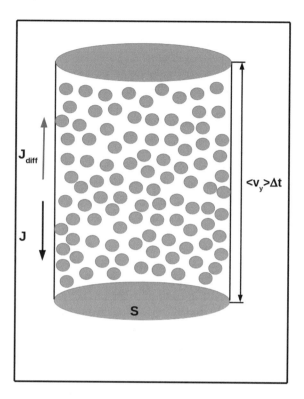

1.3 Phase Separation in Living Cells

The cell nucleus was the first cellular organelle observed under the microscope and was named by Robert Brown in 1831. Since then two types of organelles have been described, organelles bound by a membrane such as mitochondria and Golgi apparatuses, and membraneless organelles like the nucleolus and Cajal bodies. More than a hundred years ago, Edmund Wilson observed different molecular assemblies in cells that behaved like liquid droplets.

The physics and chemistry of these droplet-like structures during the past decade have become a major focus of research and although many properties of these droplets are now known, there are many unanswered questions regarding the kinetics, dynamics, and regulation of their assembly and disassembly. Currently, it is thought that these structures are the result of reversible liquid–liquid phase separation (LLPS) driven by multivalent molecular interactions between the molecules forming the droplets [7, 8]. The liquid–liquid phase transition proposed as a mechanism of droplet formation in cells is a transition between two states of species of the same molecular composition, rather than compositional separation like water and oil. These phase transitions are thought to be analogous to the transitions between different amorphous states in a solid, an example of which is the transition between the two amorphous states of ice observed under pressure. Water is also thought, in its super-cooled state, to undergo such phase separation into a mixture of high and low-density liquids. Double-well interaction potentials between the molecules in a liquid have been shown to favor liquid–liquid phase separation theoretically and the complex multivalent interactions between the proteins in cells may favor such polyliquid states.

In contrast to many non-biological systems, where the transition between the two states can be driven by pressure and temperature, in cells, the transitions are driven by changes in the concentration of the molecular species. In biological systems, the LLPS is the transition from a monophasic liquid state to a multiphase liquid state, where the monophasic state spontaneously separates into dense and diffuse states. Recent studies suggest that LLPS is involved in multiple cellular functions including cell compartmentalization [9–11], genome architecture [12–15], gene expression [16–18], and cell signaling [19, 20]. Therefore, LLPS and its regulation have emerged as a fundamental topic in cellular biology. Both nucleation and spinodal decomposition have been implicated as the mechanism behind membranelles organelle formation. In many of the systems in which LLPS have been observed, multiple stable droplets exist in contrast to the predictions of the classical nucleation theory in which minimization of the surface energy leads to Ostwald ripening (droplet coarsening through diffusion of molecules from small droplets to larger droplets) as well as coalescence (the merger of two droplets in contact with one another) the result of which is the formation of one large droplet and the disappearance of small droplets. It was recently suggested (Hyman) that this deviation from the classical nucleation behavior may be due to the existence of reaction sites in cells that would lead to a stable multi-droplet state. A number of experimental [21–24] and computational studies [25–28] aimed at understanding the physical properties of liquid–liquid phase separation have shown that factors known to influence LLPS are the valence of interacting domains [29], electrical charge content [30], physical dimension [31, 32], etc. However, the mechanism by which both nucleation and equilibrium/nonequilibrium conditions favor LLPS is still unclear.

The existence of intracellular molecular aggregates has been known since the eighteenth century, but only recently detailed experimental studies have indicated that many membraneless organelles represent a different phase of the intracellular molecules. Examples are the nucleoli, Cajal bodies, P-bodies, stress granules, promyelocytic leukemia protein (PML) nuclear bodies, DNA repair foci, etc. One of the functions of the Cajal bodies is to assemble telomerase. Telomerase consists of two components, hTR, the RNA component that serves as a template for telomere synthesis, which is then catalyzed by the telomerase reverse transcriptase, hTERT. The PML bodies are protein assemblies named after the PML genes, whose product the PML protein is necessary for the formation of the organelle. The function of the PML nuclear bodies is not clear, but the mutation of the PML gene has been linked to cancer. In addition, PML bodies have been implicated in the mysterious alternative lengthening of telomeres (ALT) mechanism discussed earlier. There are 1−30 bodies in the nucleus.

The mechanisms of formation of these organelles are not yet clear, but they are clearly molecular assemblies different from the surrounding molecular liquid and we can start describing their formation using a master equation. The formation of the clusters is through the aggregation of molecules and it proceeds through the attachment or de-attachment of a molecule to the cluster. Initially, the cluster grows due to work performed on the cluster, which is the change of the Gibbs free energy. This is the early phase of growth. As the cluster grows, it reduces the molecular concentration in its surroundings and this depletion has to be taken into account in a self-consistent theory of cluster formation. Once reaching critical size it is expected that the system becomes unstable and the cluster continues to grow on its own. This is the nucleation phase.

We consider a molecular liquid at constant temperature and pressure, which is close to a phase transition. The master equation for the cluster size distribution function is

$$\frac{\partial p(n,t)}{\partial t} = k_{on}^{n-1\to n} p(n-1,t) + k_{off}^{n+1\to n} p(n+1,t) - k_{off}^{n\to n-1} p(n.t) - k_{on}^{n\to n+1} p(n,t).$$

(1.276)

A major goal of nucleation theory is to determine the on and off rates in this equation. The stationary solution, $\frac{\partial p(n,t)}{\partial t} = 0$ determines the stationary cluster size distribution. One way to obtain this distribution is to consider the equilibrium process of cluster formation. The equilibrium cluster size distribution function is

$$p_{eq}(n) = p_0 e^{-W(n)/kT},$$

(1.277)

where $W(n)$ is the work done on the system required to form a cluster with n molecules. At constant temperature and pressure, it is equal to the change of the Gibbs free energy ΔG. The change of the Gibbs free energy is the difference between the Gibbs free energy of the bulk liquid with a cluster with n molecules and the bulk liquid without a cluster. This difference can be written as the difference in the Gibbs free energy of the cluster of size n and the Gibbs free energy of the bulk phase, i.e.,

$$\Delta G(n) = G_c(n) - n\mu_b(P,T).$$

(1.278)

We assume here that each cluster equilibrates to its equilibrium structure very fast compared to the processes of cluster formation and from now on we will use the word drop to denote this equilibrium state of the cluster. The master equation can be written using the fluxes

$$J_{n-1} = k_{on}^{n-1\to n} p(n-1,t) - k_{off}^{n\to n-1} p(n.t)$$

(1.279)

$$J_n = k_{on}^{n\to n+1} p(n,t) - k_{off}^{n+1\to n} p(n+1,t)$$

(1.280)

as

$$\frac{\partial p(n,t)}{\partial t} = J_{n-1} - J_n.$$

(1.281)

At equilibrium, the detailed balance leads to $J_n = 0$ for $n > 1$, i.e.,

$$\frac{k_{on}^{n \to n+1}}{k_{off}^{n+1 \to n}} = \frac{p_{eq}(n+1,t)}{p_{eq}(n,t)}. \tag{1.282}$$

Using (5.3), we can write for the ratio of the on and off rates

$$\frac{k_{on}^{n \to n+1}}{k_{off}^{n+1 \to n}} = e^{-\frac{\Delta G(n+1) - \Delta G(n)}{kT}}. \tag{1.283}$$

In the so-called capillary approximation, the change in the Gibbs free energy is given by

$$\Delta G(n) = n(\mu_c(P,T) - \mu_b(P,T)) + 4\pi\sigma \left(\frac{3}{4\pi} v_c n\right)^{2/3}. \tag{1.284}$$

The second term on the right-hand side is the free energy related to the formation of a spherical surface with area A. This energy is σA, where σ is the surface tension. The surface area of the spherical drop is $A = 4\pi R^2$, where R is the radius of the drop and since the volume of the drop is $n v_c V = (4/3)\pi R^3$ follows that

$$R = \left(\frac{n v_c}{\frac{4\pi}{3}}\right)^{1/3}. \tag{1.285}$$

From (5.3) follows that

$$-(\Delta G(n+1) - \Delta G(n)) = \mu_b(P,T) - \frac{\partial G_c(n)}{\partial n}\Big|_{n=n+1} = \mu_b - \mu_c(n+1). \tag{1.286}$$

The chemical potential of the cluster is defined as

$$\mu_c(n) = \frac{\partial G_c(n)}{\partial n}. \tag{1.287}$$

In the capillary approximation, we obtain

$$\mu_c(n) = \mu_c(P,T) + \frac{8\pi\sigma}{3} \left(\frac{3}{4\pi} v_c\right)^{2/3} n^{-1/3}. \tag{1.288}$$

So, we can write using the expression (5.11)

$$\frac{k_{on}^{n \to n+1}}{k_{off}^{n+1 \to n}} = e^{(\mu_b - \mu_c(n+1))/kT}. \tag{1.289}$$

1.3.1 The Szilard Model

In this model, it is assumed that, once a drop of critical size, say with n_{cr} number of molecules is formed, it is removed from the system and n_{cr} number of molecules are added to the system, keeping the total number of molecules fixed. Starting with no clusters and N particles eventually, a steady state of clusters is reached and these clusters have some steady-state distribution. The stationary cluster-size distribution in this model is

$$p^{steady\ state}(n) = p(1)e^{-\frac{\Delta G(n)}{kT}}. \tag{1.290}$$

1.3.2 Nucleation, Growth, Coarsening, and Coalescence in Oversaturated Solutions

The work required to change the volume and the surface area of a system by dV and dA, respectively, is

$$dW = -p_{out}dV_{out} - p_{in}dV_{in} + \gamma dA, \tag{1.291}$$

where γ is the surface tension and $p_{out/in}$ is the pressure outside and inside, respectively. The increase of the enclosed liquid's volume is equal to the decrease of the volume outside so $dV_{in} = -dV_{out} = dV > 0$. So, we have

$$dW = -(p_{in} - p_{out})dV + \gamma dA. \tag{1.292}$$

In equilibrium, $dW = 0$ and

$$(p_{in} - p_{out})dV = \gamma dA \tag{1.293}$$

or

$$\Delta p = (p_{in} - p_{out}) = \frac{\gamma dA}{dV} > 0. \tag{1.294}$$

For a sphere, $dA = d(4\pi R^2) = 8\pi RdR$ and $dV = d(4/3\pi R^3) = 4\pi R^2 dR$ and thus the Laplace pressure is

$$\Delta p = \frac{2\gamma}{R}. \tag{1.295}$$

The pressure on both sides of an equilibrium flat surface, e.g., gas/liquid interface is the same when the surface is at equilibrium. And thus, the Helmholtz free energy F has a minimum

$$0 = dF = -PdV - SdT + \gamma dA. \tag{1.296}$$

Consider a molecular liquid in two or three dimensions with a uniform concentration in the absence of drops, c, which is slightly above the saturation concentration c_{sat}. Let the distance from saturation be $\Delta = c - c_{sat} << 0$. At the origin, we consider a spherical (circular) drop with radius R. We want to derive the kinetic equations of the drop size. When the drop forms the concentration is now $c(r)$. The concentration at the drop surface is related to the saturation concentration in the absence of a drop (sometimes this concentration is denoted by c_∞ to signify that this is the concentration in the case of infinite surface), c_{sat} by the Gibbs–Thomson formula:

$$c_R = c_{sat} + \frac{\alpha}{R}, \tag{1.297}$$

where $\alpha = \frac{2\sigma c_{sat}\nu}{kT}$ with σ the surface tension, ν the molecular weight, k the Boltzmann constant, and T the temperature. The change in volume of the drop in the absence of drop–drop interactions is due to the flux out or in the drop of molecules, i.e.,

$$\frac{dV}{dt} = 3\frac{4\pi R^2}{3}\frac{dR}{dt} = -4\pi R^2 J, \tag{1.298}$$

where Δ is the Laplace operator.and the flux of molecules per unit surface is

$$J = -D\frac{\partial c}{\partial r}(r = R). \tag{1.299}$$

From these relations, we obtain

$$\frac{dR}{dt} = D\frac{\partial c}{\partial r}(r = R). \tag{1.300}$$

The right-hand side can be obtained by solving the diffusion equation

$$\frac{\partial c}{\partial t} = D\Delta c, \tag{1.301}$$

where Δ is the Laplace operator. Subject to the boundary conditions

$$c(R) = c_R \tag{1.302}$$
$$c(r >> R) = c. \tag{1.303}$$

In the static limit and for a spherical drop the radial Laplace equation is

$$\frac{d^2 c}{dr^2} + \frac{2}{r}\frac{dc}{dr} = 0. \tag{1.304}$$

We look for a solution in the form:

$$c(r) = A + \frac{B}{r}. \tag{1.305}$$

Applying the boundary conditions, we obtain the solution

$$c(r) = c + \frac{R(c_R - c)}{r} \tag{1.306}$$

and therefore the

$$\frac{dc}{dr}(r = R) = \frac{c - c_R}{R}. \tag{1.307}$$

Using the Gibbs–Thomson relation, we obtain

$$\frac{dc}{dr}(r = R) = \frac{1}{R}\left(\Delta - \frac{\alpha}{R}\right). \tag{1.308}$$

Substituting in (1.300), we obtain the equation for the drop size

$$\frac{dR}{dt} = \frac{D}{R}\left(\Delta - \frac{\alpha}{R}\right). \tag{1.309}$$

The drop is stable for $R = R_c = \frac{\alpha}{\Delta}$. For $R < R_c$, the drop dissolves and, for $R > R_c$, the drop grows. The drop size distribution function and the time dependence of the drop size have also been obtained [33, 34].

It is custom to separate the coarsening of the system due to Ostwald ripening and the coarsening due to the coalescence of droplets into bigger ones. The process of Ostwald ripening is called coarsening, while the merger of droplets—coalescence. Coarsening in the LSW theory is diffusion-limited and so is coalescence in general. If droplets diffuse very slowly or don't diffuse, they would not coalesce. The reason for this might be the small transport coefficient of the droplets or the exhaustion of molecules in the extra-droplet space. There are two processes leading to coarsening of the system, i.e., the disappearance of small drops and the growth of large drops (these with a radius bigger than the critical radius). One of them is molecules leaving small drops and being captured by large drops and the second is coalescence, where two drops merge. In biological systems at biologically relevant time scales, small drops are stabilized by a currently unknown mechanism. In order for this to occur, the diffusional field established in the system has to be disrupted by some process. One proposal that leads to small droplet stabilization is the existence of chemical reaction centers in the system [35]. Indeed such centers stabilize small droplets as shown in the cited work, but it is unclear if such droplets are solid clusters or liquid structures. In biological systems, the molecules phase separate into liquid-like droplets and these droplets are reversible. In disease states, e.g., Alzheimers disease, the FUS protein aggregates in solid-like clusters, and such clusters become dysfunctional. The mechanism of cluster formation is still being currently debated (Fig. 1.3).

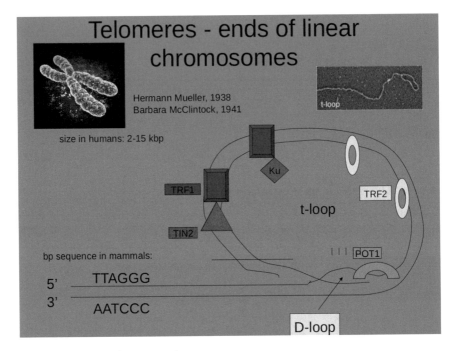

Fig. 1.3 The human telomere complex

1.4 A Biophysical Example: Telomere Homeostasis

Telomeres are repetitive DNA sequences at the end of the linear chromosome in eukaryotic cells. In humans, the repetitive sequence is 5'-TTAGGG-3' with the complementary CCCTAA sequence. Human telomeres vary between 2500 and 50 bp. Budding yeast, Saccharomyces cerevisiae, telomeres contain about 60–100 copies of the sequence TGTGGGTGTGGTG. Telomeres are protected from being recognized as double-strand breaks that must be repaired by homologoues recombination (HR, rare in humans) or non-homologous end joining (NHEJ) by a group of proteins and by the formation of a loop structure called t-loop shown in the figure.

In the early 1970s, Alexey Olovnikov in the Soviet Union and James Watson in the United States realized that during replication the lagging strand cannot be replicated all the way to the 5' end of the DNA, because the synthesis requires an RNA primer to proceed and the DNA-polymerase can only extend in the direction $5' - 3'$, while it is moving in the opposite direction for this strand, i.e., $3'-5'$. As a consequence, telomeres lose $50-200$ base pairs per cell division. When a critically short telomere emerges p53-related checkpoint response is activated and the cell is taken from the proliferation pool into the senescent cellular pool. In fibroblasts, one in a million cells bypasses this mechanism and continues to divide, while in epithelial cells, one in a hundred thousand bypasses the checkpoint response. Telomeres at the end of the chromosome form a loop, called t-loop, to protect the telomere ends from p53 response to double-strand breaks and to protect chromosomes from fusing with each other, a highly mutagenic event, although the telomere-like sequences of the centromeres indicate that during the evolution of life, chromosome fusion has been common. The formation of the t-loop requires around 2000 base pairs, below which a critically short telomeres form. Human somatic cells have telomeres around 10 kbp at birth, and the 50−200 bp loss at each cell division leads to the conclusion that human cells have between 160 and 40 cell divisions during human life. The difference between the cell death, the cell senescence, and the growth rate determines how many cells the cells at birth can produce during the lifetime of an organism. However, organ maintenance is facilitated by a small number of adult stem cells, which divide infrequently producing progenitor cells with finite proliferative potential.

The end replication problem has been solved by nature by two independent mechanisms. During embryogenesis, and in most cancers, an enzyme called telomerase can extend telomeres by using its own RNA template. A second mechanism called alternative lengthening of telomeres (ALT) is active in approximately 10% of cancers. ALT is believed to be facilitated by homologous recombination between sister chromatids. FISH experiments have shown the transfer of fluorescent signal from a labeled chromatid to its sister unlabeled chromatid showing the telomere exchange. Below, we will give a quantitative description of these two mechanisms.

1.4.1 Telomerase Control of Telomere Length

In this section, we will give a short exposition of optimal control theory and its relation to variational methods and then apply it to a model of telomere maintenance by telomerase as observed in budding yeast.

The discussion will be restricted to scalar functions, but it can be extended to vector and tensor optimization problems. Consider a system whose state is described by a time-dependent variable $x(t)$. In our case, this will be the telomere length, but it can be the position or velocity of a car or the water level in a tank, etc. Next, we introduce the control u, which in our case will be the telomerase adding base pairs, but it can be the break applied to a train or a system of water sink and source that maintain the water level in a tank around a given level. The nonequilibrium dynamics of the system is described by a differential equation:

$$\dot{x}(t) = f(x(t), u(x, t), t). \tag{1.310}$$

Given an initial state $x(t_0)$, in control theory, the goal is to bring the system to a state $x(t)$ at time t. In optimal control theory in addition to the dynamics described by the above equation, one would also like to minimize some functional (energy, cost function, etc.)

$$S[x, u] = \int_{t_0}^{t} L(x, u, t)dt. \tag{1.311}$$

Thus, the variational problem is to find a function $u(t)$ that will optimize the functional S and satisfy the initial and final state values of x as well as the non-holonomic constraint $\dot{x}(t) = f(x, u, t)$.

At each cell division, telomere length regulation consists of basal telomere loss and telomerase facilitated telomere gain. In short, this can be expressed as

(change of telomere length) = −(basal loss) + (telomere gain)*(extension probability).

The extension probability in Saccharomyces cerevisiae [36–38], human cancer cells, and in telomerase positive, normal human fibroblasts has been quantified recently. The data suggests that the extension probability or the extension frequency is a sigmoid type of curve and was well fitted by logistic regression. In wild-type cells with sufficient telomerase expression for maintaining telomere homeostasis, telomeres are maintained at an equilibrium length. In S. cerevisiae, this equilibrium length is approximately 300 base pairs (bp) [37], while in immortalized human cells, it is between 5000 and 15000 bp [39]. The basal telomere loss in S. cerevisiae is 3 nucleotides (nt) per generation [40], while in human cells, it is between 50 and 200 bp [39]. Larger telomere rapid deletions (T-RD) may occur as well, due to DNA double-strand breaks or errors during DNA replication [41]. The number of telomere repeats added by telomerase in a single cell cycle in vivo varies from few to more than a hundred nucleotides in S. cerevisiae [37] and up to 800 in human super-telomerase cells [42]. Telomerase adds nucleotides to S. cerevisiae telomeres in the late S phase but does not replenish all telomeres at each cell replication either because it might not be available at all telomeres during that time or because when available at a telomere it may not be able to extend the telomere. Recent data suggests that in budding yeast [37] telomeres switch back and forth between two states: extendible or open state, which allows telomerase to associate with the telomere and a nonextendible or closed state, which prevents telomerase from associating with the telomere. This binary response suggested in this study is consistent with the sigmoid function used to fit the data. The oscillation frequency between these two states is higher for shorter telomeres and this leads to a higher probability for a telomerase complex to associate with these telomeres. Even when telomerase associates with a telomere it might not extend that telomere. Whether or not a telomerase associated with a telomere processes it or not depends on its length and perhaps on the state of the shelterin complex. In S. cerevisiae, the repeat addition processivity (the number of telomere repeats added per round of DNA replication) is higher at shorter telomeres but is lower than in human cells. In human cells, telomerase concentrations correlate with increased repeat addition processivity [43]. In cells in which telomerase is partly inhibited, a new equilibrium length is established by a feedback control mechanism as shown in S. cerevisiae. Recently, telomerase in human cells was expressed beyond the physiological limit and, in these super-telomerase cells, the telomere extension dynamics did not seem to slow down, continuing with the same average rate for more than 60 population doublings. This constant rate of elongation suggests that in these cells the combined probability for a telomere to be in open state times the number of base pairs added to the telomere during an elongation event is a constant that is larger than the basal telomere loss. In the HEK-293 human cancer cell line, the number of telomerase complexes, 50 [44], is approximately two times smaller than the number of telomeres, suggesting that the telomerase concentrations are limiting [45, 46].

When the telomere control occurs on a time scale much larger than the cell division time, the experimental observations can be summarized by the following adaptive nonlinear control dynamics:

$$\frac{dE(t)}{dt} == b(t) + \frac{cD}{1 + e^{(E-\mu)/T}}, \tag{1.312}$$

where E is the telomere length, b is the basal telomere loss, c is the probability that a telomerase is at the site of the telomere or equivalently the telomerase concentration, D is the number of base pairs added at each elongation event, and μ and T are control parameters to be fitted to the experimental data. The logistic function, shown in the figure specifies the type of feedback control that the telomere length exercises on its elongation. The telomerase control of the telomere length using this logistic control is shown in Fig. 1.5. This differential equation cannot be solved explicitly, but several limits on the parameters of the logistic function can be set from it. When the telomere length is equal to the steady state length ($E_{eq} = 6000$), the left-hand side of (1.312) is zero. This leads to the following relation between the telomerase concentration, c and the parameters μ and T:

$$6000 = \mu + T ln \left(\frac{cD}{b} - 1 \right). \tag{1.313}$$

At $c = 50/92$, $b = 100$ bp/cell division, and $D = 250$ bp/cell division, the relation between μ and T is $\mu = 6000 + 1.025T$. On the other hand (1.313), sets a limit on the parameter T because c is a concentration with a maximum value of 1 when $L > 0$. The relation is

$$c = 0.4 + e^{-\frac{\mu}{T}} \tag{1.314}$$

leading to

$$e^{-\frac{\mu}{T}} \le 0.6 \tag{1.315}$$

or equivalently

$$T \le \frac{\mu}{5.12} = 6000 + 1.025T. \tag{1.316}$$

From the last relation follows that the maximum value of T and μ, which will guarantee that the steady-state length of the telomeres on average is 6000 bp are $T_{max} = 1467$ bp and $\mu_{max} = 7504$ bp, respectively. The parameter T sets the slope of the sigmoid curve displayed in Fig. 1.4 and μ sets the value at which the probability for an open state is equal to K. This result means that to maintain telomere length at

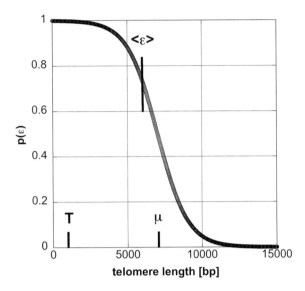

Fig. 1.4 The feedback control function, maintaining the telomere length, is shown. In this figure, the dependence of the probability (or the frequency), p, for occurrence of the open state on the telomere length e (in base pairs) is shown. The parameter m in this function is the telomere length at which the probability for the extending state is K. The parameter $T = 1000$ determines the slope of the sigmoid. The choice of T is close to the maximum possible value for telomere homeostasis at a telomere length of 6000bp (see the text for an explanation)

6000 the slope of the curve describing the frequency extendible state cannot be larger than a certain number. This slope is given by the derivative of the c with respect to L at $L = \mu$ and is equal to $0.25/T$. Here, the value $T = 1000$ is assumed, which is close to the maximum value and the results presented here are a lower limit to the speed with which telomerase extends a telomere, i.e., the worst-case scenario for treatment and more quantitative data with different human cells will be useful to better quantify the values of the parameter in the theory.

In Fig. 1.5, the probabilistic and deterministic length control dynamics are shown for two telomeres with different initial lengths: one shorter and one longer than the steady-state length. The feedback control steadily increases the length of the shorter and decreases the length of the longer telomere. The speed of telomere elongation or depletion is larger the farther a telomere is from the steady state and becomes zero for telomeres with the steady-state length. For fixed steady-state length, this speed is controlled by the parameter T in (1.313). The speed at which the telomeres approach the steady length depends on the parameters T and μ in the logistic response (Fig. 1.4). For small $T = 1, \mu = 6001$ and the speed approaches the maximum speed determined by the basal loss (100 bp/cell division in this model). This is because the logistic function is zero (one) for longer (shorter) than the steady length telomeres most of the time. Therefore, the curves in Fig. 1.5, represent the slowest telomere length dynamics. The choice of parameters in the logistic probability function is

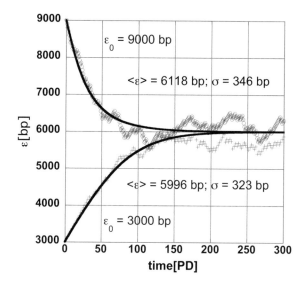

Fig. 1.5 In this figure, the time dependence (in cell divisions) of the average telomere length e is shown. Because the initial length of the two representative telomeres shown in the figure is longer (shorter) than the homeostatic length, the telomeres gradually approach the homeostatic length. Adaptive control at the telomere guarantees that short telomeres are extended and long telomeres are shortened to the average homeostatic length. The initial telomere length, the average steady-state telomere length, and the square root of the variance are shown for two telomeres in the presence of telomerase, while at the steady state, the square root of the variance is small compared to the average telomere length. As the telomere shortens below 1000 bp these two quantities will become comparable and large fluctuations may produce a sub-critically short telomere and trigger p53-independent checkpoint response. The continuous model is shown with continuous lines

consistent with current data for S. cerevisiae [33], but for human, the telomere length is not known. Lowering the telomerase concentration leads to a faster decrease in telomere length (the fastest decrease is set by the basal loss rate). The first telomere reaching a critical length is expected to trigger a p53-dependent checkpoint cellular response. In human cells, this critical length is 2000–4000bp. However, in most cancers, the p53 pathway is inactivated and the telomeres continue to decrease in length until the crisis point is reached. At the crisis point, the short telomeres are only a few hundred base pairs in length [42] and therefore telomerase inhibition is more effective in cancers with an intact p53 response. In Fig. 1.5, the telomere erosion is shown for telomeres with an initial length of 6000 bp, corresponding to cells with different degrees of telomerase suppression. In this figure, it is assumed that, when the telomerase–telomere fraction is 50/92, the telomeres are maintained at their steady-state length of 6000 bp. The critical telomerase concentration below which telomeres reach zero length is 73.6% of 50/92. The time that it takes to reach crisis (0 in this figure) and thus the PP decreases with decreasing telomerase concentration.

In S. cerevisiae, the nonlinear adaptive control that telomeres perform through a feedback loop on their extension rate is performed by the frequency at which an extendible state occurs in other organisms the mechanism is currently unknown. As the steady-state telomere length in human cells is much longer than in S. cerevesiae, the length control in this case might be accomplished by the adaptive extension length of the added telomere repeats by telomerase and not by the frequency of occurrence of the extendible state. This alternative mechanism is also captured by the model introduced here, because the term

$$\frac{D}{1 + e^{(E-\mu)/T}} \tag{1.317}$$

in (1.312) can be interpreted as the length of the added telomere repeats by the telomerase complex. This means that the model is more general than the specific mechanism of telomere length homeostasis. This paper is concerned with the dynamics of telomere attrition caused by telomerase depletion. In this case, the speed at which telomerase replenishes a telomere is related to the probability with which the extending state occurs and the probability of a telomerase complex being at the telomere. Below a critical telomerase concentration, the telomere cannot be sufficiently replenished and will become eventually critically short, despite the increasing frequency, with which it is in an extendible state. Lower telomerase concentration will lead to less frequent extension events and consequently to slower extension. For cells with an abundance of telomerase, the telomeres are extended beyond the predicted by the theory steady-state length [37]. It is possible that the telomere length control system has an upper limit on counting telomere length above which it is incapable of tracking the length of a telomere and therefore the probability for an open state above this limit is a nonzero constant [37] explaining the results in human super-telomerase cells. This effect can be incorporated into the theory, but more studies are needed to understand how does the extension rate in super-telomerase cells relate to their telomerase concentration. Currently, available data is inconclusive regarding the relation between the telomerase concentration and the rate of telomere extension. One of the conclusions of this paper is that in clinical settings, telomerase inhibitors need to be administered robustly for a long time: more than 250 cell divisions, if telomerase is suppressed only 60%. However, low levels of the RNA component of telomerase lead to a deleterious condition, dyskeratosis congenital, characterized by short telomeres and premature death from bone marrow failure [46]. A question that can be addressed by the model presented here is the optimal drug administration schedule that could limit the effect of telomerase inhibition on the replicative capacity of the hematopoietic cells and maximize the suppression of tumor proliferation. In clinical settings, the grade of a tumoris often quantified histologically in terms of the mitotic index [47, 48], i.e., the fraction of mitotic spreads visible under a microscope in a given slice of tissue. Most cells in real tumors do not divide at all times, because otherwise, a single cell after 60 divisions will create approximately 10–20 cells, more than the number of cells in the human body. How is the telomere-derived proliferative potential related to the mitotic index? The mitotic index of a cell population is between the smallest and the largest proliferative potential. The lowest proliferative potential of a dividing cell after n cell divisions is log 2 (n+1), when only one of the two daughter cells divides at each cell division. This will correspond to a stem cell renewing itself and a non-dividing progenitor cell. The maximum is n, when every daughter cell divides at each cell division. The mitotic index depends on many factors that limit and stimulate cell proliferation and telomere length is one of them. In a tumor in the presence of telomerase inhibition, different proliferating cells will most likely be exposed to a different level of telomerase inhibition and will require different times to stop dividing. Several studies have shown positive correlations between the median survival time of cancer patients and the average telomere length of the tumor cells [49–51]. Telomerase expression levels are also negatively correlated with the median survival time [52]. Therefore, the telomere-related proliferation potential is positively correlated with the mitotic index and it will be interesting to know how the dynamics of the mitotic index is related to the telomerase inhibitor concentration and duration. This will allow the quantification of the relation between the tumor malignancy and the proliferative capacity in cancer.

1.4.2 Telomere Sister Chromatid Exchange and Biased Diffusion

A small percent of cancers do not express telomerase, but the telomeres of the cancer cells are maintained at finite length by a mechanism called alternative lengthening of telomeres (ALT). There is evidence that the ALT is related to homologous recombination between sister chromatids during replication. In this section we will study a model in which one or two telomeres are present in a cell. This is highly idealized model for human cells where 92 telomeres are present, but it shows the nonequilibrium dynamics of telomere loss with exchange. It is worth mentioning that the lowest number of chromosomes and thus telomeres in eaukaryotic cells has been observed in *Myrmeciapilosula*, the Jack jumper ant, in which the males have one chromosome and the females a pair of chromosomes [47]. Recently a single chromosome budding yeast has been been created by fusing 16 linear chromosomes [48], with similar transcriptome and phenotype, but reduced fitness. The simple model described below might be applicable to these organisms.

We consider a population of dividing cells with a telomere in each of them, which at each cell division shortens by some fixed length. At each cell division also part of a telomere is exchanged between the two sister cells. If cells reach critically short telomeres, they are removed from the dividing pool. Thus the senescent state is an absorbing state for the cell population. In Fig. 1.6 these processes are shown and these processes lead to drift-diffusion stochastic process defined as.

We consider a toy model of cell division. Each cell has a finite number of telomeres each of which has a specified lifetime T. When a cell divides, each of the telomeres in each daughter cell runs down by a fixed amount ΔT plus some fluctuation. That is,

Fig. 1.6 Chromatid
exchange processes

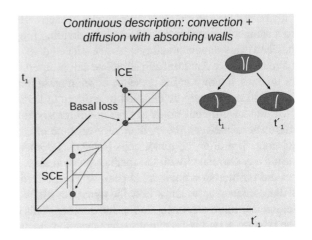

$$T \rightarrow T - \Delta T + \eta, \tag{1.318}$$

where η is a stochastic variable with mean zero, and without correlations. When the value of T for any one of the internal telomeres reaches zero, the cell stops dividing and is removed'.

The system can be modeled as a biased towards the origin random walk with absorbing boundary condition at the origin, while in the continuum limit, the evolution of each telomere length T can be viewed as biased towards the origin diffusion with an absorbing boundary condition at the origin. An analogy can be made with Brownian particles in a river which end at a waterfall. Cell division then would be a process in which the Brownian particles divide at some rate. The divisions will lead to exponential growth of the population that have not yet died. This growth competes with the bias associated with the running down of the internal telomeres of the cell (specified by ΔT) or analogously with the disappearance of the Brownian particles in the waterfall. As a result, there is a transition in the time dependence of the cell density as the growth rate and the bias are varied. For n telomeres per cell the resulting equation is

$$\frac{\partial c(\mathbf{x}, t)}{\partial t} = -\mathbf{v} \cdot \nabla c(\mathbf{x}, t) + \nabla D(\mathbf{x}, t) \cdot \nabla c(\mathbf{x}, t) + kc(\mathbf{x}, t) \tag{1.319}$$

subject to an absorbing boundary condition at $x = 0$. The growth of the cell population by division is given by k. Here k is assumed independent of the length of the telomeres inside the cell, except, when a critically short telomere appears. In this description, the values of $\mathbf{x}$ is equivalent to the current clock value of a particular cell, while the drift velocity $\mathbf{v}$ is proportional to $-\Delta T_i$, and D_i is proportional to $\langle \eta^2 \rangle$. Here we have included different drift and diffusion and terms for each telomere.

One can solve the resulting equations for the case of one and two telomeres per cell. How to solve the problem for a single telomere is shown here and we refer the reader to the literature where the two telomeres per cell problem has been solved.

1.4.2.1 Diffusion and the Method of Images

The problem of diffusing particle with an absorbing state is closely related to the calculation of the probability of first passage through the origin, because once the diffusing particle is absorbed at the origin it is removed from the system. In our case a cell that reaches critically short telomere, which in this case is assumed to be of zero length becomes senescent and is removed from the pool of dividing cells. The method of images from electrostatics can be generalized to the case of particles diffusing on the positive x-axis and being absorbed at $x = 0$ once they reach this point. The reason the method of images can be used here is that both electrostatics and the diffusion equation are governed by a Laplacian. The generalization of the method of images to include time dependence we notice that at each time point the electrostatic method of images is satisfied. We consider a diffusing particle in one dimension that starts at $x_0 > 0$ and then place a second (antiparticle) at $x = -x_0$. Both particles in this approach diffuse on the $[-\infty, \infty]$ interval with the same diffusion coefficient, but with opposite signs of the concentration. Since the antiparticle propagator is anti-Gaussian the boundary condition for the combined particle-antiparticle system at $x = 0$ is satisfied by default and therefore the particles diffusing in the $x > 0$ axis are absorbed at the origin. The solution to the diffusion equations for the concentration $c(x, t)$ with $c(0, t) = 0$ is:

$$c(x, t) = \frac{1}{\sqrt{4\pi Dt}} \left(e^{-(x-x_0)^2/4Dt} - e^{-(x+x_0)^2/4Dt} \right) \qquad (1.320)$$

To include biased diffusion to this method we notice that now the particle will drift towards the origin with velocity $-v$ and the antiparticle with the same drift velocity $-v$ towards $x = 0$. If $c(x, t)$ is the concentration at x at time t with the initial condition $c(x, t = 0) = \delta(x - x_0)$, corresponding to a single particle starting at x_0, the concentration at later times is [49]

$$c(x, t) = \frac{1}{\sqrt{4\pi Dt}} \left[e^{-(x-x_0-vt)^2/4Dt} - e^{-vx_0/D} e^{-(x+x_0-vt)^2/4Dt} \right]. \qquad (1.321)$$

The second term represents the "image" contribution and the bias velocity of the image is in the same direction as that of the initial particle and the exponential pre-factor ensures that the absorbing boundary condition at any point of time, $c(x = 0, t) = 0$ is satisfied.

From this concentration profile the probability for a diffusing particle to hit the origin for the first time at time t, is

$$F(t) = D \frac{\partial c}{\partial x}\bigg|_{x=0} = \frac{x_0}{\sqrt{4\pi Dt^3}} \, e^{-(x_0+vt)^2/4Dt}. \tag{1.322}$$

and the survival probability, that is, the probability that the diffusing particle has not hit the origin by time t, is

$$S(t) = 1 - \int_0^t F(t') \, dt'$$

$$\sim \sqrt{\frac{4}{\pi} \frac{x_0 \sqrt{Dt}}{(vt)^2}} \, e^{-(vt)^2/4Dt}. \tag{1.323}$$

For one telomere per cell the equation is the diffusion equation with a growth and drift terms,

$$\frac{\partial c}{\partial t} = -v \frac{\partial c}{\partial x} + D \frac{\partial^2 c}{\partial x^2} + kc \tag{1.324}$$

subject to an absorbing boundary condition at $x = 0$. Here subscripts denote partial differentiation and the growth of the cell population by division is given by k. In this mathematical description, the value of x is equivalent to the current clock of a particular cell, while the drift velocity v is proportional to $-\Delta T$, and D is proportional to $\langle \eta^2 \rangle$.

We represent the density as a product of the growth solution (exponential growth) and an unknown function u, i.e. $c(x, t) = e^{kt} u(x, t)$, $u(x, t)$. Thus the cell density is given by

$$c(x, t) = \frac{e^{kt}}{\sqrt{4\pi Dt}} \left[e^{-(x-x_0-vt)^2/4Dt} - e^{-vx_0/D} \, e^{-(x+x_0-vt)^2/4Dt} \right]. \tag{1.325}$$

Let $w = -v$. Then, using Eq. (1.325), the total number of living cells is given by

$$N(t) = \int_0^\infty c(x, t) \, dx,$$

$$= \frac{e^{kt}}{\sqrt{\pi}} \left[\int_{\frac{wt-x_0}{\sqrt{4Dt}}}^\infty e^{-z^2} \, dz - e^{wx_0/D} \int_{\frac{wt+x_0}{\sqrt{4Dt}}}^\infty e^{-z^2} \, dz \right]$$

$$= \frac{1}{2} e^{kt} \left[\text{erfc}\left(\frac{wt - x_0}{\sqrt{4Dt}} \right) - e^{wx_0/D} \text{erfc}\left(\frac{wt + x_0}{\sqrt{4Dt}} \right) \right], \tag{1.326}$$

where $\text{erfc}(z)$ is the complementary error function. From the asymptotics of the error function [50], $\text{erfc}(z) \sim e^{-z^2}/\sqrt{\pi} \, z$, the long-time behavior of $N(t)$ is given by

$$N(t) \sim \frac{e^{kt}}{2\sqrt{\pi}} \left[\frac{\sqrt{4Dt}}{wt - x_0} e^{(wt-x_0)^2/4Dt} - \frac{\sqrt{4Dt}}{wt + x_0} e^{wx_0/D} e^{(wt+x_0)^2/4Dt} \right]$$

$$\sim \sqrt{\frac{Dt}{\pi}} \frac{2x_0}{(wt)^2 - x_0^2} e^{kt} e^{(wt-x_0)^2/4Dt}$$

$$\sim \sqrt{\frac{Dt}{\pi}} \frac{2x_0}{(wt)^2} e^{wx_0/2D} e^{(k-w^2/4D)t}. \tag{1.327}$$

Thus the fundamental parameter of the system is $\epsilon_1 \equiv 4Dk/w^2$. For $\epsilon_1 < 1$, cell division is insufficient to overcome the rate of death due to decreasing length of the telomeres and the population of living cells decays exponentially in time.

The cell life time (as $t \to \infty$) dependence of the density of living cells can be obtained by taking the limit $t \to \infty$ in Eq. (1.325), we find

$$c(x, t \to \infty) \sim \frac{xx_0}{\sqrt{4\pi Dt^3}} e^{-(k-w^2/4D)t} e^{-w(x-x_0)/2D}. \tag{1.328}$$

and apart from an overall time-dependent factor, the cell density decays exponentially, with a characteristic lifetime $2D/w$ that is independent of whether the total cell population is increasing or decreasing in time.

1.4.2.2 Mean Lifetime

We can evaluate the integrals in Eq. (1.328) to determine the mean lifetime of a cell population. Using Eq. (1.322), the denominator in Eq. (1.328) is

$$\mathcal{D} = \frac{x_0 e^{wx_0/2D}}{\sqrt{4\pi D}} \int_0^\infty \frac{dt}{t^{3/2}} e^{-x_0^2/4Dt} e^{(k-w^2/4D)t}. \tag{1.329}$$

Making the substitution $y = 1/\sqrt{z}$, we transform the integral into the standard form (see (7.4.3) in Ref. [51]):

$$\int_0^\infty e^{-az^2 - b/z^2} \, dz = \sqrt{\frac{\pi}{4a}} e^{-\sqrt{4ab}}.$$

Using this result, Eq. (1.329) gives

$$\mathcal{D} = e^{wx_0/2D[1-\sqrt{1-4Dk/w^2}]}.$$

Similarly, the numerator in Eq. (1.328) is given by

$$\mathcal{N} = \frac{x_0 e^{wx_0/2D}}{\sqrt{4\pi D}} \int_0^\infty \frac{dt}{t^{1/2}} e^{-x_0^2/4Dt} e^{(k-w^2/4D)t}. \tag{1.330}$$

For this integral, we make the substitution $y = \sqrt{z}$ to again transform the integral into (7.4.3) in Ref. [51] and obtain

$$\mathcal{N} = \frac{x_0}{w} \frac{e^{wx_0/2D[1-\sqrt{1-4Dk/w^2}]}}{\sqrt{1-4Dk/w^2}}.$$

Taking the ratio $\mathcal{N}/D$ then give the result for the mean lifetime quoted in Eq. (1.328).

References

1. L.D. Landau, E.M. Lifshitz, *Statistical Physics* (Butterworth-Heinemann, Oxford, 1980)
2. M. Elenius, M. Dzugutov, J. Chem. Phys. **131**(10), 104502 (2009)
3. D.D. Holm, V. Putkaradze, C. Tronci, Kinet. Relat. Models **6**(2), 429 (2013)
4. A.M. Bloch, J.E. Marsden, D.V. Zenkov, Not. Amer. Math. Soc. **52**(3), 324 (2005)
5. G.S. Ezra, J. Math. Chem. **35**(1), 29 (2004)
6. J.P. Hansen, I.R. McDonald, *Theory of Simple Liquids: With Applications to Soft Matter* (Academic, Cambridge, 2013)
7. C.P. Brangwynne, C.R. Eckmann, D.S. Courson, A. Rybarska, C. Hoege, J. Gharakhani, F. Jülicher, A.A. Hyman, Science **324**(5935), 1729 (2009)
8. Y. Shin, C.P. Brangwynne, Science **357**(6357) (2017)
9. B. Bonev, G. Cavalli, Nat. Rev. Genet. **17**(11), 661 (2016)
10. M. Falk, Y. Feodorova, N. Naumova, M. Imakaev, B.R. Lajoie, H. Leonhardt, B. Joffe, J. Dekker, G. Fudenberg, I. Solovei et al., Nature **570**(7761), 395 (2019)
11. A.R. Strom, A.V. Emelyanov, M. Mir, D.V. Fyodorov, X. Darzacq, G.H. Karpen, Nature **547**(7662), 241 (2017)
12. Y. Shin, Y.C. Chang, D.S. Lee, J. Berry, D.W. Sanders, P. Ronceray, N.S. Wingreen, M. Haataja, C.P. Brangwynne, Cell **175**(6), 1481 (2018)
13. A.G. Larson, D. Elnatan, M.M. Keenen, M.J. Trnka, J.B. Johnston, A.L. Burlingame, D.A. Agard, S. Redding, G.J. Narlikar, Nature **547**(7662), 236 (2017)
14. S.F. Banani, H.O. Lee, A.A. Hyman, M.K. Rosen, Nat. Rev. Mol. Cell Biol. **18**(5), 285 (2017)
15. M. Feric, N. Vaidya, T.S. Harmon, D.M. Mitrea, L. Zhu, T.M. Richardson, R.W. Kriwacki, R.V. Pappu, C.P. Brangwynne, Cell **165**(7), 1686 (2016)
16. A. Boija, I.A. Klein, B.R. Sabari, A. Dall'Agnese, E.L. Coffey, A.V. Zamudio, C.H. Li, K. Shrinivas, J.C. Manteiga, N.M. Hannett et al., Cell **175**(7), 1842 (2018)
17. B.R. Sabari, A. Dall'Agnese, A. Boija, I.A. Klein, E.L. Coffey, K. Shrinivas, B.J. Abraham, N.M. Hannett, A.V. Zamudio, J.C. Manteiga et al., Science **361**(6400) (2018)
18. D. Hnisz, K. Shrinivas, R.A. Young, A.K. Chakraborty, P.A. Sharp, Cell **169**(1), 13 (2017)
19. X. Su, J.A. Ditlev, E. Hui, W. Xing, S. Banjade, J. Okrut, D.S. King, J. Taunton, M.K. Rosen, R.D. Vale, Science **352**(6285), 595 (2016)
20. P. Li, S. Banjade, H.C. Cheng, S. Kim, B. Chen, L. Guo, M. Llaguno, J.V. Hollingsworth, D.S. King, S.F. Banani et al., Nature **483**(7389), 336 (2012)
21. C.P. Brangwynne, P. Tompa, R.V. Pappu, Nat. Phys. **11**(11), 899 (2015)
22. J. Wang, J.M. Choi, A.S. Holehouse, H.O. Lee, X. Zhang, M. Jahnel, S. Maharana, R. Lemaitre, A. Pozniakovsky, D. Drechsel et al., Cell **174**(3), 688 (2018)
23. S. Maharana, J. Wang, D.K. Papadopoulos, D. Richter, A. Pozniakovsky, I. Poser, M. Bickle, S. Rizk, J. Guillén-Boixet, T.M. Franzmann et al., Science **360**(6391), 918 (2018)
24. A. Patel, H.O. Lee, L. Jawerth, S. Maharana, M. Jahnel, M.Y. Hein, S. Stoynov, J. Mahamid, S. Saha, T.M. Franzmann et al., Cell **162**(5), 1066 (2015)
25. H. Kang, J. Yoo, B.K. Sohn, S.W. Lee, H.S. Lee, W. Ma, J.M. Kee, A. Aksimentiev, H. Kim, Nucleic Acids Res. **46**(18), 9401 (2018)

26. T. Sun, A. Mirzoev, V. Minhas, N. Korolev, A.P. Lyubartsev, L. Nordenskiöld, Nucleic Acids Res. **47**(11), 5550 (2019)
27. R.V. Pappu, FASEB J. **32**, 102 (2018)
28. A. Molliex, J. Temirov, J. Lee, M. Coughlin, A.P. Kanagaraj, H.J. Kim, T. Mittag, J.P. Taylor, Cell **163**(1), 123 (2015)
29. S. Banjade, M.K. Rosen, Elife **3**, e04123 (2014)
30. S. Alberti, Curr. Biol. **27**(20), R1097 (2017)
31. I. Koltover, K. Wagner, C.R. Safinya, Proc. Natl. Acad. Sci. **97**(26), 14046 (2000)
32. J. Elf, M. Ehrenberg, Syst. Biol. **1**(2), 230 (2004)
33. I.M. Lifshitz, V.V. Slyozov, J. Phys. Chem. Solids **19**(1–2), 35 (1961)
34. C. Wagner, Zeitschrift für Elektrochemie, Berichte der Bunsengesellschaft für physikalische. Chemie **65**(7–8), 581 (1961)
35. D. Zwicker, A.A. Hyman, F. Jülicher, Phys. Rev. E **92**(1), 012317 (2015)
36. M. Chang, M. Arneric, J. Lingner, Genes & Devel. **21**(19), 2485 (2007)
37. M.T. Teixeira, M. Arneric, P. Sperisen, J. Lingner, Cell **117**(3), 323 (2004)
38. A.D. Mozdy, T.R. Cech, RNA **12**(9), 1721 (2006)
39. U.M. Martens, E.A. Chavez, S.S. Poon, C. Schmoor, P.M. Lansdorp, Exp. Cell Res. **256**(1), 291 (2000)
40. S. Marcand, V. Brevet, E. Gilson, EMBO J. **18**(12), 3509 (1999)
41. B. Li, A.J. Lustig, Genes & Devel. **10**(11), 1310 (1996)
42. G. Cristofari, J. Lingner, EMBO J. **25**(3), 565 (2006)
43. R.H. Datar, W.Y. Naritoku, P. Li, D. Tsao-Wei, S. Groshen, C.R. Taylor, S.A. Imam, Gynecol. Oncol. **74**(3), 338 (1999)
44. S.B. Cohen, M.E. Graham, G.O. Lovrecz, N. Bache, P.J. Robinson, R.R. Reddel, Science **315**(5820), 1850 (2007)
45. L.Y. Hao, M. Armanios, M.A. Strong, B. Karim, D.M. Feldser, D. Huso, C.W. Greider, Cell **123**(6), 1121 (2005)
46. C. Greider, in *Cold Spring Harbor Symposia on Quantitative Biology*, vol. 71 (Cold Spring Harbor Laboratory Press, New York, 2006), pp. 225–229
47. M.W. Crosland, R.H. Crozier, Science **231**(4743), 1278 (1986)
48. Y. Shao, N. Lu, Z. Wu, C. Cai, S. Wang, L.L. Zhang, F. Zhou, S. Xiao, L. Liu, X. Zeng et al., Nature **560**(7718), 331 (2018)
49. S. Redner, *A Guide to First-Passage Processes* (Cambridge University Press, Cambridge, 2001)
50. M. Abramowitz, I. Stegun, *Handbook of Mathematical Functions* (1970)
51. I.S. Gradshteyn, I.M. Ryzhik, Y.V. Geronimus, M.Y. Tseytlin, A. Jeffrey, D. Zwillinger, V.H. Moll, Inc.(Academic, New York, 1965)
52. https://www.nature.com/articles/bjc2012602

Chapter 2
Probing the Energy Landscapes of Biomolecular Folding and Function

Paul Charles Whitford and José N. Onuchic

Abstract This chapter focuses on the development and application of energy land-scape principles to study biomolecular dynamics. Over the last several decades, our understanding of biomolecular folding has transitioned from traditional biochemical ideas to a statistical-mechanical approach to studying dynamics. These fundamental concepts, which were developed for protein folding, are now ready to be extended and applied to investigate emergent phenomena related to the functional dynamics of biomolecules.

2.1 Energy Landscape Theory: The Interface of Physics and Molecular Biology

From a physical standpoint, protein folding is the most well-understood area of molecular biophysics. Historically, the principal challenge was to determine how the energetics of biomolecules can give rise to rapid folding kinetics. That is, during fold-ing, a protein chain must be able to search the expansive phase space associated with the unfolded ensemble and then rapidly form a complex network of interactions that defines the native configuration (Fig. 2.1). Interestingly, the folding process does not occur in a piece-wise fashion. Instead, folding is typically cooperative, where there is collective folding/unfolding of the entire chain. The development and application of energy landscape principles has been central to understanding folding in quantitative terms. Rather than attempting to describe interconversion events between a large number of discrete states, it is common to describe folding as a diffusive process along a continuous energy landscape.

While originally developed to study protein folding, energy landscape concepts may be applied to any number of molecular processes. For example, during biologi-

P. C. Whitford (✉)
Northeastern University, Boston, MA, USA
e-mail: p.whitford@northeastern.edu

J. N. Onuchic
Rice University, Houston, TX, USA
e-mail: jonuchic@rice.edu

© Springer Nature Switzerland AG 2022
K. B. Blagoev and H. Levine (eds.), *Physics of Molecular and Cellular Processes*,
Graduate Texts in Physics, https://doi.org/10.1007/978-3-030-98606-3_2

Fig. 2.1 Protein folding
Proteins are linear
polypeptides that are
composed of distinct
sequences of amino acids.
When initially translated, a
protein may adopt an
extended, unfolded ensemble
of configurations (left). From
this unfolded ensemble, most
proteins will rapidly
(sub-second timescales) fold
to well-defined low-energy
"native" configurations
(right)

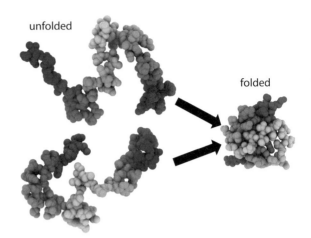

cal function, molecules may associate/dissociate, undergo conformational rearrangements, transiently adopt partially disordered states, as well as utilize energy that is
released by chemical reactions. While each serves a distinct biological purpose, this
range of dynamics is governed by a common energy landscape, which has evolved
to respond to changes in cellular conditions.

In this chapter, we will describe principles that have helped shape the energy
landscapes associated with biomolecular dynamics. We will begin with a discussion
of energy landscape theory, as developed for the study of protein folding. We then
provide a variety of examples where folding-inspired concepts have been extended to
investigate functional dynamics. As will become evident, an emerging theme is that
biomolecules frequently exploit molecular disorder to regulate biological function.

2.2 The Landscapes of Protein Folding

To provide a context for describing how folding concepts and tools have been
extended to study functional dynamics, we begin with an overview of the general
character of folding landscapes and modern theoretical models that are available for
their study.

2.2.1 Principle of Minimal Frustration

Energy landscape theory [1–4] provides a general framework that can allow for qualitative arguments and ideas to be transformed into quantitative models. A striking
example involved the quantitative inspection of the so-called "Levinthal's Paradox".
That is, it appeared paradoxical that a protein would be able to search the immense

unfolded configuration space and rapidly find the lowest energy conformation. However, this type of search problem could only manifest if the energy landscape was relatively random and/or flat. When traversing a rough landscape, the protein would have to undergo a nearly-unguided search, where the native configuration would be encountered by chance. A crude estimate of the number of possible disordered states reveals that such a search would be prohibitively slow. For example, if each residue in a 100-residue protein chain could sample two isoenergetic conformations, there would be $O(2^{100})$ accessible disordered states. Since local reconfiguration of a chain may occur on the timescale of picoseconds, then a single folding event should require approximately $2^{100} * 1ps \approx 10^{12}$ years. This simple argument makes it clear that the underlying landscape must not be rough.

To establish a quantitative relationship between landscapes and timescales, Bryngelson and Wolynes used a random energy model to study the balance between energetic roughness (ΔE) and stability of the folded ensemble (δE; Fig. 2.2). They showed that, in order for a protein to fold on physiological timescales, the stabilizing energy gap must be large, relative to the scale of the energetic roughness (i.e., $\Delta E/\delta E$ is small). Adopting the terminology of glass-forming liquids, δE is analogous to frustration [3, 5–9]. Thus, it was postulated that nature has selected for protein sequences that minimize frustration (i.e., the Principle of Minimal Frustration), where the overall energy landscape is funnel-like [10]. According to this description, each configuration is locally connected to states that have slightly more/less native content. In addition, the stabilizing energy is highly correlated with the degree of native content. The system may then circumvent Levinthal's paradox, and the dynamics can be described as a diffusive walk on an energetically "smooth" landscape.

The Principle of Minimal Frustration has been largely validated through comparison with experimental measures of stability and kinetics. A typical measure of stability is the folding temperature T_f. In order to define T_f, one should note that the folding of many proteins may be described as a pseudo-first-order[1] phase transition [3, 11], where there is an abrupt transition between the disordered (unfolded) ensemble and the ordered (folded) ensemble. At low temperatures, proteins will predominantly adopt folded conformations, whereas the unfolded ensembles are sampled at higher temperatures. T_f is then defined as the temperature at which the system is equally likely to be folded or unfolded. An alternate definition of T_f would be the temperature at which the specific heat reaches a maximum value. Since folding is a pseudo-phase transition, the sharp increase in energy associated with unfolding will lead to a spike in the specific heat. With regard to energetic roughness, one may define a temperature at which protein dynamics will become extremely slow, i.e., the "glass"-transition temperature T_g. Using a random energy model [1], one may relate the T_f and T_g. First, the free energy as a function of a reaction coordinate may be expressed as

$$F(\rho) = \bar{E}(\rho) - \frac{\Delta E^2(\rho)}{2k_B T} - T S_0(\rho), \tag{2.1}$$

---

[1] Technically, folding is not a pure phase transition, due to the finite size of a single protein.

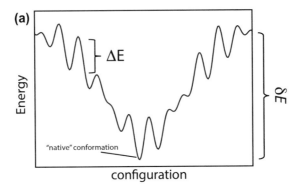

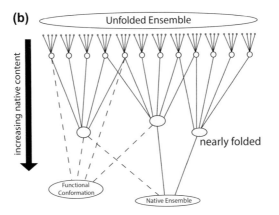

Fig. 2.2 Energy landscapes of biomolecular folding a In order for a biomolecule to rapidly find the native configuration, the global character of the energy landscape must be funnel-like. In these minimally frustrated landscapes, the energy gap between the unfolded and folded ensemble δE is large, relative to the scale of the energetic traps ΔE. **b** One may describe folding as the process in which a molecule initially can access large numbers of unfolded configurations. As the protein folds, there is an increase in native content, which is accompanied by a gradual reduction in the number of accessible states. Within these funneled landscapes, relatively low-energy conformations that are required for function are often structurally distinct from the native conformation. There are many ways in which a molecule may reach these functional configurations. In some cases, some degree of unfolding (dashed lines) is required

where $\bar{E}$, S_0, and ΔE are the average energy, configurational entropy, and scale of roughness for a given value of the coordinate ρ. Assuming the roughness is uniform across the landscape, the folding temperature is then given by

$$F(\rho_{\text{unfolded}}) - F(\rho_{\text{folded}}) = 0 = \delta\bar{E} - \frac{\Delta E^2}{2k_B T_f} - T_f \delta S_0. \tag{2.2}$$

In addition, the random energy model yields the following expression for the glass temperature:

$$T_g = \frac{\Delta E}{\sqrt{2\delta S_0}}.$$

(2.3)

Together, (2.2) and (2.3) allow one to estimate the ratio $\frac{T_f}{T_g}$. Based on early studies using lattice models, minimally frustrated protein sequences were predicted to have a ratio of approximately 1.6 [12]. Similarity of this value to experimental estimates [13] further bolsters the idea that protein folding landscapes are generally funnel-like.

2.2.2 Landscape-Inspired Models for the Study of Folding

The funnel-like character of protein folding energy landscapes suggests that some broad approximations may be applied when studying protein dynamics. Specifically, the Principle of Minimal Frustration implies that interactions formed in the native configuration are effectively more stable than non-native interactions. That is, native interactions ensure a large energy gap δE, while the weakness of non-native interactions ensures that the roughness ΔE is small. One may take this notion to a logical extreme, where there is no roughness, or the landscape is "unfrustrated". As described below, while this represents a limiting scenario, the study of unfrustrated landscapes has provided a range of insights into the dynamics of folding.

One class of models that describes the landscape as unfrustrated are structure-based models [14–16]. The defining feature of these models is that the energetics is defined based on knowledge of an experimentally obtained structure. In this representation, every bond, angle, and dihedral is assigned a potential energy, where the minimum of each interaction corresponds to the known configuration. In addition, interactions between any atoms/residues that are near one another in the folded conformation are defined as stabilizing, while all other interactions are repulsive.

Even though pure structure-based models lack roughness arising from non-native interactions, the landscape can be severely limited by molecular structure. For example, repulsive excluded-volume interactions impose limitations on the accessible phase space, which will lead to configurational entropy contributions that depend on how close the protein is to the native conformation. Non-trivial changes in configurational entropy are then determined, in part, by the molecular structure of the chain. In addition to non-linear/non-monotonic changes in configurational entropy, the connectivity of interactions in the native configuration also has the potential to introduce energetic traps. Due to steric interactions, it may not be possible to form contacts in an arbitrary order. Rather, it is possible that premature folding of a particular element could be associated with formation of a stabilizing interaction, though this early-forming component may need to break (i.e., "backtrack" [17, 18]) in order to continue toward the fully folded configuration. Thus, the system may encounter sterically induced energetic traps, even in the absence of stabilizing non-native inter-actions.

A powerful method for probing the energy landscape of folding has been ϕ-value analysis [19]. The objective of ϕ-value analysis is to infer the energetic and structural composition of the transition-state ensemble associated with protein folding. This inference is based on measured changes in folding rates and stability upon introduction of a point mutation. To calculate the ϕ value, one describes folding as a two-state transition between the unfolded (U) and folded (F) ensembles:

$$U \underset{k_u}{\overset{k_f}{\rightleftharpoons}} F \tag{2.4}$$

where k_f and k_u are the folding and unfolding rates. The ϕ value is then defined as follows:

$$\phi = \frac{\Delta \Delta F_{\text{TSE-U}}}{\Delta \Delta F_{\text{F-U}}} = \frac{\Delta F_{\text{TSE}} - \Delta F_{\text{U}}}{\Delta F_{\text{F}} - \Delta F_{\text{U}}}. \tag{2.5}$$

ΔF_{TSE}, ΔF_{U}, and ΔF_{F} are the changes in free energy of the transition-state ensemble (TSE), the unfolded ensemble, and folded ensemble (See Fig. 2.3) upon mutation. ϕ is equal to 1 if a residue contributes equally to the free energy of the TSE and the folded state. If the mutation does not impact the free-energy barrier, then $\phi = 0$. Since the Principle of Minimal Frustration indicates that native interactions should dominate folding energetics, the degree of native-structure formation should correlate with energy. Accordingly, if a residue is associated with a ϕ value of 1, then it is appropriate to infer that the majority of the native content around that residue is already formed by the time the system has reached the top of the rate-limiting free-energy barrier. From an experimental standpoint, one may measure the relative changes in stability through calorimetric measurements, while changes in barrier height may be inferred from changes in kinetics. Recent discussion also suggests single-molecule techniques may be used to measure ϕ values [20]. Specifically, if the rates of interconversion and relative populations of the endpoints (folded and unfolded) can be measured, then within the two-state description, the numerator and denominator in (2.5) may be calculated according to

$$\Delta F_{\text{TSE}} - \Delta F_{\text{U}} = -k_B T \ln\left(\frac{k_f^{\text{mut}}}{k_f^{\text{wt}}}\right), \tag{2.6}$$

and

$$\Delta F_{\text{F}} - \Delta F_{\text{U}} = -k_B T \ln\left(\frac{P_{\text{F}}^{\text{mut}}}{P_{\text{U}}^{\text{mut}}} \Big/ \frac{P_{\text{F}}^{\text{wt}}}{P_{\text{U}}^{\text{wt}}}\right), \tag{2.7}$$

where the superscripts "wt" and "mut" indicate quantities corresponding to the wild-type and mutant sequence.

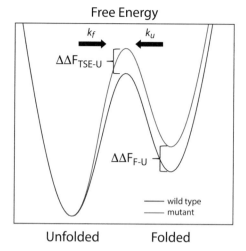

Fig. 2.3 Quantifying landscapes One widely used technique for characterizing protein folding is to calculate/measure ϕ values. The general idea is to introduce a point mutation and then determine the extent to which the free-energy barrier and the free energy of the folded ensemble are altered. Specifically, the ϕ value is defined as $\frac{\Delta\Delta F_{\text{TSE}-\text{U}}}{\Delta\Delta F_{\text{F}-\text{U}}} = \frac{\Delta F_{\text{TSE}}-\Delta F_{\text{U}}}{\Delta F_{\text{F}}-\Delta F_{\text{U}}}$. When a residue has a ϕ value of 1, it is appropriate to infer that all of the native content local to that residue is formed prior to reaching the top of the free-energy barrier

From a theoretical standpoint, ϕ values may also be obtained through free-energy perturbation techniques [21]. Rather than simulating the mutant system, one may calculate the change in free-energy upon exclusion of stabilizing interactions with a residue [14]. The change in free energy of each state is then calculated according to

$$\Delta F_n = -k_B T \langle \exp\left(-\Delta U / k_B T\right)\rangle_{0,n}, \tag{2.8}$$

where ΔU is the change in potential energy upon removal of all stabilizing interactions with a residue. The subscript 0,n indicates that the average is calculated over the unperturbed ensemble n (i.e., TSE, U, or F).

Through comparison of experimental folding dynamics and predictions from simplified models, there has been growing support for the funneled/unfrustrated landscape paradigm. While there are aspects of folding that are not well described by unfrustrated models, the points of agreement indicate general properties of proteins. The utility of these models has largely resulted from the fact that the accessibility of native contacts and the constraints imposed by excluded volume are dominant contributors to folding mechanisms. The study of circularly permutated protein sequences provides an excellent example of the pronounced effect that contact connectivity can have on dynamics. In a circularly permuted protein sequence, the tails of the protein are ligated, and a chemical bond is introduced at a different point along the sequence [18, 22]. This can leave the protein with nearly the same global configuration, though the connectivity of interactions is dramatically altered (Fig. 2.4).

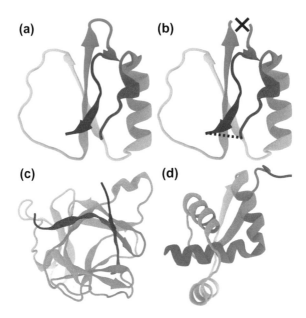

Fig. 2.4 Complexity of protein folds a Native configuration of the protein CI2. **b** In circular permutant studies, the tails are ligated (dashed line) and a peptide bond is broken elsewhere in the chain (X). **c** Complex folds such at the β-trefoil (IL-1β shown) can allow for premature folding of particular regions, which leads to partial unfolding (i.e., backtracking) during the folding process. **d** The most complicated folding motif is the knotted protein. If one were to pull the terminal tails of a knotted protein, the chain would not fully extend. Rather, a tight knotted structure would form

For example, contacts near the tails become short range (in sequence), which alters the entropic changes associated with folding this region. For the case of circular permutants, unfrustrated models were found to predict structural characteristics of the TSE that are consistent with experimental measurements [18, 22]. Since the permuted model only differed in connectivity, the consistency across systems may only be rationalized in terms of excluded-volume contributions. Another example of how structural complexity can guide dynamics is the cytokine IL-1β, which forms a complex beta-trefoil native structure (Fig. 2.4c). For this system, the application of simplified models has shown that early folding of the trefoil 3 region can help collapse the protein [17, 23]. However, in order to continue folding, this region must unfold (i.e., backtrack) while other regions of the protein simultaneously fold. As a consequence of this en-route unfolding, the kinetics of folding is extremely slow. Using simplified models, it was later shown that the rate of folding could be significantly increased by removing a loop that is essential for receptor binding [24]. Thus, it appears that nature has selected a protein sequence for which folding is sub-optimal, in order to preserve the functional properties of the protein. As a final example, an emerging topic of interest is the dynamics of protein folding in "knotted" (Fig. 2.4d) proteins [25]. Initially, it was unclear whether proteins with knotted topologies would be able to fold reversibly without the assistance of chaperones. However, consistent with simulations using structure-based models, experiments showed that reversible folding is accessible. Further, these simplified models predicted approximately a one order-of-magnitude change in folding rates when a knot was introduced for the protein 1ouf [26]. The consistent effect of introducing a knot further supports the general observation that folding rates are predominantly determined by fold complexity, rather than precise energetic details.

2.2.3 All-Atom Explicit-Solvent Models

For any physical phenomenon, a range of models may be applied, where each is defined to address distinct questions. In the above discussion, we described how simplified models, which were motivated by the Principle of Minimal Frustration, have been used to understand how a protein's native structure can influence folding dynamics. While those models are intended to provide approximate descriptions to the underlying landscape, one may also study folding through the use of more highly detailed representations, such as semi-empirical all-atom explicit-solvent models [27, 28]. In contrast to structure-based models, semi-empirical models assign partial charges to each atom (to approximate the local electron density), and every atom is assigned non-specific interactions with the remaining atoms in the system. The interactions are "non-specific," in the sense that the energetic parameters are assigned based on atom/residue type, without consideration of a known configuration. When adopting this approach, the assumption is that one may calibrate a model that is sufficiently general in functional form, such that any polypeptide may be simulated without possessing knowledge of the native (lowest energy) configuration. If the sequence corresponds to that of a stable protein, then one would expect that any reasonably accurate model will lead to spontaneous folding.

There have been two persistent challenges when trying to study protein folding with all-atom explicit-solvent simulations. First, by including a high level of energetic and structural detail, simulations with these models can be very computationally demanding. For example, when using a high-end compute node that is accelerated with a GPGPU processor, one is often limited to 10-100 nanoseconds per day for a typical single-domain protein. Since fast-folding proteins fold on timescales of tens of microseconds [29], obtaining a single folding event can require months, or years, of computing. This high requirement is exacerbated by the fact that quantifying statistical characteristics requires many folding events to be simulated. A more fundamental challenge when using these models is that the force fields are typically refined through comparison of experimental measures for smaller model systems (protein fragments). Thus, it is not known, a priori, whether the selected benchmarks will yield parameters that are sufficient to accurately simulate folding.

While accuracy and computational limitations have historically limited the study of folding with these models, there has been dramatic progress in recent years. With regard to computational demand, there is now a line of customized supercomputers [30] that were designed specifically for simulations with these models. By developing hardware algorithms and advanced I/O strategies, these machines can generate microseconds of simulated time per day, for smaller proteins. Unfortunately, since this represents a scaling performance of approximately 1 atom per compute core, it is not expected that this meteoric increase in performance will continue in the coming years. Nonetheless, this notable technical advance has allowed for millisecond-scale simulations of small protein folding with highly detailed models [31]. This scale of computational speed can now enable more direct comparison of folding simulations and experimental measurements [32], which may prove to be superior to comparisons based on model compounds.

Since it is now possible to simulate folding of fast-folding proteins using explicit-solvent models, it is necessary to consider what insights this may provide into protein energetics. In experiments, one typically introduces mutations to identify which residues most strongly influence the kinetics (e.g., ϕ values). Experiments are then repeated under different conditions (changes in pH, etc.) in order to determine which types of interactions are dominant. While highly detailed models can be used to simulate folding, it is not tractable to completely repeat folding simulations under varying conditions, and free-energy perturbation techniques have technical limitations when applied to explicit-solvent models. It is also not trivial to represent arbitrary conditions, or properly introduce non-aqueous solvent molecules, which can limit direct comparisons with experiments.

Despite the challenges when interpreting results with explicit-solvent models, some general trends have been demonstrated. A major observation has been that naturally occurring fast-folding proteins exhibit folding mechanisms [31] that are remarkably similar to predictions from simplified models. To specifically ask whether non-native interactions tend to contribute systematically to folding, the conditional probability of folding was calculated, given that a non-native interaction was formed [33]. That is, if a non-native contact is formed, what is the probability that the protein will continue toward the folded ensemble prior to reaching the unfolded ensemble? For naturally occurring proteins, these models indicate that formation of a non-native contact is not predictive of a folding event. The lack of a dependence on non-native contacts is in accord with the construction of simpler structure-based models, where the landscape is explicitly defined to be unfrustrated and funnel-like.

2.3 Models for Studying Biomolecular Functional Dynamics

After a protein folds, it may execute any number of cellular functions, which can include catalyzing chemical reactions, coordinating the positions of cellular components, or facilitating biomolecular recognition, transmembrane transport, degradation, and recycling. To fulfill these many roles, biomolecules must be adaptable entities that can undergo small-scale (few residue) fluctuations, as well as large-scale collective rearrangements. This range of mechanistic phenomena highlights the diverse and dynamic character of proteins, which is determined by the underlying energy landscape.

Biomolecules often are able to interconvert between putative "native" conformations and alternative (typically higher energy, Fig. 2.2b) states that are competent for function (i.e., "on" state in a biochemical pathway). Since it is necessary to fold, as well as function, conformational dynamics must be encoded within a generally funnel-like landscape. In this context, the native and functional conformations may be connected in configuration space in a variety of ways. For example, simple deformations may result in apparent rigid-body-like morphing from one structure to the other, or a biomolecule may partially (or fully) unfold in order to reach the functional conformation (dashed lines). The latter case would be in contrast to macroscopic machines, where rigid components work in lock-step with one another. Instead, since biomolecules are marginally stable, they can utilize a broader range of mechanistic possibilities, such as partial unfolding.

From a physical perspective, partial disordering may appear quite intuitive. However, the prediction of this type of behavior [34] represented a turning point for the field, where it had to move beyond rigid-rod-like descriptions of the dynamics [35]. In support of this prediction, the last 15 years have provided a long list of examples where partial/complete disordering of biomolecules is directly linked to functional dynamics. Below, we will describe some of the theoretical techniques that have been instrumental in elucidating the influence of disorder during biomolecular functional dynamics.

2.3.1 Normal Mode Analysis

Since detailed descriptions of normal mode analysis may be found in any introductory mechanics text, as well as in Chap. 4 of this volume, we will only briefly summarize the approach. Normal mode analysis (NMA) provides a low-order description of structural fluctuations about a potential energy minimum. For small oscillations, one may expand the Lagrangian to second order and apply a linear transformation of the coordinates from the original coordinate system (e.g., Cartesian coordinates or internal dihedral coordinates) to the "normal coordinate" system. In this transformed coordinate system, small displacements may be described in terms of uncoupled collective oscillations. However, since large-scale biomolecular dynamics occur in an overdamped environment, the term "low frequency" is used to describe the curvature of the landscape, rather than kinetics. Specifically, low-frequency modes will correspond to directions of motion that are associated with slowly increasing potential energy. Accordingly, these modes represent "soft" deformations of the system, along which fluctuations are energetically accessible.

When applying NMA, there is a number of factors that one should be aware. First, NMA is not specific to a single potential energy function (i.e., model). In principle, one may calculate the Hessian matrix using an all-atom explicit-solvent force field, coarse-grained models, or all-atom models with varying levels of energetic detail.

While there can be advantages to selecting different models, studies have found that the lowest frequency (largest amplitude) modes are generally insensitive to the precise energetic details [36]. This robustness may be understood in terms of the separation of energy scales that are associated with bonded, non-bonded, and excluded-volume interactions. Since bonded and excluded-volume interactions are strong, any mode that would be associated with their deformation would necessarily correspond to high-energy motions. In addition, according to the equipartition theorem, each mode should have the same average energy ($\frac{1}{2}k_B T$). Accordingly, high-frequency motions, which are associated with larger effective spring constants, will correspond to smaller amplitude motions. In contrast, attractive non-bonded terms (e.g., vdW) are generally weak; however, the density of interactions will vary significantly within a molecule. Further, since all non-bonded interactions are generally on a comparable energetic scale, the energy required to deform a molecule along a specific mode is expected to be correlated with the number of non-bonded interactions that are perturbed. As the density of non-bonded interactions is governed by the structure of the molecule, even crude energetic models (e.g., uniform non-bonded interaction strength) are likely to distinguish between stiff and soft regions.

In addition to using normal modes to describe fluctuations about a minimum, normal coordinates may be used to predict possible pathways associated with large-scale deformations [37, 38]. However, when attempting to study large-scale motions, it is common that the modes are iteratively evaluated [34, 39], to ensure that the basis set preserves the stereochemistry of the molecule. Even though the application of NMA has become less frequently in recent years, it was central to the development of early ideas regarding the energetics of functional dynamics.

2.3.2 Multi-basin Effective Potential Energy Models

While normal mode analysis describes the energetics up to second order, more complete descriptions will account for anharmonic interactions that can spontaneously break/reform during biomolecular motions. One strategy that has had utility in exploring such events are so-called "multi-basin" structure-based models [40]. These models are built upon the Principle of Minimal Frustration. However, in contrast to folding models, multi-basin models encode potential energy minima that are defined based on multiple known conformations. The idea behind these models is that, since both conformations are found (experimentally) to be stable, the interactions unique to each are effectively stable. That is, the potentials of mean force associated with interatomic interactions have minima that correspond to the distances (or angles) that are found in the endpoint configurations.

To clarify the construction of a multi-basin model, it is necessary to first describe the functional form of a single-basin structure-based model. The general functional form of a "SMOG" type of structure-based model [41, 42] may be expressed as follows:

$$U(\mathbf{x}, \mathbf{x}^0) = \sum_{ij \in \text{bonds}} \frac{\epsilon_b}{2}(r_{ij} - r_{ij}^0)^2 + \sum_{ijk \in \text{angles}} \frac{\epsilon_\theta}{2}(\theta - \theta_{ijk}^0)^2$$

$$+ \sum_{ijkl \in \text{dihedrals}} \epsilon_D F_D(\phi_{ijkl} - \phi_{ijkl}^0)$$

$$+ \sum_{ij \in \text{contacts}} \epsilon_C \left[\left(\frac{r_{ij}^0}{r_{ij}}\right)^{12} - 2\left(\frac{r_{ij}^0}{r_{ij}}\right)^6 \right] + \sum_{ij \notin \text{contacts}} \epsilon_{NC} \left(\frac{\sigma_{NC}}{r_{ij}}\right)^{12} \quad (2.9)$$

where

$$F_D(\delta\phi) = [1 - \cos(\delta\phi)] + \frac{1}{2}[1 - \cos(3\delta\phi)].$$

While this generic functional form is common to most classical mechanical models, the distinguishing feature of a structure-based model is that the lowest energy value for each bond, angle, dihedral, and non-local interactions is defined by a pre-assigned structure. That is, r_{ij}^0, θ_{ijk}^0, and ϕ_{ijkl}^0 are given the values found in the native configuration $\mathbf{x}^0$. As discussed above, this explicitly defines an unfrustrated energy landscape where the native structure is the lowest energy conformation.

When describing conformational transitions, one may combine structure-based models that are defined based on multiple endpoints. First, one defines single-basin potentials for a number of structures: U_i. The most straightforward way of merging these energy functions would be to define the combined energy as the minimum of the set of potentials:

$$U_{\text{comb}}(\mathbf{x}) = \min(\{U_i(\mathbf{x})\}). \quad (2.10)$$

In this case, the system would be described as transitioning between the various energy surfaces, where the adopted surface is always the one that is lowest in energy. While this is a reasonable initial representation, which may be utilized in Monte Carlo simulations, this will introduce discontinuities in force that will lead to artificial noise/heating when using molecular dynamics protocols. To circumvent the discontinuous nature of the min function, one may alternatively combine potential energy functions through use of a Boltzmann-style weighting [43]:

$$e^{-U_{\text{comb}}/k_B T_{\text{mixing}}} = \sum_i e^{-(U_i + \Delta_i)/k_B T_{\text{mixing}}}, \quad (2.11)$$

where the relative difference in the input potential energy minima can be shifted by Δ_i. In this representation, one may adjust how sharp the transition is between potentials by adjusting the value of T_{mixing}. In the limit $T_{\text{mixing}} = 0$, for a given configuration $\mathbf{x}$, U_{comb} will simply adopt the value of the smallest term $U_i + \Delta_i$ (2.10). In contrast, when T_{mixing} is larger, the potential energy will smoothly transition between the surfaces (see Fig. 2.5). For very large value of T_{mixing}, U_{comb} is the mean of the U_i values.

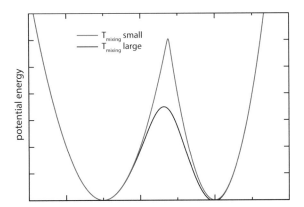

Fig. 2.5 Multi-minima landscapes To study large-scale conformational transitions, one often aims to construct multi-basin models, where multiple known structures are defined as minima. This schematic illustrates how one may use a Boltzmann-style method (2.11) to combine two harmonic wells to form a single, continuous landscape. When using the Boltzmann approach, a critical choice is the variable T_{mixing}. If T_{mixing} is small, then the landscape abruptly transitions between the two wells (red). When T_{mixing} is large, the barrier may be artificially reduced (black)

An alternate approach to define a multi-basin model is to combine the single-basin potentials through a superposition strategy [44, 45]. For this, a smooth potential energy function, U_{mixed}, is defined as the eigenvalue of the characteristic equation:

$$\begin{pmatrix} U_1 & \Delta \\ \Delta & U_2 + C_2 \end{pmatrix} \begin{pmatrix} c_1 \\ c_2 \end{pmatrix} = U_{\mathrm{mixed}}^{S} \begin{pmatrix} c_1 \\ c_2 \end{pmatrix}, \tag{2.12}$$

where Δ is a constant that is introduced to couple the potentials U_1 and U_2, and $(c1, c2)$ is the corresponding eigenvector. In this approach, similar to the Boltzmann weighting strategy, the barrier height may be modulated by adjusting the value of Δ. In addition, similar to the role of Δ_i in the Boltzmann approach, the constant C_2 is introduced to shift the relative weights of the potentials U_1 and U_2. The lowest energy solution is then obtained by solving the secular equation:

$$U_{mixed} = \frac{U_1 + U_2 + C_2}{2} - \sqrt{\left(\frac{U_1 - U_2 - C_2}{2}\right)^2 + \Delta^2}. \tag{2.13}$$

Similar to the Boltzmann-based approach, the superposition strategy also ensures a smooth transition between the potentials, making it suitable for performing MD simulations of conformational rearrangements.

Another broad strategy for constructing multi-basin models is to "piece together" various energetic terms from the input potentials U_i [46]. Rather than globally couple the two energy surfaces (e.g., Boltzmann or Superposition approaches), one may exploit the fact that many interactions will be common, or similar, between the two energy functions. For example, in the absence of a chemical reaction, the covalent geometry (i.e., bond lengths and angles) is preserved throughout a conformational rearrangement. Further, since conformational rearrangements often involve structurally conserved elements (e.g., domains), many of the contacts and dihedral angles will also be nearly identical in both endpoint configurations. However, there will necessarily be non-bonded contacts that are unique to each configuration. By constructing a model in which contacts from both endpoints are simultaneously defined, one may describe conformational rearrangements as resulting from energetic competition between these sets of native-like interactions.

In all of the above methods, a multi-basin landscape is constructed by merging multiple single-basin potential energy functions. Even though there is a predefined number of configurations used to define the potential energy, one does not know *a priori* how many free-energy (and potential energy) minima will be present in the combined landscape. For example, landscapes defined based on knowledge of two competing structures frequently lead to the system adopting configurations in which contacts from both endpoints are partially formed [47, 48]. While this phenomenon is more likely in models that combine interactions on a per-contact basis, it is also possible (though less likely) when using the Boltzmann and Superposition strategies. Accordingly, in the current discussion, we will refer to models that are constructed from two structures as "multi-basin," rather than "dual-basin".

2.3.3   Simulations with Semi-empirical All-Atom Models

While the above methods explicitly build upon the Principle of Minimal Frustration, semi-empirical models are often employed to study conformational rearrangements [49, 50]. As described earlier, these models do not define each interaction based on knowledge of a specific configuration (e.g., 2.9). Instead, non-specific parameters are defined for each atom. That is, the functional form of the potentials is the same (e.g., harmonic bonds, cosine-based dihedrals), however the positions of the minima $(r_{ij}^0, \theta_{ijk}^0, \phi_{ijkl}^0)$ and their weights are calibrated through comparison with experiments and quantum mechanical calculations for model compounds. If the potential is sufficiently well calibrated, one may anticipate that this representation could be used to perform a simulation of any system (given sufficient computational resources are available). However, similar to studies of folding, it is not typically possible to simulate spontaneous large-scale rearrangements with these models. Rather, it is generally necessary to employ targeting techniques [51] or enhanced sampling protocols [52]. While targeting methods can suggest mechanistic features that may serve as useful guides when analyzing dynamics, these far-from-equilibrium descriptions are limited. Fortunately, recent advances in computing hardware and software development [30] are allowing for larger conformational changes to be simulated with explicit-solvent models [50]. With these advances, the field is now beginning to more clearly compare the dynamics predicted by semi-empirical and energy-landscape-inspired models [53].

2.4 How Disorder Guides Biomolecular Function

While traditional biochemical thought would suggest that conformational rearrangements are best described in terms of well-defined transitions between distinct free-energy minima, the marginal stability of biomolecules also allows for transient disorder to be exploited as a mechanism for controlling dynamics. Here, we will highlight several well-documented examples for how disorder is central to biomolecular processes.

2.4.1 Partial Unfolding During Function: Cracking

Normal mode-based analysis first predicted that Adenylate Kinase [34, 54, 55] may partially unfold and refold during large-scale domain rearrangements. In particular, in the harmonic regime, it was predicted Adk may accumulate internal "strain" energy that is non-uniformly distributed throughout the protein. In fact, the total level of strain energy was found to be sufficient to unfold a typical protein that is under physiological conditions (>20 kCal/Mol). In addition to suggesting that there is sufficient energy to unfold the entire protein, the predicted free-energy barriers would correlate with non-biologically relevant timescales for domain rearrangements (i.e., hours versus milliseconds). Together, these observations led to the suggestion that highly strained regions will transiently unfold as the domains transition, and then refold upon reaching the alternate endpoint (i.e., "crack") [34]. Breaking a number of local interactions would allow for a higher entropy partially unfolded state to be adopted, which would increase the rate by decreasing the free-energy barrier [56].

Subsequent studies have applied more complex modeling approaches to isolate the precise character of cracking in Adk. In our studies, we developed a model in which a union of native contacts was introduced [47]. In simulations with this model, we found evidence of cracking in the same residues predicted by normal mode analysis. However, another group applied a Boltzmann-mixed potential, which predicted a lower degree of cracking [57]. One explanation for these differential predictions is that the Boltzmann-mixing approach ensures that the landscape is "dual-funnel" in character. In contrast, introducing a union of contacts can allow for subsets of "open" and "closed" contacts to be formed simultaneously. It should be noted that these simulations employed coarse-grained models, where each residue was represented as a single bead. Accordingly, cracking measures were based on changes in backbone dynamics. Complementary to these energy-landscape-inspired approaches, explicit-solvent simulations also revealed signatures of side chain disordering [49]. Instantaneous normal mode (INM) calculations [58] found that immediately before domain rearrangement, where there were anomalous changes in the INMs (i.e., dissimilar from either of the end-states). While indirect, this observation suggests that low-energy (i.e., rigid-body-like) rearrangements cannot fully account for the functional dynamics of Adk. Finally, various experiments have also uncovered signatures of cracking in Adenylate Kinase [59, 60]. Together, these studies are revealing a general picture in which cracking may be described as a balancing act between order and disorder in the protein backbone and sidechains.

Since there are many proteins that undergo large-scale domain rearrangements, it is perhaps unsurprising that cracking has been implicated in other systems. For example, kinesin molecules are responsible for "walking" on microtubules [61], where each "step" involves ATP binding and an ≈ 80 Å displacement of the domains [62]. By employing a time-dependent Hamiltonian, it was shown how the "power stroke" may be facilitated by cracking [63, 64]. In terms of a physical picture, these studies assumed that ATP binding shifts the potential energy surface of the leading head, which results in the displacement of the trailing domain. Since the trailing head may also bind the microtubule, there is an internal energetic conflict that can induce disorder within the linker region. If the inherent flexibility of the linker was too low, it would not be possible to transfer energy between the domains. In contrast, if the domain linker was too rigid, then large-scale displacements would not be accessible, which would hinder the walking process. Cracking was also predicted by a dual-basin statistical-mechanical model applied to the C-terminal domain of Calmodulin [65]. Similarly, a variational model introduced by Tripathi and Portman predicted non-monotonic changes in flexibility of the N-terminal domain of Calmodulin [66]. A coarse-grained model of Protein Kinase A also implicated cracking during functional rearrangements [67]. Finally, in a survey of multi-domain proteins, Okazaki et al. showed a pervasive role of cracking during domain rearrangements in proteins [45]. More recently, as computing resources have continued to improve, long-timescale simulations using explicit-solvent models are also beginning to reveal cracking-like behavior [50, 53].

2.4.2 Biomolecular Association: Fly-Casting

Cellular dynamics involves a balance between many kinetically competitive processes [68, 69]. Due to this highly non-equilibrium character of the cell, there has been evolutionary pressure to develop precisely tunable signaling processes. At the heart of gene expression is the ability of transcription factors (TFs) to bind DNA and signal the expression of the genetic message. If TF-DNA binding event was governed merely by diffusive random walks, association kinetics could only be modulated by introducing large changes in a protein's shape and size. However, producing a much larger protein in order to adjust binding kinetics would not be an energetically viable approach. Accordingly, there are likely more sophisticated principles that exploit molecular disorder in order to enable broad adaptability of protein binding dynamics.

To explain the rapid kinetics of protein-DNA binding, Shoemaker et al. proposed the "fly-casting" mechanism [70], where a protein may allow for a disordered arm-like extension to search for the target binding site. Based on free-energy functional calculations, it was demonstrated that the rate of association could be increased if this disordered region has a minimal affinity for the target. That is, even though the affinity of the disordered region would be low, the vast increase in available phase space would exponentially increase the rate of substrate capture. Molecular simulations [71, 72] using coarse-grained structure-based models later demonstrated, at a molecular level, how changes in flexibility may increase bimolecular association kinetics. In addition, bioinformatic analysis has indicated that the electrostatic composition of disordered regions may help guide fly-casting dynamics [73].

Later studies have elaborated on the influence of disorder during the process of protein-DNA searching and binding. The molecular flexibility that facilitates fly-casting to a specific site can also allow for sliding-like displacements along the DNA [74]. During this process, the protein may "scan" proximal regions of the DNA as it seeks to find the lowest energy site [75]. In addition to scanning, the protein may also rapidly dissociate and reassociate at a binding site that is far (in sequence), a process known as "hopping". One reason that this is a kinetically viable process is that the reassociation step can utilize fly-casting-like dynamics. In addition, it is possible for extended and disordered regions to search distant DNA sequences before the protein dissociates from DNA [76]. In this case, the ability of a disordered region to scan potential association sites can allow the protein to completely circumvent the need for free diffusion. Rather, a protein may transiently bridge distant DNA sequences in a manner that resembles a child in a playground, where each hand is sequentially displaced across monkey bars. Together, these examples illustrate how the degree of protein flexibility can govern the balance between disorder and energetic interactions during gene expression.

2.4.3 Molecular Machines: Entropically Guided Rearrangements

A final example of the influence of molecular disorder in biology may be found in the context of elongation by the ribosome. The ribosome is responsible for recruiting transfer RNA (tRNA) molecules in order to read the message that is encoded in messenger RNA (mRNA). To accomplish this, aminoacyl-tRNA (aa-tRNA) molecules form interactions with the three-residue mRNA "codon". Since each codon is composed of only four possible nucleic acids (A, U, C, or G), there are 64 possible codon messages. Accordingly, the differences between correct and incorrect codon-tRNA pairing only lead to small differences in energetics, which would lead to error rates of approximately 1:100. However, since the error rates in bacteria are $\sim 1 : 10^3 - 10^4$, additional proofreading mechanisms are required [77]. To explain the high-fidelity of translation, Hopfield proposed the "kinetic proofreading" mechanism [78]. According to this framework, the ribosome could amplify the level of fidelity if each tRNA molecule was initially loaded into a marginally stable intermediate, in a GTP-hydrolysis-dependent manner. EF-Tu could then deliver the tRNA in an energetically stained "A/T" configuration [79]. Upon release of EF-Tu, the tRNA can fully relax into the A site of the ribosome, a process known as accommodation. Consistent with accommodation being central to proofreading, modeling efforts have implicated a trade-off between fidelity and elongation rates [80]. If accommodation was too fast for all tRNA molecules, then large energetic differences would be required to distinguish correct and incorrect molecules. In contrast, if accommodation was too slow, then nearly all tRNA molecules would be rejected, including correct molecules.

The influence of accommodation rates on the fidelity of gene expression has inspired many investigations into accommodation kinetics [81], structural properties [82–84], and large-scale conformational rearrangements [51, 85–90]. One striking feature that has been revealed is the critical role of disorder during accommodation. In the first simulation that employed a single-basin structure-based model, it was noted that large changes in flexibility of the 3'-CCA end of the tRNA may facilitate a multi-step accommodation mechanism [86]. In subsequent simulations, it was shown that, if aa-tRNA is released by EF-Tu, and EF-Tu was to not dissociate rapidly, then it may confine the mobility of the 3'-CCA end [88]. This confinement of the 3'-CCA tail could stabilize the fully bound conformation, analogous to the potential influence of molecular crowders on protein stability [91]. Similarly, simulations were employed where the flexibility of the L11 stalk was modified [89]. These calculations revealed that L11 may act in a manner that can amplify the confinement effect of EF-Tu, which raises the possibility of cooperative effects of EF-Tu and L11. Finally, during the process of EF-Tu dissociation, simulations with structure-based models predicted that domain 1 may enter a transient disordered state [90], where there is separation [52] of the domains.

While the first targeted simulations of accommodation revealed a need for 3'-CCA end flexibility during accommodation [51], simulations with simplified models demonstrated how many molecular factors may serve to modulate disorder in order to control accommodation dynamics. Interestingly, in addition to simulations, there have been many experimental signatures that illustrate the dynamic character of the 3'-CCA end. For example, the 3'-CCA tail is often unresolved in crystallographic structures [92], or the Debye-Waller factors are large [93]. Similarly, cryogenic electron microscopy methods often only provide a low-density for the 3'-CCA tail [94], which suggests it may rapidly sample a range of configurations. In addition, the many binding partners of the 3'-CCA tail [95–98] necessitate that there be a significant degree of flexibility in this region.

2.5 Concluding Remarks

As we look forward to the investigation of any number of emergent phenomena in nature, the energy landscape principles developed to study protein folding will prove invaluable. Here, we provided an overview of a few examples where concepts from protein folding have been adapted and extended to understand conformational processes that are related to biological function. In these efforts, one of the most pronounced and exciting themes to emerge is the critical role of biomolecular disorder. That is, rather than operating as sets of rigid rods and pivots, molecular assemblies can exploit a much broader range of conformational space, where some (or all) of the native content is transiently lost. By broadening our perspective on the motions accessible to function, there are now many new avenues for the investigation of the balance between order, disorder, and functional dynamics.

Acknowledgements This work was supported by the Center for Theoretical Biological Physics sponsored by the NSF (Grant PHY-2019745). PCW was supported by a National Science Foundation Grants MCB-1350312 and MCB-1915843. JNO was also supported by the NSF- CHE 1614101. JNO is a CPRIT Scholar in Cancer Research sponsored by the Cancer Prevention and Research Institute of Texas.

References

1. J.D. Bryngelson, P.G. Wolynes, J. Phys. Chem.-Us **93**(19), 6902 (1989)
2. J. Bryngelson, P. Wolynes, Biopolymers **30**(1–2), 177 (1990)
3. J.D. Bryngelson, J.N. Onuchic, N.D. Socci, P.G. Wolynes, Proteins **21**(3), 167 (1995)
4. H. Frauenfelder, S. Sligar, P. Wolynes, Science **254**(5038), 1598 (1991)
5. J. Onuchic, Z. Luthey-Schulten, P. Wolynes, Annu. Rev. Phys. Chem. **48**, 545 (1997)
6. J. Shea, J. Onuchic, C. Brooks, Proc. Nat. Acad. Sci. USA **96**(22), 12512 (1999)
7. J. Shea, J. Onuchic, C. Brooks, J. Chem. Phys. **113**(17), 7663 (2000)
8. S. Cho, Y. Levy, J. Onuchic, P. Wolynes, Phys. Biol. S **2**(2), 44 (2005)
9. D. Ferreiro, J. Hegler, E. Komives, P. Wolynes, Proc. Nat. Acad. Sci. USA **104**(50), 19819 (2007)
10. P. Leopold, M. Montal, J. Onuchic, Proc. Nat. Acad. Sci. USA **89**(18), 8721 (1992)
11. N. Socci, J. Onuchic, J. Chem. Phys. **103**(11), 4732 (1995)
12. J. Onuchic, P. Wolynes, Z. Luthey-Schulten, N. Socci, Proc. Natl. Acad. Sci. USA **92**(8), 3626 (1995)
13. H. Frauenfelder, N.A. Alberding, A. Ansari, D. Braunstein, B.R. Cowen, M.K. Hong, I.E.T. Iben, J.B. Johnson, S. Luck, J. Phys. Chem. **94**(3), 1024 (1990)
14. C. Clementi, P. Jennings, J. Onuchic, Proc. Nat. Acad. Sci. USA **97**(11), 5871 (2000)
15. J. Karanicolas, C.L. Brooks, Prot. Struct. Func. Bioinfo. **53**(3), 740 (2003)
16. P. Whitford, J. Noel, S. Gosavi, A. Schug, K. Sanbonmatsu, J. Onuchic, Prot. Struct. Func. Bioinfo. **75**(2), 430 (2009)
17. S. Gosavi, L. Chavez, P. Jennings, J. Onuchic, J. Mol. Biol. **357**(3), 986 (2006)
18. D. Capraro, M. Roy, J. Onuchic, P. Jennings, Proc. Nat. Acad. Sci. USA **105**(39), 14844 (2008)
19. A. Fersht, Curr. Opin. Struct. Biol. **5**, 79 (1995)
20. M. Levi, P.C. Whitford, J. Phys. Chem. B **123**, 2812 (2019)
21. B. Roux, Comput. Phys. Commun. **91**(1–3), 275 (1995)
22. C. Clementi, P. Jennings, J. Onuchic, J. Mol. Biol. **311**(4), 879 (2001)
23. L.L. Chavez, S. Gosavi, P.A. Jennings, J.N. Onuchic, Proc. Nat. Acad. Sci. USA **103**(27), 10254 (2006)
24. S. Gosavi, P. Whitford, P. Jennings, J. Onuchic, Proc. Nat. Acad. Sci. USA **105**(30), 10384 (2008)
25. N. King, A. Jacobitz, M. Sawaya, L. Goldschmidt, T. Yeates, Proc Nat. Acad. Sci. USA **107**(48), 20732 (2010)
26. J.I. Sulkowska, J.K. Noel, J.N. Onuchic, Proc. Nat. Acad. Sci. **109**(44), 17783 (2012)

27. B. Brooks, C. Brooks, A. Mackerell, L. Nilsson, R. Petrella, B. Roux, Y. Won, G. Archontis, C. Bartels, S. Boresch, A. Caflisch, L. Caves, Q. Cui, A. Dinner, M. Feig, S. Fischer, J. Gao, M. Hodoscek, W. Im, K. Kuczera, T. Lazaridis, J. Ma, V. Ovchinnikov, E. Paci, R. Pastor, C. Post, J. Pu, M. Schaefer, B. Tidor, R. Venable, H. Woodcock, X. Wu, W. Yang, D. York, M. Karplus, J. Comput. Chem. **30**, 1545 (2009)
28. D. Case, T.C. III, T. Darden, H. Gohlke, R. Luo, K.M. Jr., A. Onufriev, C. Simmerling, B. Wang, R. Woods, J. Comput. Chem. **26**(16), 1668 (2005)
29. W.A. Eaton, V. Munoz, S.J. Hagen, G.S. Jas, L.J. Lapidus, E.R. Henry, J. Hofrichter, Annu. Rev. Biophys. Biomol. Struct. **29**(1), 327 (2000)
30. D.E. Shaw, J.P. Grossman, J.A. Bank, B. Batson, J.A. Butts, J.C. Chao, M.M. Deneroff, R.O. Dror, A. Even, C.H. Fenton, A. Forte, J. Gagliardo, G. Gill, B. Greskamp, C.R. Ho, D.J. Ierardi, L. Iserovich, J.S. Kuskin, R.H. Larson, T. Layman, L.S. Lee, A.K. Lerer, C. Li, D. Killebrew, K.M. Mackenzie, S.Y.H. Mok, M.A. Moraes, R. Mueller, L.J. Nociolo, J.L. Peticolas, T. Quan, D. Ramot, J.K. Salmon, D.P. Scarpazza, U.B. Schafer, N. Siddique, C.W. Snyder, J. Spengler, P.T.P. Tang, M. Theobald, H. Toma, B. Towles, B. Vitale, S.C. Wang, C. Young, in *SC14: International Conference for High Performance Computing Networking, Storage and Analysis* (IEEE, 2014), pp. 41–53
31. K. Lindorff-Larsen, S. Piana, R. Dror, D. Shaw, Science **334**(6055), 517 (2011)
32. S. Piana, K. Lindorff-Larsen, D.E. Shaw, Biophys. J. L **100**(9), 47 (2011)
33. R.B. Best, G. Hummer, W.A. Eaton, Proc. Natl. Acad. Sci. USA **110**(44), 17874 (2013)
34. O. Miyashita, J. Onuchic, P. Wolynes, Proc. Nat. Acad. Sci. USA **100**(22), 12570 (2003)
35. J.A. McCammon, B.R. Gelin, M. Karplus, P.G. Wolynes, Nature **262**(5566), 325 (1976)
36. C. Chennubhotla, A.J. Rader, L.W. Yang, I. Bahar, Phys. Biol. S **2**(4), 173 (2005)
37. F. Tama, Y.H. Sanejouand, Protein Eng. **14**(1), 1 (2001)
38. P. Petrone, V. Pande, Biophys. J. (2006)
39. P. Whitford, S. Gosavi, J. Onuchic, J. Biol. Chem. **283**(4), 2042 (2008)
40. P.C. Whitford, K.Y. Sanbonmatsu, J.N. Onuchic, Rep. Prog. Phys. **75**, 076601 (2012)
41. P.C. Whitford, J.K. Noel, S. Gosavi, A. Schug, K.Y. Sanbonmatsu, J.N. Onuchic, Prot. Struct. Func. Bioinfo. **75**(2), 430 (2009)
42. J.K. Noel, M. Levi, M. Raghunathan, H. Lammert, R.L. Hayes, J.N. Onuchic, P.C. Whitford, PLoS Comput. Biol. **12**(3), e1004794 (2016)
43. R.B. Best, Y.G. Chen, G. Hummer, Structure (London, England: 1993) **13**(12), 1755 (2005)
44. K. Okazaki, S. Takada, Proc. Nat. Acad. Sci. USA **105**(32), 11182 (2008)
45. K. Okazaki, N. Koga, S. Takada, J.N. Onuchic, P.G. Wolynes, Proc. Nat. Acad. Sci. USA **103**(32), 11844 (2006)
46. M. Levi, J.K. Noel, P.C. Whitford, Methods (2019)
47. P. Whitford, O. Miyashita, Y. Levy, J. Onuchic, J. Mol. Biol. **366**(5), 1661 (2007)
48. K. Nguyen, P.C. Whitford, Nat. Commun. **7**, 10586 (2016)
49. J. Brokaw, J.W. Chu, Biophys. J. **99**(10), 3420 (2010)
50. Y. Shan, A. Arkhipov, E.T. Kim, A.C. Pan, D.E. Shaw, Proc. Natl. Acad. Sci. USA **110**(18), 7270 (2013)
51. K.Y. Sanbonmatsu, S. Joseph, C.S. Tung, Proc. Natl. Acad. Sci. USA **102**(44), 15854 (2005)
52. J. Lai, Z. Ghaemi, Z. Luthey-Schulten, Biochemistry **56**(45), 5972 (2017)
53. P.C. Whitford, Proc. Nat. Acad. Sci. USA **110**(18), 7114 (2013)
54. O. Miyashita, P. Wolynes, J. Onuchic, J. Phys. Chem. B **109**(5), 1959 (2005)
55. P.C. Whitford, S. Gosavi, J.N. Onuchic, J. Biol. Chem. **283**(4), 2042 (2008)
56. P.C. Whitford, J. Onuchic, P. Wolynes, HFSP J. **2**(2), 61 (2008)
57. M.D. Daily, G. Phillips, Q. Cui, J. Mol. Biol. **400**(3), 618 (2010)
58. C. Peng, L. Zhang, T. Head-Gordon, Biophys. J. **98**(10), 2356 (2010)
59. U. Olsson, M. Wolf-Watz, Nat. Commun. **1**(8), 111 (2010)
60. L. Rundqvist, J. Adén, T. Sparrman, M. Wallgren, U. Olsson, M. Wolf-Watz, Biochemistry **48**(9), 1911 (2009)
61. R. Vale, R. Milligan, Science **288**(5463), 88 (2000)
62. S. Block, Biophys. J. **92**(9), 2986 (2007)

63. C. Hyeon, J. Onuchic, Proc. Natl. Acad. Sci. USA **104**(7), 2175 (2007)
64. C. Hyeon, J.N. Onuchic, Proc. Natl. Acad. Sci. USA **104**(44), 17382 (2007)
65. K. Itoh, M. Sasai, J. Chem. Phys. **134**(12), 125102 (2011)
66. S. Tripathi, J.J. Portman, Proc. Nat. Acad. Sci. USA **106**(7), 2104 (2009)
67. C. Hyeon, P. Jennings, J. Adams, J. Onuchic, Proc. Natl. Acad. Sci. USA **106**(9), 3023 (2009)
68. A. Walczak, J. Onuchic, P. Wolynes, Proc. Nat. Acad. Sci. USA **102**(52), 18926 (2005)
69. D. Schultz, E.B. Jacob, J. Onuchic, P.G. Wolynes, Proc. Nat. Acad. Sci. USA **104**(45), 17582 (2007)
70. B.A. Shoemaker, J.J. Portman, P.G. Wolynes, Proc. Nat. Acad. Sci. USA **97**(16), 8868 (2000)
71. Y. Levy, S. Cho, J. Onuchic, P. Wolynes, J. Mol. Biol. **346**(4), 1121 (2005)
72. Y. Levy, J.N. Onuchic, P.G. Wolynes, J. Am. Chem. Soc. **129**(4), 738 (2007)
73. K. Chen, J. Eargle, K. Sarkar, M. Gruebele, Z. Luthey-Schulten, Biophys. J. **99**, 3930 (2010)
74. O. Givaty, Y. Levy, J. Mol. Biol. **385**, 1087 (2009)
75. D. Vuzman, Y. Levy, Proc. Nat. Acad. Sci. USA **107**(49), 21004 (2010)
76. D. Vuzman, Y. Levy, Mol. BioSys. **8**, 47 (2012)
77. R. Green, H.F. Noller, Annu. Rev. Biochem. **66**, 679 (1997)
78. J. Hopfield, Proc. Nat. Acad. Sci. USA **71**(10), 4135 (1974)
79. J. Frank, J. Sengupta, H. Gao, W. Li, M. Valle, A. Zavialov, M. Ehrenberg, FEBS Lett. **579**(4), 959 (2005)
80. M. Johansson, M. Lovmar, M. Ehrenberg, Curr. Opin. Microbio. **11**(2), 141 (2008)
81. I. Wohlgemuth, C. Pohl, J. Mittelstaet, A.L. Konevega, M.V. Rodnina, Phil. Trans. R. Soc. B: Biol. Sci **366**(1580), 2979 (2011)
82. J. Frank, C. Spahn, Rep. Prog. Phys. **69**(5), 1383 (2006)
83. T.M. Schmeing, R.M. Voorhees, A.C. Kelley, Y.G. Gao, F.V. Murphy, J.R. Weir, V. Ramakrishnan, Science **326**(5953), 688 (2009)
84. A. Korostelev, D.N. Ermolenko, H.F. Noller, Curr. Opin. Chem. Biol. **12**(6), 674 (2008)
85. S.C. Blanchard, R.L. Gonzalez, H.D. Kim, S. Chu, J.D. Puglisi, Nat. Struct. Mol. Biol. **11**(10), 1008 (2004)
86. P.C. Whitford, P. Geggier, R.B. Altman, S.C. Blanchard, J.N. Onuchic, K.Y. Sanbonmatsu, RNA **16**, 1196 (2010)
87. J.K. Noel, J. Chahine, V.B.P. Leite, P.C. Whitford, Biophys. J. **107**(12), 2872 (2014)
88. J.K. Noel, P.C. Whitford, Nat. Commun. **7**, 13314 (2016)
89. H. Yang, J.K. Noel, P.C. Whitford, J. Phys. Chem. B **121**, 2777 (2017)
90. H. Yang, J. Perrier, P.C. Whitford, Prot. Struct. Func. Bioinfo. **86**, 1037 (2018)
91. A. Samiotakis, M. Cheung, J. Chem. Phys. **135**(17), 175101 (2011)
92. R.T. Byrne, A.L. Konevega, M.V. Rodnina, A.A. Antson, Nucleic. Acids Res. **38**(12), 4154 (2010)
93. H. Shi, P.B. Moore, RNA **6**(8), 1091 (2000)
94. N. Fischer, A.L. Konevega, W. Wintermeyer, M.V. Rodnina, H. Stark, Nature **466**(7304), 329 (2010)
95. S. Chladek, M. Sprinzl, Angew. Chem. Int. Edit. **24**(5), 371, 175101 (1985)
96. W.A. Rebecca, J. Eargle, Z. Luthey-Schulten, FEBS Lett. **584**(2), 376 (2010)
97. J. Eargle, A.A. Black, A. Sethi, L.G. Trabuco, Z. Luthey-Schulten, J. Mol. Biol. **377**(5), 1382 (2008)
98. A. Bashan, I. Agmon, R. Zarivach, F. Schluenzen, J. Harms, R. Berisio, H. Bartels, F. Franceschi, T. Auerbach, H.A.S. Hansen, E. Kossoy, M. Kessler, A. Yonath, Molecular Cell **11**(1), 91 (2003)

Chapter 3
Energetic and Structural Properties of Macromolecular Assemblies

Paul Charles Whitford

Abstract This chapter addresses the structural and energetic properties of macro-molecular assemblies. We first describe the chemical composition of biomolecules and then elaborate on the dynamics of two classes of well-studied complexes: ribosomes and viruses. The biological context of each system is first presented, which is followed by a discussion of physical considerations and questions that have been posed, as well as an overview of experimental and theoretical approaches available for their study.

3.1 Chemical Composition of Macromolecular Assemblies

Before introducing specific physical questions, it is important to first review the chemical composition of biomolecular assemblies. While these complexes may be composed of a variety of types of biomolecules, the current focus will be on two of the most prevalent molecular constituents: amino acid (protein) and nucleic acid (RNA, DNA) polymer chains.

Proteins and nucleic acids are described as linear chains of residues. Each residue (Fig. 3.1a) contains a set of backbone atoms that defines the class of molecule. Protein residues have identical backbone atoms (C, C_α, N and O, in addition to the associated H atoms), whereas each RNA/DNA residue has a common sugar-phosphate backbone. With regard to molecular structure, the only difference in the backbone of RNA and DNA is that there is an additional oxygen at the 2' position in RNA. While the backbone atoms define the class of molecule, within each class there is a range of possible side-chain atoms, which define the specific amino acid or nucleic acid. In proteins, there are 20 types of residues that are genetically encoded, where each contains a unique set of side chain atoms. In RNA, there are four residue types (C, G, A, and U). Similarly, there are four types of DNA residues, three of which have side chains that are identical to those of RNA (C, G, and A). The backbone atoms of each residue form covalent bonds with the preceding and subsequent residues in

P. C. Whitford (✉)
Northeastern University, Boston, MA, USA
e-mail: p.whitford@northeastern.edu

© Springer Nature Switzerland AG 2022
K. B. Blagoev and H. Levine (eds.), *Physics of Molecular and Cellular Processes*,
Graduate Texts in Physics, https://doi.org/10.1007/978-3-030-98606-3_3

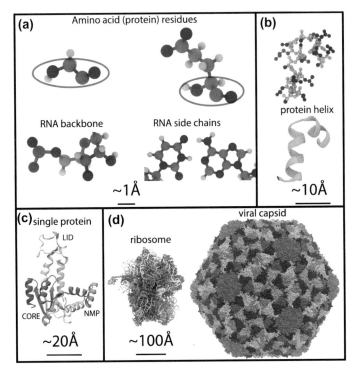

Fig. 3.1 Hierarchy of structure in biomolecular systems a A biopolymer may be described as a linear chain of residues. Each residue has a common set of backbone atoms (circled) and a specific set of side-chain atoms. **b** The backbone atoms form covalent bonds with adjacent residues, which lead to an overall linear chemical structure. Since the backbone atoms of proteins are common, as are the backbone atoms of RNA, or DNA, the functional properties unique to each molecule are determined by the composition of the side chains. When viewing larger assemblies, one typically does not represent every atom, but rather uses a cartoon depiction (bottom). **c** Individual polymer chains can fold to low-energy "native" configurations. Here, Adenylate Kinase (214-residue protein) is shown. Within this molecule, there are three globular regions called "domains" (LID, NMP, CORE). **d** Biomolecular assemblies are formed by collections of biopolymers. For example, a ribosome is composed of several large RNA molecules (cyan, white, pink) and approximately 50 protein chains (purple, gray). Viral capsids are often composed entirely of proteins, and they can include hundreds to thousands of individual proteins. The structures shown here correspond to PDB entries 4AKE (Adenylate Kinase [1]), 5E7K (ribosome [2]), and 3J7X (T7 bacteriophage [3])

the chain. Protein and nucleic acid chains are typically composed of at least 10–20 residues, though some have as many as thousands of residues.

Many protein, DNA and RNA sequences can spontaneously adopt well-defined three-dimensional structures. These low-energy configurations are often referred to as "native" structures. Figure 3.1b shows the native configuration of a short (~ 15 residue) segment of a protein chain, where the backbone atoms are displayed in yellow and side-chain atoms are shown in red. For ease of visualization, cartoon representations are frequently used (Fig. 3.1b, bottom). While the helical structure of the backbone is not obvious when every atom is shown, automated algorithms allow for simple cartoon depictions to reveal the overall arrangement of the chain. A complete protein is shown in Fig. 3.1c. This protein, called Adenylate Kinase, contains 214 residues that spontaneously organize into three "domains" (LID, CORE, NMP), where each is comprised of a globular collection of residues.

Multiple biomolecules may associate and form higher order assemblies (Fig. 3.1d), which are the focus of the current chapter. The size and complexity of these assemblies vary widely, where the smallest assemblies are composed of only two chains (i.e., dimers), and the largest contain thousands of individual polymers. One extensively studied assembly is the ribosome [4–10]. Ribosomes exist in all living cells, and they are composed of collections of RNA and protein molecules. In bacteria, the 70S ribosome has three RNA chains and roughly 50 protein chains. Together, they form an asymmetric ∼20-nanometer assembly that converts mRNA sequences into proteins. Another well-studied class of assemblies are viral capsids, which are responsible for transporting viral DNA/RNA. Capsids typically have spatial dimensions of 50-100 nanometers, and they are composed of hundreds (e.g., 420 proteins in the T7 capsid, 3.1D) to more than a thousand (e.g., ≈1600 proteins in the HIV-1 capsid [11]). Capsids may also be enveloped in lipids and they may contain additional coat proteins, though the current chapter will focus on the simplest systems, bacteriophage.

3.2   The Ribosome

Ribosomes are multicomponent assemblies that synthesize proteins in all organisms [4–7, 9, 10, 12]. From a structural and biochemical perspective, the ribosome is one of the most well-studied large-scale assemblies. Combined with its critical role within the cell, this makes the ribosome a prime candidate for identifying physical principles that govern biomolecular assemblies.

3.2.1   Biological Role and Mechanistic Characteristics

In bacteria, the 70S ribosome (2.4 MDa) is described in terms of two subunits: the "large" (i.e., 50S subunit) and the "small" (30S) subunit (Fig. 3.2). The 50S subunit is composed of ribosomal RNA (rRNA) molecules (23S and 5S), in addition to numerous proteins. The 30S subunit contains one rRNA molecule (16S) and roughly 20 proteins. Both subunits have three binding sites (A, P, and E) to which transfer RNA (tRNA) molecules transiently associate. In bacteria, aminoacyl-tRNA (aa-tRNA) molecules are delivered to the ribosome by elongation factor thermally unstable (EF-Tu). When aa-tRNA·EF-Tu·GTP (i.e., the ternary complex) initially associates with the ribosome, codon–anticodon interactions are formed between aa-tRNA and mRNA. The aminoacyl-tRNA (aa-tRNA) molecule can then fully bind to the A-site,

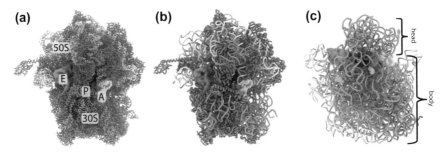

Fig. 3.2 Ribosome structure and dynamics a The 70S ribosome is described in terms of two subunits: the "large" 50S and "small" 30S, where each subunit is associated with three tRNA binding sites (A, P, and E). During the synthesis of proteins, the small subunit allows incoming tRNA molecules to read the messenger RNA at the A-site. After the incoming tRNA is partially delivered to the A-site, it must undergo an $\approx 100 Å$ conformational change, where potentially incorrect aa-tRNA molecules may be rejected (i.e., accommodation). After the aa-tRNA binds the A-site, the nascent protein chain is transferred from the P-site tRNA to the aa-tRNA, thereby extending the chain by one residue. Following peptide bond formation, the tRNA molecules must be displaced by one binding site (A to P and P to E), which involves multiple conformational rearrangements of the tRNA and ribosome (hybrid-state formation and mRNA-tRNA translocation). **b** Same structure as panel A, using tube representation for the ribosome. **c** Same as panel B (protein not shown), view rotated $\approx 90°$ about the horizontal axis. As the tRNA molecules move between binding sites, the 30S body domain undergoes a relative rotation of $\approx 7°$, and the 30S head domain undergoes a rotation of $\approx 20°$. Together, the balance of tRNA distortions and collective rearrangements in the ribosome allow the tRNA molecules to move from the A- and P-sites to the P- and E-sites, which enables the ribosome to have a new aa-tRNA molecule read the next mRNA codon

a process known as accommodation. Subsequent to accommodation, the growing peptide chain is transferred to the aa-tRNA in the A-site from the P-site tRNA, which extends the peptide chain in length by one amino acid. The A- and P-site tRNAs and mRNAs are then displaced to the P- and E-sites. This movement of tRNA-mRNA is called translocation, and it leads to an empty A-site. By vacating the A-site, the next mRNA codon may be read by an incoming aa-tRNA molecule.

Biochemical, structural, and single-molecule measurements have provided many mechanistic insights into the process of tRNA selection. For example, biochemical data has shown that tRNA selection occurs in two steps (i.e., initial selection and proofreading) [13, 14]. Initial selection involves the aa-tRNA (in the form of ternary complex) forming codon–anticodon interactions with mRNA. This signals GTPase activation and GTP hydrolysis [15–17]. After GTP hydrolysis, inorganic phosphate is released, EF-Tu undergoes a large-scale conformational rearrangement [18], and aa-tRNA accommodates into the A-site (i.e., the proofreading step). Experiments have shown that proofreading can increase the accuracy of tRNA selection at least 15-fold [13, 14, 16, 17, 19–23]. Cryoelectron microscopy (cryo-EM) and X-ray crystallography have revealed the structural properties of biochemically isolated states that precede and follow accommodation [24–30]. Single-molecule FRET (smFRET) studies have been able to follow individual aa-tRNA molecules throughout the selection process [31, 32], which has helped elucidate the kinetics and timing of individual substeps. Together, this immense body of experimental data provides a structural and biochemical foundation, upon which it may be possible to establish a detailed physicochemical understanding of ribosome function.

tRNA–mRNA translocation is typically described in terms of two general steps: tRNA hybrid-state formation and movement of the mRNA–tRNA codon–anticodon pair [4, 7, 33, 34]. During hybrid formation, tRNAs are displaced relative to the large subunit. In the second step, the tRNA anticodons and mRNA move through the small subunit. Experiments have shown that tRNA molecules can spontaneously adopt hybrid configurations [35], whereas movement along the small subunit is largely influenced by the action of elongation factor G (EF-G) [36]. To further complicate the description of this process, tRNA translocation also involves collective rotary motions ($\approx 6 - 20°$) of the ribosomal domains/subunits. For example, the so-called body domain of the small subunit rotates relative to the large subunit during hybrid-state formation [33, 37], and the head domain of the small subunit can undergo multiple rotations ("head swivel" and "head tilting") as the mRNA and tRNA move along the small subunit [38–40].

3.2.2 Physical Considerations

There are many physical factors that one must account for when describing ribosome dynamics. For example, the overall ribosome structure may favor motions along functionally related directions. In addition to structure, flexibility of the constituents may lead to changes in configurational entropy that can limit (or facilitate) individual functional steps. Further, the ribosome is composed of highly charged polymer chains. That is, bacterial ribosomes contain over 4000 RNA residues, each of which carries a net negative charge about the PO_4^-. To partially compensate for the large negative charge introduced by the RNA backbone, most ribosomal proteins carry a net positive charge. However, positively charged counterions (K^+, Mg^{2+}) are also necessary for the ribosome to remain structurally stable. As a natural consequence, changes in ion concentration can alter the relative stabilities of different functional conformations, and thereby modulate the kinetics/accuracy of elongation [22, 41]. Finally, as described above, the elongation cycle is facilitated by two GTPase proteins, EF-Tu and EF-G. These multi-domain proteins are responsible for hydrolyzing GTP and facilitating tRNA selection and translocation. Together, this range of physical contributors must balance, in order for the ribosome to function efficiently and accurately.

With the aim of establishing a physical framework for understanding ribosome function, there has been an effort by the ribosome field to adopt an energy landscape perspective. This shift has been largely inspired by the study of protein folding, where energy landscape approaches have allowed for systematic analysis of the principles that guide folding dynamics [42–46]. Within the energy landscape framework, a central objective has been to identify the precise locations and heights of rate-limiting free-energy barriers, as well as understand how these barriers are impacted by environmental conditions.

When adopting an energy landscape perspective, it is necessary to first identify appropriate reaction coordinates for describing the motion. Similarly, one may seek to understand the degree to which motion along distinct degrees of freedom are energetically coupled/correlated. Before addressing specific examples in the ribosome, it is important to discuss some general aspects of reaction coordinate analysis. In many cases, including protein folding, it is possible to accurately describe multi-dimensional motion in terms of diffusive movement along one-dimensional projections of the dynamics [47–50]. If the underlying free-energy barrier is described well by a given reaction coordinate, then one may relate the mean first-passage time between distinct conformations via the relation [42, 51]:

$$\langle t_{\mathrm{fp}} \rangle = \int_{\rho_{\mathrm{initial}}}^{\rho_{\mathrm{final}}} d\rho \int_0^\rho d\rho' \frac{\exp\left\{\left[F(\rho) - F(\rho')\right]/k_{\mathrm{B}}T\right\}}{D_\rho(\rho)}, \qquad (3.1)$$

where ρ is an appropriately chosen reaction coordinate, $D_\rho(\rho)$ is the effective diffusion coefficient along ρ, and ρ_{initial} and ρ_{final} are the values of the coordinate at the endpoint configurations of the system.[1] Perhaps surprising is that this relation is analogous to what one obtains by inverting the Smoluchowski equation (Fokker–Planck equation in the overdamped regime) in one dimension, where the potential energy and diffusion coefficient are replaced by the free energy and effective diffusion coefficient. However, a clear distinction is that (3.1) can only be applied to multi-dimensional systems if an appropriate one-dimensional coordinate exists.

To illustrate the importance of the choice of reaction coordinate, it is instructive to consider a hypothetic multi-dimensional system described by two reaction coordinates (ρ_1, ρ_2). Here, the underlying free-energy surface is given by the sum of two Gaussian functions of depth $10k_{\mathrm{B}}T$ (Fig. 3.3a). One may calculate the free energy as a function of a single coordinate, according to the relation:

$$\Delta F(\rho_{i,0}) = -k_{\mathrm{B}}T \ln \int d\rho_1 d\rho_2 \delta(\rho_i - \rho_{i,0}) \exp(-\Delta F(\rho_1, \rho_2)/k_{\mathrm{B}}T). \qquad (3.2)$$

Here, the notation ΔF is used, since we are only describing relative free energies. For the example system, the free energy along ρ_1 and ρ_2 is shown in Fig. 3.3b. In this simple case, the free-energy barriers obtained with the two coordinates differ by several $k_{\mathrm{B}}T$. That is, the barrier obtained with ρ_1 is similar to the pre-defined multi-dimensional barrier, whereas ρ_2 significantly underestimates the barrier height. The free energy is also shown using a collective coordinate (ρ_{comb}) constructed from a linear combination of ρ_1 and ρ_2. When using the collective coordinate, the exact height of the original two-dimensional barrier is recovered. For this hypothetical system, it is possible to compare each one-dimensional projection to the known multi-

---

[1] This form of the equation assumes that $\rho_{\mathrm{final}} > \rho_{\mathrm{initial}}$, where ρ may only adopt positive values.

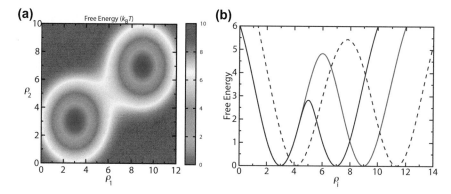

Fig. 3.3 Apparent barrier height depends on the choice of coordinate. When studying multi-dimensional process, the calculated/measured free-energy barrier can be significantly influenced by which coordinate is used. **a** Hypothetical two-dimensional free-energy surface $F(\rho_1, \rho_2)$. **b** Apparent free energy along each individual coordinate. While ρ_1 (red) provides a larger estimate of the barrier than ρ_2 (solid black), the correct height is only recovered when a linear combination of coordinates is used $\rho_{comb} = C_1\rho_1 + C_2\rho_2$ (dashed black)

dimensional barrier height. However, in molecular assemblies, the motion occurs in $3N$ dimensions, where N is on the order of 10,000–1,000,000. Accordingly, it can be challenging to determine which coordinates provide suitable one-dimensional descriptions of the collective process.

3.2.3 Methods for Probing Ribosome Energetics

There are numerous ongoing efforts to elucidate the energetics of large-scale conformational changes in the ribosome. Here, we will specifically discuss methods that aim to provide descriptions of individual molecules, rather than bulk properties. The reason for focusing on single-molecule approaches is that, while biochemical measurements provide insights into the kinetics of ribosome function, bulk experiments only provide indirect measures of the underlying free-energy barriers. However, since rates may be calculated from free energies (3.1), bulk kinetic measurements represent a much needed point of comparison when characterizing the energy landscape.

There have been three primary tools available to infer, or directly probe, free-energy barriers in the ribosome. Single-molecule FRET measurements have allowed for the apparent dynamics to be measured along putative reaction coordinates. Similarly, recent cryo-EM techniques have aimed to utilize ensembles of two-dimensional images of ribosomes, in order to project the distribution along specific structural metrics. To complement these experimental efforts, a number of molecular simulation studies have implicated the use of specific reaction coordinates, and they have provided predictions of the energetic characteristics of large-scale rearrangements. As described below, these techniques are beginning to provide a cohesive understanding of ribosome energetics.

3.2.3.1 Single-Molecule Spectroscopy

Over the last two decades, single-molecule techniques have continued to be refined and extended to study increasingly complex molecular assemblies. In the context of the ribosome, the most widely used single-molecule method has been single-molecule FRET (smFRET). In these experiments, individual fluorescently labeled ribosomes are immobilized to a surface and then illuminated by an intense monochromatic light source. In these constructs, two fluorescent dyes are ligated to specific residues on the ribosome. The first dye is called the donor, and it is excited by the light source. The second dye, the acceptor, is then excited through non-radiative energy transfer from the donor. The acceptor then releases energy in the form of photons, which are detected by an additional sensor. By simultaneously measuring the emission of donor and acceptor photons, one can calculate the efficiency of energy transfer, which is related to the inter-dye distance R according to

$$E = \frac{1}{1 + (R/R_0)^6}. \tag{3.3}$$

R_0 is the Förster radius, which is defined as the distance at which energy transfer efficiency is 0.5 for the given donor–acceptor pair. In applications to the ribosome, R_0 is typically 50–80 Å. With this approach, it is possible to collect real-time trajectories of ribosomes as they undergo functional processes (reviewed in [52–54]).

With regard to probing the energetics of the ribosome, a significant challenge in single-molecule studies is time averaging. That is, in order to prolong the lifetime of the dyes and prevent photobleaching, the light intensity is reduced. This reduces the rate of photon absorption and emission, which makes it necessary for the observation window of the camera to be long (10–100 ms), relative to biomolecular timescales.[2] Accordingly, the obtained time traces represent discretized trajectories after finite-time averaging. Thus, one cannot directly calculate the free energy from the measured probability distribution, since averaging will reduce/eliminate detection of the underlying barrier.[3] Finally, as discussed above, it is not known *a priori* which reaction coordinates (i.e., pairs of labeled sites) most closely follow the lowest free-energy path.

While technical limits of current smFRET technology have prevented the direct measurement of free-energy barriers for the ribosome, it has been possible to infer the relative free energies of stable conformations (i.e., free-energy minima). Specifically,

---

[2] As a note, alternate photon detection methods that allow for higher time resolution have been applied to study protein dynamics [55], but such approaches are not typically utilized in ribosome experiments.

[3] Deconvolution techniques [56] represent a potential avenue by which one may extract the potential of mean force, though applications to the ribosome have not been reported.

when the free-energy barrier is sufficiently large (i.e., slow interconversion times), it is possible to extract the populations of distinct states, from which the relative free energy may be calculated ($\Delta G_{21} = G_2 - G_1 = -k_B T \ln \left(\frac{P_2}{P_1} \right)$). To this end, smFRET methods of the ribosome allow one to obtain estimates of the differences in free energy for competing tRNA conformations during accommodation [32], hybrid-state formation, and translocation [57, 58], as well as conformational rearrangements of the ribosome [37], movement of stalk regions [59–61], and EF-G [62].

From a physical perspective, a common insight that has emerged from smFRET studies is that the endpoints associated with individual conformational rearrangements often differ in free energy by only a few $k_B T$. For example, for a variety of pre-translocation constructs, it was inferred that rotated and unrotated ribosome configurations are both accessible [37], and that their relative populations typically differ by less than an order of magnitude (i.e., $\Delta G < -k_B T \ln(10) \sim 2.5 k_B T$). Similarly, smFRET has shown that tRNA molecules and the L1 stalk of the large subunit continuously interconvert between competing conformations during tRNA translocation [57]. In accordance with this theme, when aa-tRNA is delivered to the ribosome, it may also interconvert between accommodated-like and unaccommodated conformations [32] prior to fully binding to the ribosome.

Another feature that has been pervasive in smFRET measurements of the ribosome is that motions along putative reaction coordinates are only weakly correlated. For example, in the context of subunit rotation in the ribosome, one set of FRET probes reported spontaneous interconversion between endpoints [37], whereas a second set of probes showed much slower interconversion events [63]. Since both labeling sites were motivated by available structural data, the fact that the ribosome exhibits differential dynamics along each coordinate suggests that molecular flexibility can significantly impact the apparent dynamics. In another instance, distinct conformational processes associated with tRNA hybrid-state formation were shown to exhibit differential timescales of interconversion [64]. In addition, reversible transitions were observed along each coordinate, suggesting that the full translocation process is only completed when these seemingly independent motions simultaneously adopt the endpoint values. This type of behavior has been described as implying that the degrees of freedom are "uncoupled" or "loosely coupled" [64], though this designation has been made on phenomenological grounds. Accordingly, as described below, there is an opportunity for physical characterizations to provide systematic quantitative measures of the degree of energetic coupling between apparently weakly correlated dynamics.

3.2.3.2 Cryoelectron Microscopy

Cryoelectron microscopy was originally developed to resolve the structural details of biological complexes at distinct stages of function. In a cryo-EM experiment, ribosome complexes (commonly referred to as "particles") are prepared in aqueous solution and then rapidly frozen ("plunged"). This leads to each ribosome being localized to a particular configuration. The frozen set of particles is then imaged using an electron microscope, which provides two-dimensional projections of the system. Decades of effort have been dedicated to developing algorithms by which large number of 2D projections may be classified and then used to reconstruct a small set of 3D electron densities for the system [65].

Cryo-EM has been an incredibly effective technique for identifying long-lived structures of biological systems,[4] where more recent efforts have aimed to use these 2D images to directly calculate the free energy along collective coordinates. In these approaches, the objective is to identify the most likely conformations of the system (as measured by a particular coordinate or set of coordinates) and then calculate the probability distribution. In one application, Fischer et al. [66] directly projected the 2D images onto pre-defined coordinates through the use of an image filter. In later efforts, Frank, Ourmazd, and colleagues developed a rigorous approach by which the dimensionality of the detected dynamics is not pre-defined. Rather, in their approach, called manifold embedding [67, 68], a form of singular value decomposition is applied to the covariance of the images in order to determine the number of dominant degrees of freedom. Using the low-dimensional ($\approx$5 degrees of freedom) basis set, the collection of images are projected along these collective coordinates in order to calculate a multi-dimensional free-energy surface.

While the use of cryo-EM to determine free energies represents a promising avenue for directly quantifying ensemble properties of ribosome function, there are still some outstanding physical questions that will need to be addressed. For example, manifold embedding provides natural coordinate systems for describing differences between experimentally obtained 2D images, though it is not yet clear how these coordinates relate to specific deformations of the assembly. Similarly, it is not known whether the identified coordinates follow the underlying lowest free-energy path (see above for discussion). Finally, in order to calculate rates from these extracted free-energy surfaces (e.g., through use of 3.1), it will be necessary to have a comprehensive understanding of the diffusive properties along each coordinate, which will likely require additional theoretical developments. These open questions represent exciting new challenges as we further develop our physicochemical understanding of the ribosome.

3.2.3.3 Molecular Simulations and Calculations

Advances in computing hardware have made it possible to use theoretical models to study conformational transitions in the ribosome. The range of models includes coarse-grained representations (e.g., one pseudoparticle per residue, or group of residues [69–71]), harmonic approximations to atomic interactions [72, 73], simplified models with anharmonic interactions [74], as well as conventional explicit-solvent models [75–77]. While there are some commonalities between these models, each is suited to provide specific insights into the ribosome's energy landscape. Below

---

[4] The impact of this approach was recognized by the 2017 Nobel Prize in Chemistry.

is an overview of the utility of these techniques, as well as a discussion of some key observations. For a more detailed technical review on applications of computing to the ribosome, see [78].

Before continuing, we should note that the current chapter is focused on free-energy barriers associated with large-scale conformational transitions. In addition to the approaches described below, there are also tools available for calculating detailed energetic properties of bimolecular association events. For example, models that employ Brownian dynamics protocols have been used to predict the path of approach during EF-G binding [79]. As another example, highly detailed explicit-solvent simulations have been used to calculate differences in binding energies for different codon–anticodon pairs [80]. For a more extensive discussion on the objectives and applications of these techniques, additional resources are available [81].

Normal Mode Analysis Early normal mode calculations of the ribosome were used to identify energetically accessible directions of motion in fully assembled systems. In order to appreciate the insights gained from such calculations, it is valuable to briefly review normal mode analysis.[5] Normal mode analysis involves expanding the potential energy about a minimum to second order and then identifying energetically uncoupled degrees of freedom. That is, the Lagrangian is approximated as

$$L(\{q_i\}, \{\dot{q}_i\}) = \frac{1}{2} T_{ij} \dot{q}_i \dot{q}_j - \frac{1}{2} V_{ij} q_i q_j + O(3), \tag{3.4}$$

where fluctuations are sufficiently small that $O(3)$ terms may be neglected. One then defines a linear transformation of the coordinates by solving the eigenvalue problem:

$$[V - \lambda_i T] \xi_i = 0, \tag{3.5}$$

where ξ_i and λ_i are the corresponding eigenvectors and eigenvalues. When the eigenvectors are appropriately normalized, they define a linear transformation from the original coordinates to normal coordinates Q_i. The Lagrangian may then be written as the sum of 3N uncoupled oscillators

$$L(\{Q_i\}, \{\dot{Q}_i\}) = \frac{1}{2} \dot{Q}_i^2 - \frac{1}{2} \lambda_i Q_i^2. \tag{3.6}$$

Accordingly, $\lambda_i = \omega_i^2$ describes the curvature of the potential energy along the direction of each mode. Finally, in accordance with the equipartition theorem, the average energy in each mode will equal $\frac{1}{2} k_B T$. Therefore, under equilibrium conditions, the largest amplitude motions will be along the lowest frequency modes.

In the context of the ribosome, normal mode analysis has been used in conjunction with coarse-grained models [72, 73]. In these models, each residue is represented by a single pseudoatom, where pseudoatom pairs that are within a cutoff distance are assigned a harmonic interaction. While normal mode analysis, in general, involves

---

[5] For a more complete discussion of normal modes, the reader may consult any introductory classical mechanics text.

identifying a potential minimum and then calculating the modes, applications to the ribosome typically define the known conformation as the global minimum. The weight of each interaction is then set according to external criteria (e.g., fluctuations inferred from crystallographic data), where the weights may be homogeneously defined, or inhomogeneous values may be used. Perhaps surprising is that the low-frequency modes tend to be insensitive to the specific distribution of interaction weights. From a physical standpoint, this may be understood in terms of the distribution of residue–residue interactions. That is, the composition of the ribosome may be thought of as a roughly homogeneous mixture of nucleic acids, amino acids, water, and ions, where intramolecular interactions are on approximately the same energy scale ($k_B T$). While the strength of individual interactions does not vary greatly, the density of interactions depends significantly on the structure of the complex. For example, residues buried within the ribosome interact with many residues, which leads to a high degree of connectivity between residues that are distant in sequence. In contrast, residues in elongated RNA helices (e.g., L1 stalk) or protein helices (e.g., protein L9) may interact with as few as 2–3 nearest neighbor residues. Accordingly, one can expect that high contact-density regions will be relatively stiff, while low contact-density regions will allow for larger scale fluctuations, as predicted by normal mode calculations.

Simulations with Simplified Models While normal mode analysis has proven to be an effective tool for exploring the character of fluctuations about specific conformations, anharmonic models are necessary to study conformational transitions between minima. A class of models that has been particularly useful for simulating large-scale rearrangements in the ribosome are all-atom structure-based models [82, 83]. In a structure-based model, a specific conformation is explicitly defined as the global potential energy minimum. This assignment of the energetic minimum is similar to normal mode analysis, although the anharmonic character of structure-based models allows for the spontaneous breakage and formation of interactions. While early structure-based models employed coarse-grained representations of proteins [84, 85], these models were subsequently extended to include all-atom representations of proteins [82, 86], RNA [87] and other biomolecules [83]. These higher resolution variants of the models have been extensively applied to study the ribosome, where early simulations were shown to provide structural descriptions of aa-tRNA accommodation [88] and translocation [74] that are qualitatively consistent with available experimental data.

Recently, simplified models have been used to identify appropriate reaction coordinates for describing ribosome dynamics. As discussed above, in order to systematically study the energetics of large-scale rearrangements in the ribosome, it is critical that one identifies which coordinates most closely follow the lowest free-energy path. Further, in order to calculate kinetics from the free-energy profile (Eq. 3.1), it is necessary to demonstrate the motion is diffusive along a given degree of freedom. To this end, structure-based models have allowed for hundreds of reversible conformational transitions to the observed for aa-tRNA accommodation [89], tRNA hybrid-state formation (a substep of translocation) [90], and subunit rotation [71]. Since these simulations provide unbiased spontaneous barrier-crossing events, it is possible to analyze the statistical properties associated with the motion. For example, one may calculate the probability of being on a transition path [47], which is a common technique in the study of protein folding [91, 92]. That is, when using an arbitrary coordinate, one may identify a putative transition-state ensemble (i.e., conformations described by the highest value of the free energy). One may then use transition path analysis to determine whether the properties of the identified TSE are consistent with these configurations representing a saddle point in the free energy. Specifically, if the TSE represents a saddle point, then in the diffusive limit the system will be equally likely to continue to either of the endpoint configurations. Accordingly, if the motion is Markovian along the chosen coordinate, then when the system reaches the transition-state ensemble, there will be a 50% chance that it will continue to the alternate endpoint.

For the ribosome, transition-path-based reaction coordinate analysis has been performed for both accommodation [89] and hybrid-state formation [90]. Some notable findings were that false positives may be detected when using current labeling sites. In contrast, positioning the dyes on alternate residues can eliminate these false positives, while providing a more precise measure of when the tRNA is at the TSE. Similarly, simulations have allowed for the comparison of labeling strategies that aim to measure small subunit body rotation [71]. In the case of subunit rotation, studies that employed different labeling sites provided qualitatively different descriptions of rotation (see earlier discussion). By comparing coordinates with a simplified model, it was shown that flexibility of the ribosome naturally allows for these contradictory observations. In addition, this comparison showed the degree to which each coordinate is coupled with the underlying rotary motion. Together, these examples illustrate how the application of simplified models can be used to help design more precise experimental techniques for measuring ribosome dynamics, and ultimately energetics.

Semi-Empirical Explicit-Solvent Models Another theoretical approach that has been applied to the ribosome is the use of explicit-solvent simulations. Before discussing the insights provided by these calculations, it is important to first contrast these models with the aforementioned techniques. While the term "explicit-solvent" only implies the level of resolution in a model, this term is typically intended to indicate a specific class of energetic models (i.e., semi-empirical explicit-solvent models). In these models, a transferrable set of parameters is developed, where charges, as well as bonded/nonbonded interaction types, are assigned on a per-residue basis. That is, a C_α atom in Phe will always be assigned the same energetic parameters, irrespective of whether information is available about stable configurations. In these classical mechanical models, partial charge definitions, bond lengths, and bond angles are assigned based on previous quantum mechanical calculations. In addition, many bond stretching and bending interactions were previously calibrated through comparison with spectroscopic measurements of prototypical molecular fragments. Finally, water molecules and ions are explicitly included, allowing for the simulation of a solvated polymer.

One potential advantage of explicit-solvent models is that physical quantities may be calculated without prior knowledge of stable configurations. That is, in principle, one may initiate the simulation from any configuration and then observe where the energetic minima/maxima are located. However, there are some notable limitations that are important to mention. First, since these simulations include a very large number of explicitly represented atoms (e.g., 2–3 million for the ribosome), they are typically limited to nanoseconds [75], hundreds of nanoseconds [76, 77], or microseconds [93, 94]. While the continued increases in computing capabilities allow for many-microsecond simulations to now be performed [95], biologically relevant processes typically occur in the millisecond regime. A second point that must be considered when interpreting explicit-solvent simulations is that the accuracy of these RNA models is unclear. That is, while fully solvated ribosomes may remain stable for microseconds in simulations [94], simulations of small RNA fragments regularly suggest improvements to force field parameters [96].

While there are outstanding challenges when using explicit-solvent models, recent calculations have begun to implicate some aspects of ribosome energetics. With regard to calculating energetics with these models, the most extensive study to date involved performing 12 simulations, each for approximately 200 ns [77]. Each simulation was initiated from a distinct configuration that had been identified in time-resolved cryo-EM measurements of the ribosome during reverse translocation. As discussed above, these reconstructed images represent minima on the free-energy landscape of the ribosome. Since the simulations were restricted to hundreds of nanoseconds, spontaneous interconversion events between adjacent minima were not directly observed. However, each simulation can describe the fluctuations about a minimum. Based on this, the authors interpreted the long-time dynamics by applying a quasi-harmonic approximation to the landscape about each minimum. From this, the height of each free-energy barrier was then roughly approximated. This provided estimates of the timescale of translocation that are similar to experimental measures, demonstrating that the inferred landscape is generally consistent with known kinetics.

In addition to predicting free-energy barriers, explicit-solvent simulations of the ribosome have also provided estimates of diffusion coefficients along putative collective reaction coordinates. By performing 200ns-1μs explicit-solvent simulations, it has been possible to ask whether the dynamics along a given coordinate exhibits diffusive behavior. These types of calculations have provided estimates of diffusion coefficients for tRNA displacements [76], as well as collective rotary motions [93] of the ribosome. As described above, apparent diffusion coefficients may be used to quantitatively relate free-energy barriers and the mean first-passage time for a conformational change (3.1). Thus, this helps provide a quantitative bridge between kinetics and large-scale free-energy barriers. Further, the apparent diffusion coefficients can be used to infer the scale of short length scale energetic roughness [42, 51]. Physically, when the short-scale roughness is larger, the apparent diffusion coefficient will be reduced. While the precise distribution of short-scale roughness is not known, the theoretically calculated coefficients suggest the scale of the roughness is approximately 2–4$k_B T$ for tRNA displacements and ribosomal subunit motion.

3.3 Viruses

In contrast to bacteria, viruses are not independent living organisms. Rather, a virus propagates by utilizing the host cell's transcription and translation machinery in order to express its molecular components. During the "life cycle" of the virus, it first enters a cell, then expresses its genetic message, which produces copies of its molecular constituents. These components then assemble into new viral particles, which exit the cell and infect new cells. Depending on the type of virus, the viral genome is stored in the form of DNA or RNA. For example, in bacteriophage (i.e., viruses that infect bacterial cells), the genomic message is stored in the form of double-stranded DNA (dsDNA) molecules, whereas the HIV genome is stored in the form of RNA. Figure 3.4 shows the life cycle of a dsDNA bacteriophage. As shown, the virion first punctures the cell wall and injects its genome into the bacterial cell. The host cell's translation (ribosome) and transcription (polymerase) machinery then generates numerous copies of the viral proteins and DNA. The viral proteins assemble to form an immature procapsid, into which a copy of the viral genome is loaded. The capsid then serves as a protective shell that allows the viral genome to be transported outside of the cell and ejected into a new host cell. In the present chapter, we will focus specifically on the physical properties of bacteriophage.

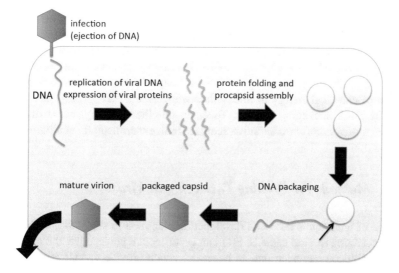

Fig. 3.4 Simplified schematic of the "life cycle" of a virus bacteriophage is a virus that can infect a bacterial cell (blue). Upon recognizing the bacterial cell, the virion punctures the cell membrane and ejects the viral DNA. After the viral proteins (orange) are expressed by the host cell, they fold and assemble to form an empty procapsid shell (green). Next, the viral DNA is packaged into the immature procapsid, which is often associated with conformational changes in the assembly (red, see Fig. 3.1d). Finally, there are typically additional biochemical modifications (e.g., addition of a tail) to the capsid, resulting in a fully mature particle that may leave the bacterial cell and then infect new host cells

3.3.1 Physical Considerations and Questions

The complexity of viral particles provides many avenues for exploring the physical principles of biological function. For example, after a viral particle recognizes a host cell, it is able to rapidly eject its DNA without utilizing additional chemical energy. Rather, the energy stored in the packaged DNA is responsible for ejection dynamics. After the viral proteins are expressed in the cell, they must self-organize into large assemblies. Next, ATPase proteins convert the chemical energy in ATP into mechanical work, which is used to load DNA into the assembled procapsid. In addition, the loading process can also be associated with conformational changes in the capsid shell. While each of these topics has been studied experimentally and theoretically, our quantitative physical understanding of packaging is most well developed.

The process of DNA packaging in viral capsids has received significant interest from the physics community due to its rich set of physical attributes. The principal objective has been to understand how a long linear polymer (e.g., $5 - 10\,\mu m$ DNA) is able to package into a compact shell with a radius of tens of nanometers. That is, how is the DNA packaging machinery able to drive the molecule from a disordered, high-entropy, ensemble to a highly ordered compact state. In addition to these dramatic changes in entropy, DNA is also a negatively charged molecule. That is, similar to RNA, there is a -1 charge on every DNA residue. Thus, not only should configurational entropy resist packaging, but charge–charge repulsion must also be counterbalanced by the local ionic environment. In terms of specific conformations that the DNA will adopt inside of the capsid, it is also possible that the chirality of the molecule will contribute to the overall ordering. Another key aspect of DNA packaging is whether the process may be described as a quasi-equilibrium process or a non-equilibrium process. Specifically, the capsid portal proteins convert a large amount of chemical energy to mechanical work, where ATP hydrolysis is used to drive the DNA packaging process. Together, this litany of physical attributes has been the motivation for quantitative analysis, both experimentally and theoretically.

3.3.2 Methods for Probing Packaging in Viruses

A range of techniques has been developed and applied to study the mechanics of DNA packaging in viral capsids. From an experimental perspective, the most informative technique has been optical tweezers, which provide precise measurements of the forces encountered during DNA packaging. From the theoretical perspective, two general classes of approaches include continuum elastic models and molecular simulations. As described below, analytical models have proven useful in providing theoretical frameworks for describing packaging, whereas molecular simulations have provided atomic (or pseudo-atomic) resolution descriptions of individual events. Through the integration of theoretical and experimental insights, it has been possible to directly test physical models, which has helped elucidate the key factors that govern viral packaging dynamics.

3.3.2.1 Continuum-Elastic Models

A broadly utilized theoretical technique for exploring the dynamics of DNA packaging has been the use of continuum elastic models. In these models, the DNA is treated as a continuous linear medium, where an elastic restoring potential energy is associated with bending. One reason for the popularity of these models is that it is possible to obtain analytical or nearly analytical solutions. In addition, by using an ideal representation of the DNA molecule, one may gain insights into the direct relationship between fundamental physical features. For example, one may consider how stiffness of the DNA (or capsid size) will change the relationship between packing forces and percent of the genome packed. This is in contrast to highly detailed simulation techniques, where the scale of the calculation limits the accessible dynamics to nano/microsecond scales. While these simplified models allow for general relationships to be established, there are some clear limitations. For example, since these models often utilize free-energy functional techniques, there is an assumption of quasi-equilibrium behavior. In addition, with the lack of molecular detail, the structural features of the packaged DNA are often assumed a priori. Despite some limitations, these models allow for broad relationships to be explored, as discussed below.

A major focus of continuum elastic model studies has been to understand the factors that give rise to load-dependent forces during packaging. That is, one may intuitively anticipate that as a capsid is filled, the force required to continue loading the DNA will increase. As mentioned earlier, there are numerous features that can give rise to this resistance [97–100]. First, the persistence length of DNA is on the same scale as the dimensions of a typical capsid, which implies bending forces will be significant. Next, as the DNA becomes confined, there will be a reduction in configuration entropy, and intramolecular charge–charge interactions will also need to be balanced. To address the relative contribution of each factor, Riemer and Bloomfield [97] experimentally showed that entropic changes should contribute roughly one order of magnitude less than the bending energy. Accordingly, continuum models often do not explicitly account for entropy, but rather focus on the role of electrostatic repulsion and bending. As an example, Purohit et al. [101] considered a model in which only bending and charge-repulsion effects were included, and they were able to approximately recover the relationship between packaging force and the percent of the genome already packed, as measured in contemporary single-molecule experiments [102]. Using a similar approach, Kindt et al. [98] considered the force as a function of percent packaged under conditions where the DNA was either self-attractive or repulsive. This comparison showed that an abrupt increase in packaging forces during the last one-third of genome packaging was not charge-dependent. That is, while forces encountered earlier in packaging do depend on charge–charge interactions, the final stage appeared to be limited by forces associated with steric compaction. In a subsequent study, the authors demonstrated that the high elastic energy stored at the late stages of packaging may also provide a mechanism by which a viral capsid can rapidly eject DNA during infection [99]. Building on these studies, Marenduzzo and Micheletti used mean-field arguments to show that the forces at late stages of packing depend strongly on the effective thickness of the DNA [103], which is consistent with a minimal role of electrostatics. As already mentioned, one assumption in continuum models has been of quasi-equilibrium dynamics where dissipative effects are negligible. To estimate the energy lost to dissipation, Purohit et al. [104] considered the viscosity of water and argued that dissipative effects are orders of magnitude smaller than the elastic energy stored in the packaged DNA. Together, this set of studies provides a general theoretical framework for understanding and describing the packaging of DNA in viral assemblies.

3.3.2.2 Molecular Dynamics Simulations

As described above, while continuum models provide a means for studying the factors that contribute to packaging energetics, the structural properties of DNA (e.g., bending energy, packaged conformation, etc.) are typically introduced as pre-defined parameters (or parameters obtained by fitting to experiments). To allow for detailed molecular features to be studied, there has been a variety of pseudo-atomistic models developed and applied through molecular dynamics techniques. These explicit representations of the polymer chain allow one to ask how specific chemical features can impact the final packaged configuration in a viral capsid. In addition, simulations represent a means by which to explore time-dependent behavior, in addition to equilibrium dynamics.

Since DNA packaging occurs on relatively large length scales, where long-range electrostatics and short-range excluded volume effects appear to be dominant at different stages of packaging, a coarse-grained representation of the DNA chain is commonly employed. That is, rather than explicitly represent every atom in a DNA/RNA residue (approximately 20 atoms, c.f. Fig. 3.1,) one represents the entire residue by 1–3 point masses, where the excluded volume and bonded interactions are tuned in order to account for the local geometry of the DNA. It should be noted that by coarse-graining one must take care to properly parameterize the effective interactions between pseudo-particles. While a detailed description of coarse-graining techniques is beyond the scope of the current chapter, additional reading is suggested for those interested in techniques for calibrating coarse-grained models [105, 106].

Molecular simulations have provided numerous insights into the relationship between packaged DNA structure and energetics. For example, Spokowitz and Wang [107] used Brownian Dynamics simulation with a CG model to explore the relationship between DNA twisting and packing. They found that if the end of the DNA chain is rotated as the chain is fed into the capsid, then there is a higher degree of order in the packaged DNA, where the DNA appears to adopt spool-like conformations. In contrast, when twist is not introduced, the chain is largely disordered, which may lead to internal barriers during ejection. In a similar series of studies, Petrov, Harvey, and colleagues [108–110] used CG models to study the structural and thermodynamic properties of DNA packaging. While cryo-EM reconstructions had implicated an averaged spool-like arrangement of DNA in epsilon15, the distribution of accessible conformations was elucidated through simulation. In a series of compelling studies, they demonstrated how an ensemble of spooled and folded toroid configurations was compatible with available cryo-EM measurements [108]. Further, in applications to ϕ29, their simulations demonstrated the critical role that chain stiffness and capsid shape have on the final packaged configuration [110]. Finally, these calculations were able to show that the portal protein may, in addition to driving packaging, influence the final configuration of the DNA [109]. With regard to ejection, Marenduzzo et al. performed simulations of DNA ejection when the DNA is packaged in different configurations [111]. These calculations showed that a well-ordered DNA (e.g., spooled) is associated with less internal friction, which can thereby allow for a more effective ejection process. In summary, simulations complement the use of continuum models by providing direct relationships between molecular structure and dynamics.

3.3.2.3 Optical Tweezers

The most quantitative experimental measures of DNA packaging energetics have been provided by optical tweezer experiments [102]. In these experiments, a partially packaged DNA-capsid complex is biochemically stalled. The unpackaged tail of the DNA is chemically linked to a large polystyrene microsphere, which is captured by an optical trap. The trap is then used to localize the bead/capsid to a second bead, which is held by a micropipette. The second bead is coated by an antibody that is reactive to the viral particle, which forms strong interactions between the bead and viral particle. The optical trap is then used to increase the distance between the beads, and thereby measure the force exerted. Alternately, one may apply a constant force between the beads, in order to study the response of the packaging assembly. When ATP is reintroduced, the capsid prohead begins to package DNA, which results in a spontaneous increase in force between the beads. Using this approach, it has been possible to measure the force exerted by the DNA packaging machinery under a range of solution conditions.

Since theoretical models were developed in parallel with optical tweezer techniques, there has been a natural synergy between the approaches. This has allowed for systematic comparison of theoretical models, which has helped shed light on the physical principles that govern DNA packaging. For example, in measurements of ϕ29, it was possible to determine the dependence of packaging forces on ionic composition [112]. Specifically, the authors found that the internal force depends significantly on the relative concentrations of Na^+, Mg^{2+}, and Co^{3+} ions. That is, forces were smallest when Mg^{2+} was the dominant ion, or when Co^{3+} was present, consistent with electrostatic screening being a critical determinant of packaging forces. However, the authors also noted that the measured forces were significantly larger than those predicted by models that assume the DNA adopts a spool-like configuration, suggesting that packaging is more structurally complex. In a subsequent study, the experimental sensitivity was improved, which allowed for more precise characterization of the early stage of packing [113]. Consistent with earlier measurements of late-stage packaging, these measurements also showed that the forces were larger than predictions obtained with inverse-spool models. As another test of theoretical methods, the relaxation kinetics of packaged DNA were studied by introducing predefined pauses during the packaging process [114]. These measurements showed that the force after pausing was lag-time dependent. Further, it was found that this time dependence would persist for several minutes. In relation to theoretical models, these findings suggest that the assumption of quasi-equilibrium is not always warranted, particularly during the late stages of packaging.

3.4 Concluding Remarks

In this chapter, we have aimed to provide an overview of our current understanding of the physical principles that govern dynamics in large-scale assemblies. In particular, we have highlighted some of the key questions that have been addressed in the study of ribosomal and viral systems. While earlier studies have made substantial progress, there are many aspects of these large assemblies that are poorly understood, from a physical perspective. Accordingly, there remain many opportunities to further our quantitative understanding of these complex biological machines.

Acknowledgements PCW was supported by National Science Foundation Grants MCB-1350312 and MCB-1915843. This work was supported by the Center for Theoretical Biological Physics sponsored by the NSF (Grant PHY-2019745).

References

1. G. Schlauderer, K. Proba, G. Schulz, Structure of a mutant adenylate kinase ligated with an ATP-analogue showing domain closure over ATP (1996)
2. A. Rozov, N. Demeshkina, I. Khusainov, E. Westhof, M. Yusupov, G. Yusupova, Nat. Commun. **7**(1), 1 (2016)
3. F. Guo, Z. Liu, P.A. Fang, Q. Zhang, E.T. Wright, W. Wu, C. Zhang, F. Vago, Y. Ren, J. Jakana et al., Proc. Nat. Acad. Sci. E **111**(43), 4606 (2014)
4. H.F. Noller, M.M. Yusupov, G.Z. Yusupova, A. Baucom, J. Cate, FEBS Lett. **514**(1), 11 (2002)
5. V. Ramakrishnan, Cell **108**(4), 557 (2002)
6. A. Bashan, I. Agmon, R. Zarivach, F. Schluenzen, J. Harms, R. Berisio, H. Bartels, F. Franceschi, T. Auerbach, H.A. Hansen et al., Mol. Cell **11**(1), 91 (2003)
7. J. Frank, H. Gao, J. Sengupta, N. Gao, D.J. Taylor, Proc. Nat. Acad. Sci. **104**(50), 19671 (2007)
8. M. Johansson, M. Lovmar, M. Ehrenberg, Current Opin. Microbiol. **11**(2), 141 (2008)
9. C.E. Aitken, A. Petrov, J.D. Puglisi, Ann. Rev. Biophys. **39**, 491 (2010)
10. H. Yamamoto, Y. Qin, J. Achenbach, C. Li, J. Kijek, C.M. Spahn, K.H. Nierhaus, Nat. Rev. Microbiol. **12**(2), 89 (2014)
11. G. Zhao, J.R. Perilla, E.L. Yufenyuy, X. Meng, B. Chen, J. Ning, J. Ahn, A.M. Gronenborn, K. Schulten, C. Aiken et al., Nature **497**(7451), 643 (2013)

12. M.V. Rodnina, A. Savelsbergh, W. Wintermeyer, FEMS Microbiol. Rev. **23**(3), 317 (1999)
13. R.C. Thompson, P.J. Stone, Proc. Nat. Acad. Sci. **74**(1), 198 (1977)
14. T. Ruusala, M. Ehrenberg, C. Kurland, EMBO J. **1**(6), 741 (1982)
15. M. Rodnina, R. Fricke, L. Kuhn, W. Wintermeyer, EMBO J. **14**(11), 2613 (1995)
16. T. Pape, W. Wintermeyer, M. Rodnina, EMBO J. **18**(13), 3800 (1999)
17. K.B. Gromadski, M.V. Rodnina, Mol. Cell **13**(2), 191 (2004)
18. U. Kothe, M.V. Rodnina, Biochemistry **45**(42), 12767 (2006)
19. T. Daviter, K. Gromadski, M. Rodnina, Biochimie **88**(8), 1001 (2006)
20. I. Wohlgemuth, C. Pohl, M.V. Rodnina, EMBO J. **29**(21), 3701 (2010)
21. H.S. Zaher, R. Green, Mol. Cell **39**(1), 110 (2010)
22. M. Johansson, J. Zhang, M. Ehrenberg, Proc. Nat. Acad. Sci. **109**(1), 131 (2012)
23. J. Zhang, K.W. Ieong, H. Mellenius, M. Ehrenberg, RNA **22**(6), 896 (2016)
24. H. Stark, M.V. Rodnina, J. Rinke-Appel, R. Brimacombe, W. Wintermeyer, M. van Heel, Nature **389**(6649), 403 (1997)
25. M. Valle, A. Zavialov, W. Li, S.M. Stagg, J. Sengupta, R.C. Nielsen, P. Nissen, S.C. Harvey, M. Ehrenberg, J. Frank, Nat. Struct. Biol. **10**(11), 899 (2003)
26. R.M. Voorhees, T.M. Schmeing, A.C. Kelley, V. Ramakrishnan, Science **330**(6005), 835 (2010)
27. X. Agirrezabala, E. Schreiner, L.G. Trabuco, J. Lei, R.F. Ortiz-Meoz, K. Schulten, R. Green, J. Frank, EMBO J. **30**(8), 1497 (2011)
28. T.M. Schmeing, R.M. Voorhees, A.C. Kelley, Y.G. Gao, F.V. Murphy, J.R. Weir, V. Ramakrishnan, Science **326**(5953), 688 (2009)
29. L.B. Jenner, N. Demeshkina, G. Yusupova, M. Yusupov, Nat. Struct. Mol. Biol. **17**(5), 555 (2010)
30. N. Fischer, P. Neumann, A.L. Konevega, L.V. Bock, R. Ficner, M.V. Rodnina, H. Stark, Nature **520**(7548), 567 (2015)
31. S.C. Blanchard, R.L. Gonzalez, H.D. Kim, S. Chu, J.D. Puglisi, Nat. Struct. Mol. Biol. **11**(10), 1008 (2004)
32. P. Geggier, R. Dave, M.B. Feldman, D.S. Terry, R.B. Altman, J.B. Munro, S.C. Blanchard, J. Mol. Biol. **399**(4), 576 (2010)
33. M. Valle, A. Zavialov, J. Sengupta, U. Rawat, M. Ehrenberg, J. Frank, Cell **114**(1), 123 (2003)
34. D.N. Ermolenko, H.F. Noller, Nat. Struct. Mol. Biol. **18**(4), 457 (2011)
35. D. Moazed, H.F. Noller, Nature **342**(6246), 142 (1989)
36. M.V. Rodnina, W. Wintermeyer, Biochem. Soc. Trans. **39**(2), 658 (2011)
37. P.V. Cornish, D.N. Ermolenko, H.F. Noller, T. Ha, Mol. Cell **30**(5), 578 (2008)
38. D.J. Ramrath, L. Lancaster, T. Sprink, T. Mielke, J. Loerke, H.F. Noller, C.M. Spahn, Proc. Nat. Acad. Sci. **110**(52), 20964 (2013)
39. J. Zhou, L. Lancaster, J.P. Donohue, H.F. Noller, Science **345**(6201), 1188 (2014)
40. A.H. Ratje, J. Loerke, A. Mikolajka, M. Brünner, P.W. Hildebrand, A.L. Starosta, A. Dönhöfer, S.R. Connell, P. Fucini, T. Mielke et al., Nature **468**(7324), 713 (2010)
41. I. Wohlgemuth, C. Pohl, J. Mittelstaet, A.L. Konevega, M.V. Rodnina, Philos. Trans. R. Soc. B: Biol. Sci. **366**(1580), 2979 (2011)
42. J.D. Bryngelson, P.G. Wolynes, J. Phys. Chem. **93**(19), 6902 (1989)
43. H. Frauenfelder, S.G. Sligar, P.G. Wolynes, Science **254**(5038), 1598 (1991)
44. J.D. Bryngelson, J.N. Onuchic, N.D. Socci, P.G. Wolynes, Proteins: structure. Fun. Bioinform. **21**(3), 167 (1995)
45. J.N. Onuchic, Z. Luthey-Schulten, P.G. Wolynes, Ann. Rev. Phys. Chem. **48**(1), 545 (1997)
46. J.N. Onuchic, P.G. Wolynes, Current Opin. Struct. Biol. **14**(1), 70 (2004)
47. G. Hummer, New J. Phys. **7**(1), 34 (2005)
48. S.S. Cho, Y. Levy, P.G. Wolynes, Proc. Nat. Acad. Sci. **103**(3), 586 (2006)
49. P. Das, M. Moll, H. Stamati, L.E. Kavraki, C. Clementi, Proc. Nat. Acad. Sci. **103**(26), 9885 (2006)
50. S.V. Krivov, M. Karplus, Proc. Nat. Acad. Sci. **105**(37), 13841 (2008)
51. R. Zwanzig, Proc. Nat. Acad. Sci. **85**(7), 2029 (1988)

52. A. Korostelev, D.N. Ermolenko, H.F. Noller, Current Opin. Chem. Biol. **12**(6), 674 (2008)
53. R.A. Marshall, C.E. Aitken, M. Dorywalska, J.D. Puglisi, Annu. Rev. Biochem. **77**, 177 (2008)
54. J. Frank, R.L. Gonzalez Jr., Ann. Rev. Biochem. **79**, 381 (2010)
55. H.S. Chung, W.A. Eaton, Nature **502**(7473), 685 (2013)
56. J.A. Hanson, K. Duderstadt, L.P. Watkins, S. Bhattacharyya, J. Brokaw, J.W. Chu, H. Yang, Proc. Nat. Acad. Sci. **104**(46), 18055 (2007)
57. J.B. Munro, R.B. Altman, N. O'Connor, S.C. Blanchard, Mol. Cell **25**(4), 505 (2007)
58. S. Adio, T. Senyushkina, F. Peske, N. Fischer, W. Wintermeyer, M.V. Rodnina, Nat. Commun. **6**(1), 1 (2015)
59. P.V. Cornish, D.N. Ermolenko, D.W. Staple, L. Hoang, R.P. Hickerson, H.F. Noller, T. Ha, Proc. Nat. Acad. Sci. **106**(8), 2571 (2009)
60. J. Fei, P. Kosuri, D.D. MacDougall, R.L. Gonzalez Jr., Mol. Cell **30**(3), 348 (2008)
61. J. Fei, J.E. Bronson, J.M. Hofman, R.L. Srinivas, C.H. Wiggins, R.L. Gonzalez, Proc. Nat. Acad. Sci. **106**(37), 15702 (2009)
62. E. Salsi, E. Farah, J. Dann, D.N. Ermolenko, Proc. Nat. Acad. Sci. **111**(42), 15060 (2014)
63. R.A. Marshall, M. Dorywalska, J.D. Puglisi, Proc. Nat. Acad. Sci. **105**(40), 15364 (2008)
64. J.B. Munro, R.B. Altman, C.S. Tung, K.Y. Sanbonmatsu, S.C. Blanchard, EMBO J. **29**(4), 770 (2010)
65. J. Frank, M. Radermacher, P. Penczek, J. Zhu, Y. Li, M. Ladjadj, A. Leith, J. Struct. Biol. **116**(1), 190 (1996)
66. N. Fischer, A.L. Konevega, W. Wintermeyer, M.V. Rodnina, H. Stark, Nature **466**(7304), 329 (2010)
67. A. Dashti, P. Schwander, R. Langlois, R. Fung, W. Li, A. Hosseinizadeh, H.Y. Liao, J. Pallesen, G. Sharma, V.A. Stupina et al., Proc. Nat. Acad. Sci. **111**(49), 17492 (2014)
68. J. Frank, A. Ourmazd, Methods **100**, 61 (2016)
69. J. Trylska, V. Tozzini, J.A. McCammon, Biophys. J. **89**(3), 1455 (2005)
70. Z. Zhang, K.Y. Sanbonmatsu, G.A. Voth, J. Am. Chem. Soc. **133**(42), 16828 (2011)
71. M. Levi, K. Nguyen, L. Dukaye, P.C. Whitford, Biophys. J. **113**(12), 2777 (2017)
72. F. Tama, M. Valle, J. Frank, C.L. Brooks, Proc. Nat. Acad. Sci. **100**(16), 9319 (2003)
73. Y. Wang, A. Rader, I. Bahar, R.L. Jernigan, J. Struct. Biol. **147**(3), 302 (2004)
74. K. Nguyen, P.C. Whitford, Nat. Commun. **7**(1), 1 (2016)
75. K.Y. Sanbonmatsu, S. Joseph, C.S. Tung, Proc. Nat. Acad. Sci. **102**(44), 15854 (2005)
76. P.C. Whitford, J.N. Onuchic, K.Y. Sanbonmatsu, J. Am. Chem. Soc. **132**(38), 13170 (2010)
77. L.V. Bock, C. Blau, G.F. Schröder, I.I. Davydov, N. Fischer, H. Stark, M.V. Rodnina, A.C. Vaiana, H. Grubmüller, Nat. Struct. Mol. Biol. **20**(12), 1390 (2013)
78. P.C. Whitford, Biophys. Rev. **7**(3), 301 (2015)
79. M. Długosz, G.A. Huber, J.A. McCammon, J. Trylska, Biopolymers **95**(9), 616 (2011)
80. P. Satpati, J. Sund, J. Åqvist, Biochemistry **53**(10), 1714 (2014)
81. J. Åqvist, C. Lind, J. Sund, G. Wallin, Current Opin. Struct. Biol. **22**(6), 815 (2012)
82. P.C. Whitford, J.K. Noel, S. Gosavi, A. Schug, K.Y. Sanbonmatsu, J.N. Onuchic, Proteins: structure. Fun. Bioinform. **75**(2), 430 (2009)
83. J.K. Noel, M. Levi, M. Raghunathan, H. Lammert, R.L. Hayes, J.N. Onuchic, P.C. Whitford, PLoS Comput. Biol. **12**(3), e1004794 (2016)
84. C. Clementi, H. Nymeyer, J.N. Onuchic, J. Mol. Biol. **298**(5), 937 (2000)
85. J. Karanicolas, C.L. Brooks III., J. Mol. Biol. **334**(2), 309 (2003)
86. C. Clementi, A.E. García, J.N. Onuchic, J. Mol. Biol. **326**(3), 933 (2003)
87. E.J. Sorin, B.J. Nakatani, Y.M. Rhee, G. Jayachandran, V. Vishal, V.S. Pande, J. Mol. Biol. **337**(4), 789 (2004)
88. P.C. Whitford, P. Geggier, R.B. Altman, S.C. Blanchard, J.N. Onuchic, K.Y. Sanbonmatsu, RNA **16**(6), 1196 (2010)
89. J.K. Noel, J. Chahine, V.B. Leite, P.C. Whitford, Biophys. J. **107**(12), 2881 (2014)
90. K. Nguyen, P.C. Whitford, J. Phys. Chem. B **120**(34), 8768 (2016)
91. K. Lindorff-Larsen, S. Piana, R.O. Dror, D.E. Shaw, Science **334**(6055), 517 (2011)
92. R.B. Best, G. Hummer, W.A. Eaton, Proc. Nat. Acad. Sci. **110**(44), 17874 (2013)

93. P.C. Whitford, S.C. Blanchard, J.H. Cate, K.Y. Sanbonmatsu, PLOS Comput. Biol. **9**(3), e1003003 (2013)
94. S. Arenz, L.V. Bock, M. Graf, C.A. Innis, R. Beckmann, H. Grubmüller, A.C. Vaiana, D.N. Wilson, Nat. Commun. **7**(1), 1 (2016)
95. D.E. Shaw, J. Grossman, J.A. Bank, B. Batson, J.A. Butts, J.C. Chao, M.M. Deneroff, R.O. Dror, A. Even, C.H. Fenton, et al., in *SC'14: Proceedings of the International Conference for High Performance Computing, Networking, Storage and Analysis* (IEEE, 2014), pp. 41–53
96. A.A. Chen, A.E. García, Proc. Nat. Acad. Sci. **110**(42), 16820 (2013)
97. S.C. Riemer, V.A. Bloomfield, Biopolym.: Orig. Res. Biomol. **17**(3), 785 (1978)
98. J. Kindt, S. Tzlil, A. Ben-Shaul, W.M. Gelbart, Proc. Nat. Acad. Sci. **98**(24), 13671 (2001)
99. S. Tzlil, J.T. Kindt, W.M. Gelbart, A. Ben-Shaul, Biophys. J. **84**(3), 1616 (2003)
100. T. Odijk, Biophys. J. **75**(3), 1223 (1998)
101. P.K. Purohit, J. Kondev, R. Phillips, Proc. Nat. Acad. Sci. **100**(6), 3173 (2003)
102. D.E. Smith, S.J. Tans, S.B. Smith, S. Grimes, D.L. Anderson, C. Bustamante, Nature **413**(6857), 748 (2001)
103. D. Marenduzzo, C. Micheletti, J. Mol. Biol. **330**(3), 485 (2003)
104. P.K. Purohit, M.M. Inamdar, P.D. Grayson, T.M. Squires, J. Kondev, R. Phillips, Biophys. J. **88**(2), 851 (2005)
105. S. Izvekov, G.A. Voth, J. Phys. Chem. B **109**(7), 2469 (2005)
106. A. Savelyev, G.A. Papoian, J. Phys. Chem. B **113**(22), 7785 (2009)
107. A.J. Spakowitz, Z.G. Wang, Biophys. J. **88**(6), 3912 (2005)
108. A.S. Petrov, K. Lim-Hing, S.C. Harvey, Structure **15**(7), 807 (2007)
109. A.S. Petrov, M.B. Boz, S.C. Harvey, J. Struct. Biol. **160**(2), 241 (2007)
110. A.S. Petrov, S.C. Harvey, Structure **15**(1), 21 (2007)
111. D. Marenduzzo, C. Micheletti, E. Orlandini et al., Proc. Nat. Acad. Sci. **110**(50), 20081 (2013)
112. D.N. Fuller, J.P. Rickgauer, P.J. Jardine, S. Grimes, D.L. Anderson, D.E. Smith, Proc. Nat. Acad. Sci. **104**(27), 11245 (2007)
113. J.P. Rickgauer, D.N. Fuller, S. Grimes, P.J. Jardine, D.L. Anderson, D.E. Smith, Biophys. J. **94**(1), 159 (2008)
114. Z.T. Berndsen, N. Keller, S. Grimes, P.J. Jardine, D.E. Smith, Proc. Nat. Acad. Sci. **111**(23), 8345 (2014)

Chapter 4
Organization of Intracellular Transport

Qian Wang and Anatoly B. Kolomeisky

Abstract Basic principles governing the organization of intracellular transport are presented from a theoretical point of view. It is argued that active diffusion processes are required in order to move particles inside the cell. The main components of the cellular transport and their functions are outlined. A special attention is devoted to explain the role of various motor proteins in moving cellular cargoes. Several open questions are introduced and discussed.

4.1 Introduction

Biological cells are dynamic systems where various species such as biomolecules, organelles, and vesicles are constantly produced, moved around, bound to other objects, dissociated from them, and finally degraded [1, 2]. All these processes are needed in order to maintain the functionality of cells. As an illustrative example, let us consider the fate of the secretory proteins, i.e., the proteins which cells secrete outside of their boundaries such as some hormones, toxins, and antimicrobial peptides [1–3]. They are synthesized in a cellular compartment known as the endoplasmic reticulum (ER), where they are packed into vesicles and moved into another cellular compartment called the Golgi apparatus. In the Golgi apparatus, these proteins are unpacked, sorted, and packed again in larger vesicles, after which they are transported to the cellular membranes for the removal to the cell exterior. This example shows the importance of fast and efficient transportation between different compartments. Generally, most cellular processes involve a very large number of such transport events that take place simultaneously at different locations [1, 2]. Clearly, cellular transport is crucial for keeping biological cells alive and functional. But what are the main elements of the transport system in the cell? How are they regulated? What

Q. Wang
University of Science and Technology of China, Hefei, Anhui, China
e-mail: wqq@ustc.edu.cn

A. B. Kolomeisky (✉)
Rice University, Houston, TX, USA
e-mail: tolya@rice.edu

© Springer Nature Switzerland AG 2022
K. B. Blagoev and H. Levine (eds.), *Physics of Molecular and Cellular Processes*,
Graduate Texts in Physics, https://doi.org/10.1007/978-3-030-98606-3_4

are the mechanisms that govern the intracellular motion of particles? These are the fundamental questions that are still not fully answered. However, many properties of the cellular transport are now reasonably well accounted for. In this chapter, we will discuss some features of the cellular transport. Our goal is not to present a comprehensive description on the intracellular transport, especially because multiple excellent reviews are already available [4–7], but rather to discuss some important theoretical concepts on which the transport phenomena are based.

4.2 Why Intracellular Transport Requires Active Processes?

Because in cells the locations where the biomolecules are produced and where they are eventually utilized are usually separated by large distances, an efficient transportation system is required. One could suggest that such motion can be accomplished by a simple diffusion. At living temperatures of typical organisms, the molecules in cytosol randomly move in all directions, and this is known as a Brownian motion. In principle, one biomolecule can indeed reach its destination simply by passive diffusion. However, as detailed below, the unbiased diffusion is an inefficient and unrealistic way of transportation for most cellular processes [8]. Let us estimate the time t for a biomolecule to move a distance r. For the unbiased diffusion, it can be approximated to

$$t \simeq r^2/D, \tag{4.1}$$

where D is called a diffusion coefficient. It can be estimated from the Stokes-Einstein equation:

$$D = k_B T/6\pi \eta r, \tag{4.2}$$

where k_B is the Boltzmann constant, T is temperature, η is the viscosity of the solution, and r is the hydrodynamic radius of the biomolecule. In cells, biomolecules such as proteins are typically packed in vesicles with a radius of 100 nm. So we can estimate that at room temperature the vesicle's diffusion constant in a pure water is $D = 10^{-12}$ m^2/s using (4.2). The eukaryotic cells range in sizes from 10 to 100 μm. Therefore, the time for the vesicle to diffuse across a cell can reach up to 10^4 s, i.e., several hours, using (4.1)! In addition, this estimate is too optimistic because it assumes that vesicles freely move in a dilute water solution. However, the cytosol is a very crowded environment including a large number of macromolecules, vesicles, organelles, and a complex cytoskeletal mesh network. This will significantly lower the motility of the vesicles. There is one more argument against the transportation of cellular species using the unbiased diffusion. In this case, the particles will be equally spread over in all directions, while they are needed only at a specific location. This would mean that resources for producing biomolecules would be spent very inefficiently. Too many molecules will be synthesized while only a few will be actually used. In addition, a lot of energy will be consumed to degrade the unused biomolecules. Thus, we conclude that the unbiased diffusion cannot be the single mechanism to support the cellular transport. Active diffusion is needed to move particles inside living cells.

4.3   Components of Intracellular Transport

Our discussions above identified two conditions that should be achieved during the cellular transport. First, the transport must be fast so that other biochemical and biological processes can take place on time. Second, it must be very specific: cellular cargoes must be delivered to proper locations in the cell so that precious cellular resources are not wasted in futile production. Biological cells have developed a highly efficient and robust transport system that satisfies these requirements. It consists of two main parts: a network of cellular protein filaments known as a cytoskeleton, and several classes of active enzymatic molecules, called motor proteins, that move cargoes along the cytoskeletal tracks [1, 2, 8–10]. This is schematically shown in Sect. 4.1. If one could compare the living cell with a large city, then the cytoskeletal network plays the role of highways and streets while the motor proteins correspond to cars and trucks that carry loads and people in different directions. Together, the cytoskeleton and motor proteins create a very efficient system that ensures fast and specific delivery of particles in cells (Fig. 4.1).

The cytoskeleton is shown in Sect. 4.2, and it is known that it consists of three types of protein filaments: actin filaments, microtubules, and intermediate filaments [1, 2, 9, 10]. Together they form a dense network of linear structures that connects all compartments in the cell. The cytoskeletal filaments are polar objects, meaning that one end of each biopolymer molecule (the "plus" end) is chemically different from another one (the "minus" end). This is an important property because it specifies the direction in which the motor proteins should carry the loads. What is also interesting is

Fig. 4.1 Electron micrograph showing how multiple motor proteins work together to support the intracellular transport along the cytoskeleton filaments. K: Kinesin; M: Myosin; D: Dynein; Red: Actin; Green: Microtubule. The figure is adopted with permission from [11]

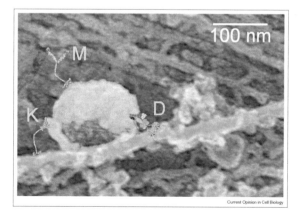

Fig. 4.2 A view of the cytoskeleton network of the eukaryotic cells obtained by microscopic studies with fluorescently labeled proteins. Actin filaments are shown in red, while microtubules are marked in green. The figure is adopted from Wikimedia Commons

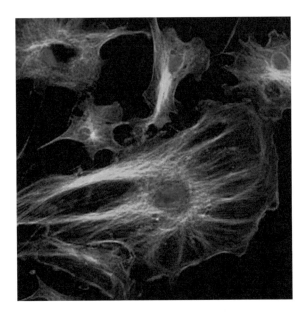

that cytoskeleton is not a frozen network structure but rather a very dynamic system that can easily adapt to changes in cellular dynamics. If there is a need to move proteins or other biomolecules to a new location in the cell, the cytoskeletal network in this direction can be quickly built. At the same time, at the locations where the cellular activities are diminished, the cytoskeletal filaments can be dismantled. This leads to a very efficient use of cellular resources. For actin filaments, it happens mostly via the so-called *treadmilling* process when the filament grows at one end and shrinks at another one [9, 10, 12]. Microtubules are also very dynamic biopolymers because of the process known as a *dynamic instability* when these filaments stochastically alternate between the periods of growth or shrinking in response to changes in cellular conditions [9, 10]. These properties make the intracellular transport very specific and robust (Fig. 4.2).

Another important component of the cellular transport is several classes of cytoskeletal motor proteins [1, 2, 8, 10, 13]. It is known that these biological molecular motors, when bound to cytoskeletal filaments, can catalyze highly exothermic chemical reactions such as the hydrolysis of adenosine triphosphate (ATP) or related compounds [1, 2, 8]. During this process, they are able to convert a fraction of the released chemical energy into a mechanical motion. This allows bound motor proteins with attached cellular cargoes to move processively in one direction at high speeds. Clearly, this ATP-driven directional one-dimensional motion on the cytoskeletal filaments is much faster than the unbiased passive diffusion process described in Sect. 4.2. Let us show this by estimating the time for a vesicle to be moved $\sim 100\,\mu$m, assuming now that this time the transport is accomplished with the help of a motor protein, kinesin. At typical cellular ATP concentrations, the velocity of kinesin is ~ 1 μm/s [6, 8, 14]. Thus, the vesicle will be moved in ~ 100 s, which is significantly faster than the estimate based on the unbiased diffusion ($\sim 10^4$ seconds). Furthermore, because vesicles and organelles are typically moved by several molecular motors, they will help to overcome various obstacles and traps on their way.

There are several families of cytoskeletal motor proteins that are involved in the cellular transport [1, 2, 8]. They differ by walking directions, processivity (i.e., how far they can travel along the filaments before dissociating into the bulk), types of cytoskeletal tracks, speeds, and forces which they exert on cellular cargoes. Some of them are presented in Sect. 4.3. In this chapter, we will concentrate on three main classes of motor proteins: kinesin-1, cytoplasmic dynein, and myosin V [8], which are responsible for the main features of the cellular transport [8].

(a) Kinesin-1, also known as a conventional kinesin, is a motor protein that walks along microtubules (MT) in the plus end. It is a dimeric molecule, i.e., it has two motor heads serving as both microtubule-binding domains and the place where ATP hydrolysis happens. Using the energy of ATP hydrolysis, its two motor heads walk on MT in a "hand-over-hand" fashion—the leading head and the trailing head alternate their steps, similar to how humans walk. Kinesin-1 is a highly processive motor protein, meaning that it can walk long distances along MT before it detaches, typically up to 1 μm. In the absence of forces, kinesin-1 walks toward the plus end of MT with a step size of ~ 8 nm, which corresponds to the distance between neighboring tubulin subunits on the microtubule. In vitro, the kinesin velocity was measured to be around 800 nm/s without external loads [8]. This velocity gradually decreases with resisting external forces and drops to zero at around 7–8 pN which is called a stall force. With forces larger than 8 pN, kinesin-1 can even walk reversely toward the minus end of MT, although quite slowly [14].

(b) Cytoplasmic dynein behaves differently from the kinesin-1, although it also moves on microtubules but mainly in the opposite "minus" direction of the filament. Cytoplasmic dynein is a massive multi-subunit complex protein composed of two heavy chains and multiple intermediate/light chains [15]. It mainly walks toward the minus end of the microtubule while the movement toward the plus end is also possible under some conditions. In addition, its step size has a wide distribution with multiple peaks at 8, 16, and −8 nm [16]. It has been proposed that this unique step-size distribution profile is the result of multiple walking patterns—"hand-over-hand" (leading head and trailing head swap) as well as "inchworm" (leading head always leads). The *in vitro* velocity of cytoplasmic dynein was measured to be less than 100 nm/s [17] while the in vivo velocity can be as high as 1 μm/s [18]. This discrepancy has not been fully understood yet. Cytoplasmic dynein is also a processive motor. Without resisting forces, its run length along MT is ~ 1.7 μm [15].

(c) Myosin V walks along a different cytoskeletal filament, actin filaments (F-actin), instead of MT. Like kinesin, it walks toward the plus end by employing a "hand-over-hand" mechanism [19]. Myosin V is also a processive motor whose run length is ~ 2.0 μm [20]. At typical cellular concentrations of ATP, the speed of myosin V is ~ 400 nm/s [21] with the stall force being slightly less than 3 pN [22]. The step size of myosins-V is equal to ~ 36 nm [19], which is significantly larger than the values for kinesins and dyneins. It is also interesting to note that myosins V are more efficient than kinesins and dyneins [8], i.e., they utilize a larger fraction of the released chemical energy of ATP hydrolysis for moving cellular cargoes. But the reason for this is still not fully understood [8] (Fig. 4.3).

4.4 Current Understanding of Mechanisms of Intracellular Transport

As a result of multiple studies, the following picture (although probably still significantly simplified in comparison with reality) of the cellular transport processes is emerging [1, 2, 4–6, 8]. Specific biomolecules (e.g., proteins) are synthesized locally at some cellular compartments (like secretory proteins which are produced in ER). Then they are packed in vesicles, which are small spherical particles that can be viewed as containers for cellular material. It is apparently more efficient to transport biomolecules in groups rather than in single species, like moving passengers in the public transportation in big cities. After that, the vesicles are picked up by the motor proteins, which start to move vesicles along the cytoskeletal filaments in all possible directions. Different types of motors are utilized depending on the nature of the cytoskeletal filaments and the required direction. Biochemical signaling networks are involved in specifying the directions of the cellular cargoes. Sometimes, vesicles with the cellular material inside are moved to intermediate compartments (like secretory proteins are moved to the Golgi apparatus), where they are unpacked, sorted, and repacked again in typically larger vesicles. This is followed by further transportation, again with the help of motor proteins, to the final destinations. Some biomolecules are also moved in the cell medium by concentration gradients, i.e., via diffusion through the cellular medium. For example, kinesin motor proteins that hop along microtubules concentrate at the periphery of the cells after unloading the cargoes and this concentration gradient moves free kinesin motors back to the center of the cell. Thus, the cellular transport is a complex combination of biased and unbiased diffusion processes coupled to multiple biochemical reactions. In most cases, it is controlled by motor proteins that run on cytoskeletal filaments.

Now let us discuss how the motor proteins carry the loads along the cytoskeletal filaments. Experiments indicate that several motor proteins transport together cellular cargoes on cytoskeletal filaments. This seems to be a reasonable choice because if one motor accidentally dissociates from the filament, other motors will continue the work. But these observations also raise a question if motors might cooperate more closely during the transportation by sharing the load and moving faster? It turns out that the degree of cooperativity depends on the nature of motors [23, 24]. Kinesins, which step along the microtubules in the direction of the cell periphery, do not collaborate much. At the same time, dyneins that move cargoes on microtubules in the minus direction and myosins V and VI that travel along actin filaments cooperate much stronger. Recent experimental and theoretical investigations were able to explain

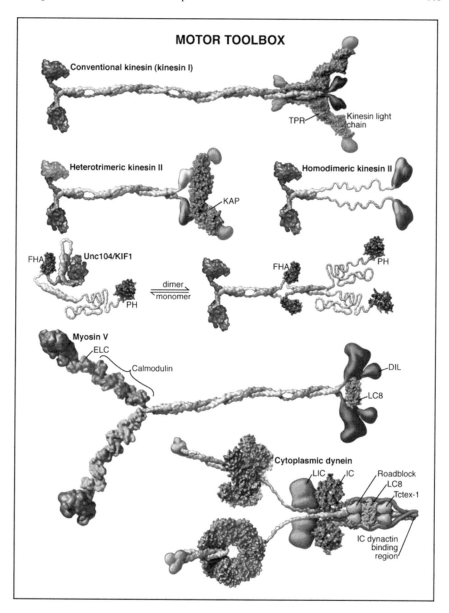

Fig. 4.3 A structural view on most important cytoskeleton motor proteins. The figure is adopted with permission from [13]

such behavior for groups of various motor proteins. It turned out that the collective motion of the team of coupled kinesins depends on the properties of individual motors [25]. To be more precise, the weak response of the single kinesin to external forces means that when several of them are coupled together, the leading motor does not slow down much, not allowing other motors to catch up and to share the external load. The leading motor will eventually dissociate before other motors can help. This effectively leads to a low cooperativity for multiple kinesin motors. The situation is different for dyneins and myosins V and VI. Each single dynein or single myosin is more sensitive to external forces, and this allows teams of motors to cooperate better by sharing the external forces.

However, the situation with the transportation of cellular cargoes is more complex because usually different types of motors are bound to them at the same time. For example, both dyneins and kinesins associate to vesicles that move along microtubules, and these motors have different driving directions. This situation is usually referred to as Tug-Of-War (TOW) phenomenon, and it is currently being intensively investigated [23, 26]. So how the directed motion of vesicles can happen in such situations? Although we still do not fully understand many details, it is believed that the following events are taking place. The biochemical regulation targets dyneins but not kinesin molecules. In the region where the cellular cargoes should be preferentially moved by kinesins, other regulating proteins probably limit the number of dyneins bound to the vesicle. But every single kinesin is stronger (i.e., can produce larger forces) than any single dynein [8], and the vesicle is moved by kinesins in the positive direction along the microtubules. In this case, probably the dyneins are staying attached to the vesicle but they are dissociated from the microtubule filament. In the region where the cargo should be moved by dyneins, the regulation increases the association of dyneins to the vesicle and the microtubule. Now, because several dyneins cooperate better than kinesins, they start to win TOW. The vesicle moves in the negative direction along the microtubule. It seems that the nature has chosen to regulate only one component of the TOW processes, namely dyneins. This is, of course, a very efficient and robust way of regulating the intracellular transport.

4.5   Open Questions and Future Directions

Although a tremendous progress in uncovering mechanisms of intracellular transport has been achieved, many aspects of these complex phenomena remain not well understood. Let us briefly discuss several of them.

Although the presented above description of the TOW phenomena has been successful in elucidating the intracellular transport by motor proteins in vitro, several recent in vivo experiments show unexpected behaviors that cannot be explained. Our arguments of TOW phenomena indicate that inhibiting one type of motor should lead to the increased motility in the opposite direction. However, in vivo experiments showed a completely different scenario: inhibiting one motor decreases the motility in both directions [27]. This unexpected phenomenon is called a *co-dependence paradox* [28].

The co-dependence paradox implies that the activation of one motor protein might require the inhibition of another motor protein. Is this true? It is still an open question at present. However, some motor proteins do have self-inhibition mechanisms. As detailed in Sect. 4.3, motor proteins utilize the chemical energy from ATP hydrolysis. If this process happens before a motor protein attaches to its cellular cargo, it would be a waste of ATP energy. In fact, motor proteins have evolved into unique structures that inhibit MT binding or ATP hydrolysis unless certain conditions meet. Let us take the kinesin-1 molecule as an example. Most in vitro experiments used a construct of kinesin-1 without its tail (kinesin-1_{-tail}). Under these circumstances, kinesin-1_{-tail} shows a high processivity along MT as discussed in Sect. 4.3. In contrast, the full-length kinesin-1 (kinesin-1_{full}) was found to be self-inhibited [29]. The mechanism of inhibition has not been fully understood yet. For kinesin-1_{full}, its tail was found to be able to directly interact with its head [30], but the molecular details of how such interactions inhibit the motility of kinesin-1_{full} are largely unknown. In addition, the binding of another protein known as c-Jun N-terminal kinase (JNK) to kinesin-1_{full} was proposed to activate kinesin-1_{full} [31]. However, whether the activation of kinesin-1_{full} requires (or is enhanced by) the existence of the other type of motor protein is unknown. For a cargo carried by both kinesins and dyneins, there are hypotheses claiming that the activation requires a direct interaction between kinesin and dynein or dynactin, a protein that binds to dynein and assists in the binding between dynein and cargo. However, previous experiments found that the co-dependence still exists even when dynein was replaced by another motor protein that also walks toward the minus end of MT [32]. Nevertheless, this critical problem remains unsolved, and if we want to understand the transport in the living cells better, it should be addressed in future studies.

Currently, our understanding of intracellular transport by motor proteins is mostly based on studies that were performed in dilute environments. However, a real cellular medium is significantly "crowded" than the dilute solution utilized in many in vitro experiments. Large numbers of macromolecules are present inside a living cell, and this should affect the dynamic of motor proteins. One of the most important effects of the cellular medium on all molecules inside the cell is known as a "macromolecular crowding effect" [33]. The free space for diffusion of motor proteins is largely limited because many large macromolecules sterically forbid them to move in specific locations. This is an entropic effect that significantly changes the thermodynamic and kinetic properties of motor proteins such as their association rate to cargo or cytoskeletal filament. It also influences the conformational changes in its enzymatic cycle. In addition, when a motor protein walks on the cytoskeleton, other macromolecules nearby will influence the traffic along the cytoskeleton. It is important to clarify the role of crowding in the intracellular crowding, and this will require significant experimental and theoretical efforts.

Investigating the mechanisms of cellular transport remains a challenging problem. It requires a combination of advanced experimental and theoretical methods in order to clarify these fundamental issues.

Acknowledgements Authors were supported by the Center for Theoretical Biological Physics at Rice University via NSF Grant PHY-1427654.

References

1. K. Roberts, B. Alberts, A. Johnson, P. Walter, T. Hunt, New York: Garland Science (2002)
2. H. Lodish, A. Berk, C.A. Kaiser, C. Kaiser, M. Krieger, M.P. Scott, A. Bretscher, H. Ploegh, P. Matsudaira, et al., *Molecular Cell Biology* (Macmillan, 2008)
3. J. Lippincott-Schwartz, T.H. Roberts, K. Hirschberg, Ann. Rev. Cell Dev. Biol. **16**(1), 557 (2000)
4. T.A. Rapoport, Nature **450**(7170), 663 (2007)
5. A.L. Jolly, V.I. Gelfand, Biochem. Soc. Trans. **39**(5), 1126 (2011)
6. N. Hirokawa, Y. Noda, Y. Tanaka, S. Niwa, Nat. Rev. Mol. Cell Biol. **10**(10), 682 (2009)
7. C. Appert-Rolland, M. Ebbinghaus, L. Santen, Phys. Rep. **593**, 1 (2015)
8. A.B. Kolomeisky, *Motor Proteins and Molecular Motors* (CRC Press, 2015)
9. D.A. Fletcher, R.D. Mullins, Nature **463**(7280), 485 (2010)
10. J. Howard et al., *Mechanics of Motor Proteins and the Cytoskeleton*, vol. 743 (Sinauer Associates Sunderland, MA, 2001)
11. E.L.F. Holzbaur, Y.E. Goldman, Current Opin. Cell Biol. **22**(1), 4 (2010)
12. M.F. Carlier, S. Shekhar, Nat. Rev. Mol. Cell Biol. **18**(6), 389 (2017)
13. R.D. Vale, Cell **112**(4), 467 (2003)
14. R.A. Cross, A. McAinsh, Nat. Rev. Mol. Cell Biol. **15**(4), 257 (2014)
15. M.A. Cianfrocco, M.E. DeSantis, A.E. Leschziner, S.L. Reck-Peterson, Ann. Rev. Cell Dev. Biol. **31**, 83 (2015)
16. M.A. DeWitt, A.Y. Chang, P.A. Combs, A. Yildiz, Science **335**(6065), 221 (2012)
17. R. Mallik, B.C. Carter, S.A. Lex, S.J. King, S.P. Gross, Nature **427**(6975), 649 (2004)
18. J.F. Presley, N.B. Cole, T.A. Schroer, K. Hirschberg, K.J. Zaal, J. Lippincott-Schwartz, Nature **389**(6646), 81 (1997)
19. A. Yildiz, J.N. Forkey, S.A. McKinney, T. Ha, Y.E. Goldman, P.R. Selvin, Science **300**(5628), 2061 (2003)
20. T. Sakamoto, F. Wang, S. Schmitz, Y. Xu, Q. Xu, J.E. Molloy, C. Veigel, J.R. Sellers, J. Biol. Chem. **278**(31), 29201 (2003)
21. J.N. Forkey, M.E. Quinlan, M.A. Shaw, J.E. Corrie, Y.E. Goldman, Nature **422**(6930), 399 (2003)
22. G. Cappello, P. Pierobon, C. Symonds, L. Busoni, J. Christof, M. Gebhardt, M. Rief, J. Prost, Proc. Nat. Acad. Sci. **104**(39), 15328 (2007)
23. R. TyleráMcLaughlin et al., Soft Matter **12**(1), 14 (2016)
24. Q. Wang, M.R. Diehl, B. Jana, M.S. Cheung, A.B. Kolomeisky, J.N. Onuchic, Proc. Nat. Acad. Sci. E **114**(41), 8611 (2017)
25. J.W. Driver, D.K. Jamison, K. Uppulury, A.R. Rogers, A.B. Kolomeisky, M.R. Diehl, Biophys. J. **101**(2), 386 (2011)
26. M.J. Müller, S. Klumpp, R. Lipowsky, Proc. Nat. Acad. Sci. **105**(12), 4609 (2008)
27. S.E. Encalada, L. Szpankowski, C.H. Xia, L.S. Goldstein, Cell **144**(4), 551 (2011)
28. W.O. Hancock, Nat. Rev. Mol. Cell Biol. **15**(9), 615 (2014)
29. D.L. Coy, W.O. Hancock, M. Wagenbach, J. Howard, Nat. Cell Biol. **1**(5), 288 (1999)
30. D. Cai, A.D. Hoppe, J.A. Swanson, K.J. Verhey, J. Cell Biol. **176**(1), 51 (2007)
31. T.L. Blasius, D. Cai, G.T. Jih, C.P. Toret, K.J. Verhey, J. Cell Biol. **176**(1), 11 (2007)
32. S. Ally, A.G. Larson, K. Barlan, S.E. Rice, V.I. Gelfand, J. Cell Biol. **187**(7), 1071 (2009)
33. A.P. Minton, Biopolym.: Orig. Res. Biomol. **20**(10), 2093 (1981)

Chapter 5
Introduction to Stochastic Kinetic Models for Molecular Motors

Mauro L. Mugnai, Ryota Takaki, and D. Thirumalai

Abstract Molecular motors such as those of kinesin, dynein, and myosin superfamilies play a critical role in various aspects of cell physiology. These motors move along linear tracks formed by cytoskeletal filaments and can transport materials, apply forces, or both. One class of motors is referred to as processive, in that the motor typically remains bound to the filament even as it steps from one site to the next. To avoid the obvious equilibrium finding that one binding site is as good as the next and hence there should be no direct preference for the motor to move in a particular direction, motors couple their stepping motion to hydrolysis of energy-rich compounds with out-of-equilibrium concentrations (such as ATP) to tailor an appropriate free energy landscape driving them in a preferred direction. This chapter introduces stochastic models of these motors that can be used to calculate their statistical properties and provide an understanding of detailed single-molecule experimental data.

5.1   Introduction

Molecular motors such as myosins, kinesins, and dyneins are enzymes that use the energy stored in the hydrolysis of ATP in order to slide, bend, or move along the cytoskeleton, that is the filamentous actin (F-actin) and microtubules (MTs). Cytoskeletal filaments may be thought of as linear and polarized, with the plus end directed towards the periphery of the cell. Most myosins move towards the plus end of F-actin, although at least one minus-end-directed exception exists. Conventional kinesins and dyneins use the MT as their "track" and are headed in opposite directions—kinesin towards the cell periphery and dyneins towards the center. Nat-

M. L. Mugnai · D. Thirumalai (✉)
The University of Texas at Austin, Austin, TX, USA
e-mail: dave.thirumalai@gmail.com

M. L. Mugnai
e-mail: mmugnai@utexas.edu

R. Takaki
Max Planck Institute, Dresden, Germany
e-mail: rt4.6692016@utexas.edu

© Springer Nature Switzerland AG 2022
K. B. Blagoev and H. Levine (eds.), *Physics of Molecular and Cellular Processes*,
Graduate Texts in Physics, https://doi.org/10.1007/978-3-030-98606-3_5

urally, there could be tug of wars between these motors, both of which walk on MTs. Molecular motors are involved in a variety of cellular processes. For instance, they power cellular motility by bending cilia, enable muscle contraction by sliding the intertwined thick and thin filaments, separate the chromosomes during cellular division, and shuttle cargo across the cell [1]. Some of these functions require the cooperation between motors belonging to different super families, as is the case for the aggregation and dispersion of the melanosome [2].

How do molecular motors perform their functions? Without entering into structural details, all these motors have a common architecture. (i) A "head" domain which binds the filament and hydrolyzes ATP (in the case of dynein, the long stalk connects the track with the "head",) (ii) a mobile unit (converter and lever arm for myosin, neck linker for kinesin, and linker for dynein) that changes conformation and enables force generation, and (iii) a tail region for interacting with other proteins. The catalytic cycle of the motor [3] (binding and hydrolyzing ATP, releasing phosphate and ADP) is associated with conformational transitions that (i) modify the affinity of the motor for the cytoskeletal track and (ii) induce a conformational transition (power stroke) of the mobile domain of the motor which exerts the force necessary for directed movement. Of course, in order to move along the track, the force-producing transition must occur while the motor is bound to the filament. Myosins and dyneins have a low affinity for the track when they bind ATP. Phosphate release is associated with the force-producing conformational transition and leads to a new motor conformation that has a high affinity for the filament. In contrast, kinesin detaches from the MT after phosphate release, and the release of ADP favors binding to the track. The force-producing transition is associated with ATP binding (or ATP hydrolysis). Of course, reverse transitions are possible, and in order to perform the cycle preferentially in one direction, some energy needs to be expended in accordance with the Second Law of Thermodynamics [4] and the theoretical physics of biological machines. The cell maintains the concentration of ATP, ADP, and phosphate far from equilibrium, so that the hydrolysis of ATP releases $\approx 20k_\mathrm{B}T$ of free energy [5]. Therefore, completing a cycle in the reverse direction is extremely unlikely.

In this section, we focus on a particular group of molecular motors, which are said to be "processive." This means that these motors, once they bind the filament, are able to repeat the enzymatic cycle numerous times without fully detaching from the track. In general, processivity requires cooperation between (at least) two heads, which form a dimer by joining their tail domains. A dimeric motor is processive because while one head is detached from the filament the other remains bound, so that the dimeric complex does not disengage from the track. In this manner, dimeric motors "walk" along the filaments by an alternating stepping mechanism in which the trailing head (TH) of the complex releases the filament and binds again along the track in front of the bound head, thereby becoming the new leading head (LH). This mechanism is termed "hand-over-hand" stepping, and it has been established for some members of the three families of motors discussed here [6]. In order to ensure that the TH detaches first, there has to be some coordination between the two heads of the motor. This can be achieved if a step in the enzymatic cycle of one of the two heads is "gated," that is it is prevented until the conformation of the other head changes. Although it is likely that the inter-head tension, reflected in the architecture of the motor, plays a critical role in dictating the inter-head coordination, it is fair to say that the structural basis of gating is still lacking.

Although there are analogies to the macroscopic world, which allows us to use phrases like "tracks," "walking," and of course "motors," for these nanoscale machines, the reader should not forget that the energies associated with the conformational transitions are not too far off from thermal energy. This is of course necessary; otherwise, the motors would be stuck in a particular state without being able to proceed along their enzymatic cycle, let alone executing their directional movement. On the other hand, these low-energy barriers enable the occurrence of backward steps, "inchworm-like" movements, and "foot-stomps" [7, 8]. In addition, it is possible and does occur that both the heads of the dimeric motor simultaneously release the track, thereby ending the processive run. These considerations illustrate a clear feature of molecular motors: they are stochastic machines, and this should be kept in mind even though we borrow some language from the deterministic, macroscopic world to characterize their stepping behavior.

How do we characterize the motility of motors? A large number of beautiful single-molecule experiments have illustrated the movement of these proteins through a variety of different methods [9]. Often, a filament is attached to a glass surface and the movement of a motor is monitored by tracking a cargo attached to the motor construct using optical tweezers. The opposite geometry has been explored as well, in which the motors are tethered to a surface or to a surface-bound bead, and the movement of the track is monitored via optical tweezers. Alternatively, probes that can be monitored by fluorescence or scattering may be attached to one or both heads of the construct or to the track, and the movement may be followed in the absence of imposed load [10, 11]. These studies enabled the measurement of quantities that characterize the motility of motors, such as the run length (L, i.e. the distance covered during a processive run), velocity (v), step size (s), and probability of backward stepping. A wealth of relevant information can be obtained by monitoring these properties under different environmental conditions. The "knobs" that experimentalists can tune in order to probe the stepping kinetics of the motors are the nucleotide concentrations (ATP and ADP) and the load applied to the motor. Examples of such experimental measurements are shown in Fig. 5.1a, b. Because the stepping of molecular motors is stochastic in nature due to thermal noise, it is important to monitor not only average properties, but also distributions (see Fig. 5.1b–d).

In order to understand how a motor executes its processive walk, characterized by data described in Fig. 5.1, a theoretical model is necessary. Such a theory should connect the macroscopic features of motility with changes in the state of the two heads—nucleotide and filament binding/release. In some instances, it may be necessary to take many factors (architecture of the motor, the number of nucleotide states sampled by the motor during a single step, and how the two heads communicate in order to preserve processivity) in order to obtain quantitative agreement with experiments. As a way of introducing the concepts leading to theories for motors, we consider simple stochastic models. In this section, we discuss the development of analytically solvable kinetic models for extracting mechanistic information from

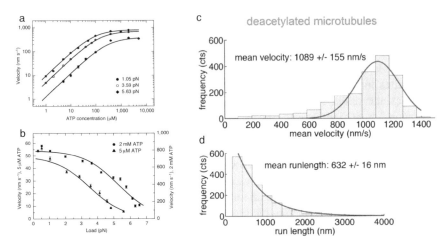

Fig. 5.1 Typical experimental observables for kinesin. **a** Average velocity as a function of ATP concentration and **b** external load. The graph is adapted from [12]. **c** Distribution of the velocity; **d** the run length. The graph is adapted from [13]. These figures are meant to give the typical stepping characteristics of kinesin. One of the objectives of a theory is to be able to calculate these observables, preferably using simple models

motility data. In particular, we focus on models that incorporate the detachment from the filament (end-of-run) as a prominent feature of the model. In the next section, we describe the key ingredients of kinetic models and why they are appropriate to study molecular motors. We then discuss two minimal models that reproduce the salient features of motor motility as a function of applied load and ATP concentration. Next, we discuss a more general framework that may be adopted in order to construct more complicated models. The introduction to stochastic kinetic models will prepare the reader to delve into more complicated models often used to analyze experimental data. We conclude by assessing the benefits and shortcomings of kinetic networks for molecular motors.

5.2   Stochastic Kinetic Models

In the framework of stochastic kinetic models, the enzymatic function of molecular motors is described in terms of a discrete set of discernible intermediate states. The chemical nature of these conformations (whether they are attached to the track, and which nucleotide is bound to the motor) can be identified using chemical kinetic studies, in which the rate k of the individual steps of the cycle can be probed. Assuming that at time $t = 0$ the system is in a particular state, the probability density that the transition away from this state occurs in a time interval dt after a dwell time t is given by the Poisson distribution,

$$f(t) = ke^{-kt}, \tag{5.1}$$

where $1/k$ is the average lifetime of the initial state $(= \int_0^\infty tf(t)dt)$. The system might leave the initial conformation via alternative pathways towards distinct states. Each one of these transitions is associated with a rate k_{ji}, where i is the initial state and j is the final state. The probability density for a $i \to j$ transition in a time interval dt after a dwell time t is

$$f_{ji}(t) = k_{ji}e^{-\sum_m k_{mi}t}, \tag{5.2}$$

where the sum at the numerator is extended over all the states accessible from i. In this case, the average lifetime of state i is $\bar{\tau}_i = \sum_j \int_0^\infty tf_{ji}(t) = (\sum_j k_{ji})^{-1}$. Moreover, $k_{ji}/\sum_{m \neq i} k_{mi}$ is the probability that after transiting from the state i the system enters the state j.

For processive molecular motors, single-molecule experiments have shown that the stepping trajectories resemble a staircase characterized by two features: stochastic dwell times and rapid jumps associated with the movement of the motor. During the dwell time, the motor changes its state in a manner that conforms to the experimental conditions, i.e. nucleotide concentrations and applied load. During the displacements, the positional changes are so fast that within the time resolution of experiments they can be characterized as almost instantaneous "jumps" along the filament.

All processive runs eventually are terminated with the detachment of the motor from the track. This process may also be included in the stochastic kinetic model by introducing a transition rate γ_i associated with the end-of-run. We assume that this state is "absorbing", meaning that there is no recovery from it; once it is visited, the system remains trapped there. Detachment is a key feature of the model and indeed is observed in all in vitro experiments. A critical quantity to calculate is $f(m, l, t)$, the joint probability density of taking m forward steps and l backward steps before detaching from the track at time t. Once this quantity is determined, experimentally measured observables such as the probability distributions and averages of the run length, the run time (time spent bound to the track before detachment), and velocity can be determined. We begin by presenting a one-state and a two-state model for which the analytical expression for the run length and velocity distributions were derived in [14, 15], respectively. The appeal of these minimal models, apart from their simplicity, is their ability to quantitatively analyze the experimental data, thus providing insights into the stepping kinetics. More importantly, these models make experimentally testable predictions. However, minimal models do not account for the details of the biochemistry of dimeric molecular motors and the potential of alternative stepping mechanisms. In order to incorporate these features, it is necessary to introduce a more general, albeit more complicated, set of models for which certain distributions and some average quantities related to experimentally measurable observables can be determined. General formulation to calculate such quantities has been recently provided [16]. For completeness, here we introduce an alternative derivation of the theoretical framework discussed elsewhere [16]. The reader will immediately appreciate the complications that arise in more realistic stochastic kinetic model descriptions of molecular motors.

5.3 One-State Model

In the minimal model shown in Fig. 5.2, the motor can be regarded as a one-state random walker taking steps of size s along a track.

The motor at site l steps forward ($l \rightarrow l + 1$) with rate k^+, takes backward steps ($l \rightarrow l - 1$) with rate k^-, and detaches from the track (i.e. reaches the state "OUT" in Fig. 5.2) with rate γ.

Following (5.2), we can construct $f(m, l, t)$ as follows:

$$
f(m, l, t) = \frac{(m + l)!}{m! l!} \int_0^t dt_{m+l} \int_0^{t_{m+l}} dt_{m+l-1} \cdots \int_0^{t_3} dt_2 \int_0^{t_2} dt_1 \\
\prod_{i=1}^{m} \left[k^+ e^{-k_T(t_i - t_{i-1})} \right] \prod_{j=m+1}^{m+l} \left[k^- e^{-k_T(t_j - t_{j-1})} \right] \left[\gamma e^{-k_T(t - t_{m+l})} \right],
$$

(5.3)

where $k_T = k^+ + k^- + \gamma$ is the sum of all the rates. Equation 5.3 can be justified with the following considerations. After each step, the motor does not "remember" (loses memory of) what was the trajectory leading to that specific location along the track, and it can step forward/backward or detach at the same rate regardless of its "history". This implies that the order of the steps does not matter, and the weight of any trajectory characterized by m forward steps, l backward steps, and detachment at time t is the same. Therefore, in order to compute the probability $f(m, l, t)$, we need (i) the total number of trajectories and (ii) the weight of a specific one (any one, they are all the same). The total number of trajectories is given by the binomial factor. We compute the probability of a trajectory in which the motor takes the first m forward steps, then l backward steps, before detaching from the track. Because

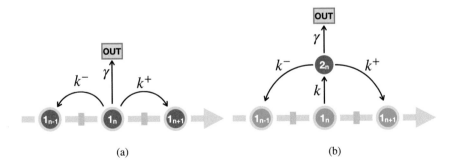

(a)   (b)

Fig. 5.2 a One-state stepping kinetics for a motor on discrete track. The motor steps in the forward direction with rate k^+ and backward direction with rate k^- and detaches with rate γ. OUT represents the absorbing state for the motor corresponding to the detachment from the track. **b** Two-state stepping kinetics for a motor on a discrete track. The motor takes mechanical steps or detaches only after the chemical transition from state 1 to state 2 with rate k

we only care about the detachment time, we integrate over all the intermediate times $t_1 \le t_2 \le \cdots \le t_{m+l} \le t$, which creates a time-ordered series of events. It is easy to carry out the integration, described in Appendix 5.11, which leads to

$$f(m, l, t) = \left(\frac{t^{m+l}}{m!l!}\right)(k^+)^m (k^-)^l \gamma e^{-k_T t}. \tag{5.4}$$

Using this expression, we will now show how to calculate the run length, run time, and velocity distributions.

Run Length Distribution. In order to obtain the run length distribution, we begin by calculating the probability $P(m, l)$ that the motor takes m forward steps and l backward steps before detachment. This can be obtained as $P(m, l) = \int_0^\infty dt f(m, l, t)$, in which we remove the dependence of $f(m, l, t)$ on the detachment time. The repeated integration by parts brings a term $(m + l)!k_T^{-m-l-1}$, from which we obtain

$$P(m, l) = \frac{(m + l)!}{m!l!} \left(\frac{k^+}{k_T}\right)^m \left(\frac{k^-}{k_T}\right)^l \left(\frac{\gamma}{k_T}\right). \tag{5.5}$$

In (5.5), k^+/k_T (k^-/k_T) is the probability of taking forward (backward) step, and γ/k_T is the probability that the motor detaches from the track. Let the net displacement on the track be $n = m - l$, so that the run length is $L = ns$. Up to a scale factor s, the run length distribution $P(n)$ is given by $P(n) = \sum_{m,l=0}^\infty P(m, l)\delta_{m-l,n}$, where $\delta_{m-l,n}$ is the Kronecker delta, which is equal to 1 if $n = m - l$ and is 0 otherwise. By carrying out the summation as detailed in Appendix 5.9, we obtain

$$P(n \gtrless 0) = \left(\frac{2k^\pm}{k_T + \sqrt{k_T^2 - 4k^+k^-}}\right)^{|n|} \frac{\gamma}{\sqrt{k_T^2 - 4k^+k^-}}. \tag{5.6}$$

Clearly, $P(n)$ indicates that the run length distribution decays exponentially with n, as illustrated in Fig. 5.3a. From (5.6), we can compute the average run length, and as we have shown in Appendix 5.9, we obtain

$$\overline{L} = s\overline{n} = s \sum_{n=-\infty}^\infty nP(n \gtrless 0) = s\frac{k^+ - k^-}{\gamma}, \tag{5.7}$$

where s is the step size of the motor. This result could have been guessed at first glance, because it is nothing more than the difference between the probability of stepping forward (k^+/k_T), minus the probability of stepping backward (k^-/k_T), divided by the probability of detaching from the track (γ/k_T).

Run Time Distribution. Next, we compute the run time distribution, $P(t)$, which is the distribution of time spent by the motor attached to the track. This

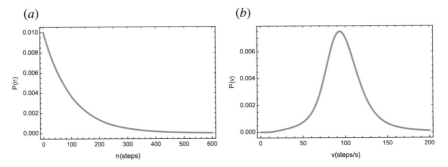

Fig. 5.3 **a** The distribution of run length of the motor stepping on a discrete track (5.6). n is the net displacement along the track from the starting position of the motor. **b** Velocity distribution of the motor stepping on a discrete track. The distribution is plotted from (5.13). Parameters for the plots are $k^+ = 100$ (s^{-1}), $k^- = 1$ (s^{-1}), and $\gamma = 1$ (s^{-1})

can be obtained from $f(m, l, t)$ in (5.4) by summing over m and l, that is $P(t) = \sum_m^\infty \sum_{l=0}^\infty f(m, l, t)$, and it equal to[1]

$$P(t) = \gamma e^{-\gamma t}, \tag{5.8}$$

from which it is easy to show that the average run time is

$$\bar{t} = \gamma^{-1}. \tag{5.9}$$

Velocity Distribution. Given the average run length and the average run time, it is tempting to define the velocity of the motor to be

$$v = \frac{\bar{L}}{\bar{t}} = s(k^+ - k^-), \tag{5.10}$$

which is quite intuitive: the velocity is given by the imbalance of the rates for stepping forward and backward. However, this expression is only approximate: it is a ratio of two averages and does not account for the correlation between run length and run time. For instance, one expects that longer times are required for longer run lengths, and this information is lost if the ratio of two averages is computed. Furthermore, (5.10) says nothing about the velocity distribution! Therefore, although appealing because it is simple, intuitive, and directly related to experimentally measurable quantities, the scope of v defined in (5.10) is somewhat limited.

In order to compute $P(v)$, we first derive $f(n, t)$, the probability distribution of net displacement n at detachment time at t, which is given by $f(n, t) = \sum_{m,l=0}^\infty f(m, l, t)\delta_{m-l,n}$. We show in Appendix 5.11 that

---

[1] Recall that $\sum_{n=0}^\infty \frac{a^n}{n!} = e^a$.

$$f(n, t) = \left(\frac{k^+}{k^-}\right)^{\frac{n}{2}} \gamma e^{-k_T t} I_n(2t\sqrt{k^+k^-}), \tag{5.11}$$

where I_n is the modified Bessel function of the first kind (5.59). The velocity distribution may be obtained by changing the variables to $v = n/t$, which is achieved by setting $P(v) = \sum_{n=-\infty}^{\infty} \int_0^{\infty} dt \delta(v - n/t) f(n, t)$, where $\delta(v - n/t)$ is Dirac's delta function. After some calculations, described in Appendix 5.11, we obtain

$$P(v > 0) = \sum_{n=0}^{\infty} \frac{n}{v^2} \left(\frac{k^+}{k^-}\right)^{\frac{n}{2}} \gamma e^{-k_T n/v} I_n\left(\frac{2n}{v}\sqrt{k^+k^-}\right);$$

$$P(v < 0) = \sum_{n=-\infty}^{0} -\frac{n}{v^2} \left(\frac{k^+}{k^-}\right)^{\frac{n}{2}} \gamma e^{-k_T n/v} I_n\left(\frac{2n}{v}\sqrt{k^+k^-}\right). \tag{5.12}$$

This expression may be written in a more compact form as

$$P(v \gtrless 0) = \frac{\gamma}{|v|} \sum_{n=0}^{\infty} \left(\frac{n}{|v|}\right)^{n+1} \frac{1}{n!} \left(k^{\pm} e^{-(k_T/|v|)}\right)^n {}_0F_1\left(; n + 1; \frac{n^2 k^+ k^-}{|v|^2}\right), \tag{5.13}$$

where ${}_0F_1$ is hypergeometric function (5.57). Note that the $P(v)$ in (5.13) is in the unit of step size s. A typical shape of $P(v)$ for the case $k^+ \gg k^-$ is plotted in Fig. 5.3b. $P(v)$ for the one-state model recovers the features of the experimentally measured velocity distribution (see Fig. 5.1).

5.4 Two-State Model

The model in the previous section is described by a single time-scale given by $\tau = 1/(k^+ + k^- + \gamma)$. However, experimental evidence suggests that the dwell time distributions between two steps are often well-described by a convolution of two exponentials. This is the case of kinesin, where careful examinations of the trajectory of a kinesin head attached to a gold-nanoparticle (GNP) [17, 18] displayed the existence of two time-scales: one associated with the transition from a state in which both heads of the dimer are bound to the MT (two-head bound, 2HB, state) to a state with one head attached to the track and the other free (one-head bound, 1HB, state), and a second transition corresponding to the completion of the step, that is 1HB→2HB. Either of these transitions could be coupled to ATP binding, with rate increasing as the the ATP concentration increases; the other transition does not depend on ATP and it is rate-limiting at large nucleotide concentrations. In order to account for the presence of two states, the minimal model presented in the previous section should be generalized as shown in Fig. 5.2b. The motor goes from state 1 (i.e. 2HB) to state 2 (i.e. 1HB) with rate k. From this new conformation, the motor can step forward with rate k^+, backward with rate k^-, and detach with rate γ. Either the rate k or at least some of the rates k^+, k^-, and γ should depend on ATP concentration.

As we did for the one-state model, we first calculate $f(m, l, t)$. Let $f_+(t)$ be the probability density for taking a forward step between t and $t + dt$, given that at $t = 0$ the motor is in the state 1. Similarly, $f_-(t)$ and $f_\gamma(t)$ are the probability densities for backward step and detachment, respectively. The three probability densities, $f_+(t)$, $f_-(t)$, and $f_\gamma(t)$, are given by

$$
\begin{aligned}
f_+(t) &= \int_0^t dt' k e^{-kt'} k^+ e^{-(k^+ + k^- + \gamma)(t - t')} \\
&= \frac{kk^+}{k^+ + k^- + \gamma - k} (e^{-kt} - e^{-(k^+ + k^- + \gamma)t})
\end{aligned}
\tag{5.14}
$$

and

$$
\begin{aligned}
f_-(t) &= \int_0^t dt' k e^{-kt'} k^- e^{-(k^+ + k^- + \gamma)(t - t')} \\
&= \frac{kk^-}{k^+ + k^- + \gamma - k} (e^{-kt} - e^{-(k^+ + k^- + \gamma)t})
\end{aligned}
\tag{5.15}
$$

$$
\begin{aligned}
f_\gamma(t) &= \int_0^t dt' k e^{-kt'} \gamma e^{-(k^+ + k^- + \gamma)(t - t')} \\
&= \frac{k\gamma}{k^+ + k^- + \gamma - k} (e^{-kt} - e^{-(k^+ + k^- + \gamma)t}).
\end{aligned}
\tag{5.16}
$$

By combining all the possible alternative pathways (i.e. all the combinations of m forward and l backward steps), $f(m, l, t)$ is given by

$$
\begin{aligned}
f(m, l, t) = \frac{(m + l)!}{m! l!} \int_0^t dt_{m+l} \int_0^{t_{m+l}} dt_{m+l-1} \cdots \int_0^{t_3} dt_2 \int_0^{t_2} dt_1 \\
\underbrace{\prod_{i=1}^{m} f_+(t_i - t_{i-1})}_{m} \underbrace{\prod_{i=m+1}^{m+l} f_-(t_i - t_{i-1}) f_\gamma(t - t_{m+l})}_{m+l},
\end{aligned}
\tag{5.17}
$$

in complete analogy with (5.3). This lengthy expression can be simplified by using the Laplace transform. Given a function $g(t)$, the Laplace transforms $\tilde{g}(s)$ is given by[2]

$$
\tilde{g}(s) = \mathcal{L}[g(t)] = \int_0^\infty dt g(t) e^{-st}.
\tag{5.18}
$$

---

[2] The reader should not confuse the Laplace variable with the step size.

It is easy to show that the Laplace transform of an exponential function $g(t) = e^{-kt}$ is $\tilde{g}(s) = 1/(s + k)$. It follows that the Laplace transforms of $f_+(t)$, $f_-(t)$, and $f_\gamma(t)$ are

$$\tilde{f}_+(s) = \frac{kk^+}{(s + k)(s + k^+ + k^- + \gamma)};$$

$$\tilde{f}_-(s) = \frac{kk^-}{(s + k)(s + k^+ + k^- + \gamma)}; \tag{5.19}$$

$$\tilde{f}_\gamma(s) = \frac{k\gamma}{(s + k)(s + k^+ + k^- + \gamma)}.$$

Importantly, the Laplace transform of the convolution of two functions $c(t) = \int_0^t g(t')h(t - t')dt'$ is given by

$$\tilde{c}(s) = \int_0^\infty dt \int_0^t dt' e^{-st} g(t')h(t - t') = \tilde{g}(s)\tilde{h}(s), \tag{5.20}$$

that is the product of the two convoluted functions. By repeatedly using this property with $f(m, l, t)$ in (5.17), $\tilde{f}(m, l, s)$ becomes the multiplication of all of the Laplace transforms of the dwell time distributions,

$$\begin{aligned}
\tilde{f}(m, l, s) &= \frac{(m + l)!}{m!l!}(\tilde{f}_+(s))^m (\tilde{f}_-(s))^l \tilde{f}_\gamma(s) \\
&= \frac{(m + l)!}{m!l!} \frac{k^{m+l+1}(k^+)^m (k^-)^l \gamma}{(s + k)^{m+l+1}(s + k^+ + k^- + \gamma)^{m+l+1}}.
\end{aligned} \tag{5.21}$$

As we show in Appendix 5.11 if we directly take the inverse Laplace transform of $\tilde{f}(m, l, s)$, we obtain

$$f(m, l, t) = \frac{\gamma \sqrt{\pi}}{m!l!} e^{-\frac{\xi_1 + \xi_2}{2}t} t^{m+l} \frac{k^{m+l+1}(k^+)^m (k^-)^l}{|\xi_2 - \xi_1|^{m+l+1}} \sqrt{|\xi_2 - \xi_1|t} \, I_{m+l+\frac{1}{2}}\left(\frac{|\xi_2 - \xi_1|}{2}t\right), \tag{5.22}$$

where I is modified Bessel function of the first kind, $\xi_1 = k$ and $\xi_2 = k^+ + k^- + \gamma$.

Run Length Distribution. In order to obtain the run length distribution, we do not need the information about the time of detachment. Therefore, we use $P(m, l) = \int_0^\infty dt f(m, l, t)$, and by noticing that $P(m, l) = \tilde{f}(m, l, s = 0)$, with (5.21), we obtain

$$P(m, l) = \frac{(m + l)!}{m!l!}\left(\frac{k^+}{k_T}\right)^m \left(\frac{k^-}{k_T}\right)^l \left(\frac{\gamma}{k_T}\right), \tag{5.23}$$

where $k_T = k^+ + k^- + \gamma$. As we did in the previous section, we compute the probability of detaching after moving along the track by $sn = s(m - l)$ by computing $\sum_{m=0}^\infty \sum_{l=0}^\infty P(m, l)\delta_{n,m-l}$. As explained in Appendix 5.9, we obtain

$$P(n \geq 0) = \left(\frac{2k^{\pm}}{k_T + \sqrt{k_T^2 - 4k^+k^-}} \right)^{|n|} \frac{\gamma}{\sqrt{k_T^2 - 4k^+k^-}}. \tag{5.24}$$

As we did before, we compute the average run length and obtain

$$\overline{L} = s \sum_{n=-\infty}^{\infty} nP(n \geq 0) = s \frac{k^+ - k^-}{\gamma}, \tag{5.25}$$

where s is the step size. That is the run length is the same as the one obtained for the one-state model.

Run Time Distribution. As with the case of the one-state model, the run time distribution is $P(t) = \sum_{m=0}^{\infty} \sum_{l=0}^{\infty} f(m, l, t)$. We can rewrite this as

$$P(t) = \mathcal{L}^{-1} \left[\sum_{m=0}^{\infty} \sum_{l=0}^{\infty} \tilde{f}(m, l, s) \right], \tag{5.26}$$

where $\mathcal{L}^{-1}$ is the inverse Laplace transform. We show in Appendix 5.10 that

$$P(t) = \frac{k\gamma}{\lambda_+ - \lambda_-} \left(e^{\lambda_+ t} - e^{\lambda_- t} \right), \tag{5.27}$$

where $\lambda_{\pm}$ are given in (5.70). The average run time is

$$\bar{t} = \int_0^{\infty} t P(t) dt = \frac{k\gamma(\lambda_+ + \lambda_-)}{(\lambda_+ \lambda_-)^2} = \frac{k_T + k}{k\gamma}. \tag{5.28}$$

Velocity Distribution. As in the case of the one-state model, we define the motor velocity to be the ratio between the average run length and the average run time, that is

$$v = \frac{\overline{L}}{\bar{t}} = s \frac{k(k^+ - k^-)}{k_T + k}. \tag{5.29}$$

Note that regardless of whether k or k^+ depends on ATP concentration, the velocity saturates at large [ATP], as expected from experiments (see Figs. 5.1 and 5.4).

By following the calculation steps outlined in the one-state model, we obtain the velocity distribution by changing the variable to v using $\int_0^{\infty} f(m, l, t) \delta(v - \frac{m-l}{t}) dt$ and summing over m and l. The calculation is somewhat lengthy, and the details are in Appendix 5.11. The final result is

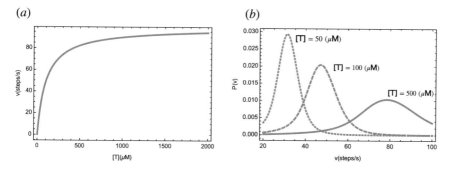

Fig. 5.4 a ATP concentration ([T]) dependence of the average velocity (v) of the motor stepping on discrete track (5.29). We set the unit for ATP concentration to be μM. **b** Velocity distribution of the motor at different ATP concentrations calculated from (5.30) with first-order ATP-dependent rate $k = k_0[T]$. Parameters for the plots are $k_0 = 1$ (μM^{-1}), $k^+ = 100$ (s^{-1}), $k^- = 1$ (s^{-1}), and $\gamma = 1$ (s^{-1})

$$P(v > 0) = \sum_{\substack{m,l \\ m>l}}^{\infty} \frac{m-l}{v^2} \frac{\gamma\sqrt{\pi}}{m!l!} e^{-\frac{\xi_1+\xi_2}{2}\frac{m-l}{v}} \left(\frac{m-l}{v}\right)^{m+l+\frac{1}{2}}$$

$$\frac{k^{m+l+1}(k^+)^m (k^-)^l}{|\xi_2-\xi_1|^{m+l+\frac{1}{2}}} I_{m+l+\frac{1}{2}}\left(\frac{|\xi_2-\xi_1|}{2}\frac{m-l}{v}\right). \tag{5.30}$$

5.5   Solution for an Arbitrary Network

The models discussed so far are very instructive because they can be solved analytically, reproduce a variety of experimental measurements, and are predictive. However, these simple kinetic networks lack a one-to-one correspondence with the plethora of states and stepping pathways that are likely to be explored by a molecular motor during a processive run. As a consequence, if one aims to use the kinetic model as a means to "look under the hood" of the motors, it may be beneficial to consider more elaborate networks. In this section, we discuss an approach to fit experimental data with a model described by an arbitrary network of states. We will show that the run length and velocity distributions are difficult to obtain in this case analytically (see Fig. 5.5 as an example of the general network which contains N internal states at site n on the track). However, averages properties and distributions of related quantities can be calculated.

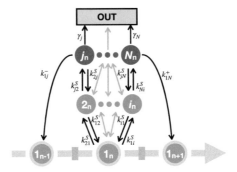

Fig. 5.5 Example of the general network. At site n on the track, there are N internal states. Some of which may lead to forward step and backward step, state N and j in this example, respectively. Red color represents the internal states leading to steps on the track and the blue color represents the states of pure chemical transitions which is not related to the movement along the track

5.5.1 Master Equation and Average Run Time

In a network with many states and edges connecting them, the changes in the probability of populating a particular state i are described by the well-known Master equation,

$$(a)\frac{dp_i(t)}{dt} = -\sum_j k_{ji} p_i(t) + \sum_j k_{ij} p_j(t),$$

$$(b)\frac{d\mathbf{p}(t)}{dt} = \hat{\mathbf{K}}\mathbf{p}(t).$$

(5.31)

Equation (a) shows that the changes in time of the probability of occupying each state i are given by the balance between the fluxes leaving state i from all other states (first term in the r.h.s.) or entering state i from elsewhere (second term in the r.h.s.). Equation 5.31(b) is a matrix form of (a), in which the entries of the vector $\mathbf{p}(t)$ correspond to the probability of occupying the N states at time t. The diagonal term of the matrix $\hat{\mathbf{K}}$ is $(\hat{\mathbf{K}})_{ii} = -\sum_j k_{ji} = -1/\tau_i$, where τ_i is the average lifetime of state i. The compact matrix representation has a number of advantages that we will exploit in the following. For instance, a formal solution of the Master equation is given by

$$\mathbf{p}(t) = e^{\hat{\mathbf{K}}t}\mathbf{p}(0),$$

(5.32)

with $e^{\hat{\mathbf{K}}0} = \hat{\mathbf{I}}$, the identity matrix. Assuming that none of the N states are absorbing and that all of the transitions k_{ji} are between the N states, the probability is conserved over time, that is $\mathbf{1} \cdot \mathbf{p}(t) = \sum_{i=1}^{N} p_i(t) = 1$, where $\mathbf{1}$ is the N-dimensional vector $(1, 1, \cdots, 1)$. This follows directly from the Master equation: $\mathbf{1} \cdot \hat{\mathbf{K}} = 0$, and therefore, $\mathbf{1} \cdot \mathbf{p}(t)$ is independent of time and equal to 1 if the initial probability was normalized.

However, in the case of molecular motors, eventually the system leaves the N filament-bound states. Let the transition from state i to the absorbing state be given by γ_i. Then the diagonal terms in the matrix $\hat{K}$ are given by $K_{ii} = 1/\tau_i = -(\sum_j k_{ji} + \gamma_i)$, and $\mathbf{1} \cdot \hat{K}$ is equal to a vector whose entries are the opposite of the γ_i rates. After a sufficient amount of time has elapsed, the probability of being absorbed approaches 1. In other words, the probability of being bound to the track $[\mathbf{1} \cdot \mathbf{p}(t)]$ is a decreasing function of time which tends to zero $[\lim_{t \to \infty} \mathbf{1} \cdot \mathbf{p}(t) = 0]$. One may then ask, what is the average time to absorption? The probability, $\phi(t)dt$, of being absorbed in a time interval dt around time t is given by

$$\phi(t)dt = \mathbf{1} \cdot [\mathbf{p}(t) - \mathbf{p}(t + dt)] = -\mathbf{1} \cdot \frac{d\mathbf{p}(t)}{dt} dt. \tag{5.33}$$

It follows that the average time to absorption is

$$\bar{t} = \int_0^\infty t\phi(t)dt = -\mathbf{1} \cdot \int_0^\infty t \frac{d\mathbf{p}(t)}{dt} dt = \mathbf{1} \cdot \int_0^\infty \mathbf{p}(t)dt = \mathbf{1} \cdot \int_0^\infty e^{\hat{K}t} \cdot \mathbf{p}(0)dt,$$

where we integrated by parts and used (5.32). Because the real part of all the eigenvalues of $\hat{K}$ must be negative,[3] we can write $\int_0^t dt' e^{\hat{K}t'} = \hat{K}^{-1}(e^{\hat{K}t} - \hat{I})$ [19]. This is a generalization of the result for a one-dimensional integral to a matrix form. It follows that

$$\bar{t} = -\mathbf{1} \cdot \hat{K}^{-1} \cdot \mathbf{p}(0), \tag{5.34}$$

which is a classic result [20].

In the case of molecular motors, at each location n along the track, there are N accessible states. For simplicity, we assume that from n only the next $(n + 1)$ and previous $(n - 1)$ filaments sites are accessible,[4] and therefore, we modify (5.31) into

$$\frac{dp_i(n, t)}{dt} = -\sum_j (k_{ji}^S + k_{ji}^+ + k_{ji}^- + \gamma_i)p_i(n, t) +$$
$$+ \sum_j k_{ij}^S p_j(n, t) + \sum_j k_{ij}^+ p_j(n - 1, t) + \sum_j k_{ij}^- p_j(n + 1, t), \tag{5.35}$$

where the first term on the r.h.s. is the out-flux towards other states at the same filament site (k_{ji}^S), towards the next filament site (k_{ji}^+), or towards the previous location along the track (k_{ji}^-). The next three terms are the in-flux to state j at site l from

---

[3] A zero eigenvalue corresponds to the stationary state. However, by construction, the stationary state is the absorbing state, which is not accounted for in the N state. Therefore, the N eigenvalues of the matrix $\hat{K}$ must be negative.

[4] In other words, the motor takes discrete steps of size s, as assumed in the case of the previous models.

states in position n, $n - 1$ (after a forward step), and $n + 1$ (after a backward step). Equation 5.35 can be simplified under two assumptions: (i) the rates do not depend on the location along the filament, and (ii) the track is infinitely long. In this case, by summing over n and defining $p_i(t) = \sum_n p_i(n, t)$, we obtain

$$\frac{dp_i(t)}{dt} = -\sum_j (k_{ji}^S + k_{ji}^+ + k_{\bar{j}i}^- + \gamma_i) p_i(t) + \sum_j (k_{ij}^S + k_{ij}^+ + k_{\bar{i}j}^-) p_j(t). \quad (5.36)$$

This expression can be written more compactly by defining the following three matrices: $\hat{K}^S$ contains all the diagonal terms $[(\hat{K}^S)_{ii} = -\sum_j (k_{ji}^S + k_{ji}^+ + k_{\bar{j}i}^- + \gamma_i)]$ and the rates associated with transitions that do not alter the location of the motor [i.e. $(\hat{K}^S)_{ji} = k_{ji}^S$ for $i \neq j$], $\hat{K}^+$ contains all the transitions associated with forward steps $[(\hat{K}^+)_{ji} = k_{ji}^+]$; the transitions leading to backward steps $[(\hat{K}^-)_{ji} = k_{\bar{j}i}^-]$ populate the matrix $\hat{K}^-$. Using these expressions, the Master equation for a molecular motor can be written

$$\frac{d\mathbf{p}(t)}{dt} = (\hat{K}^S + \hat{K}^+ + \hat{K}^-)\mathbf{p}(t). \quad (5.37)$$

5.5.2 Distributions

We first extract the probability distribution $P(\sigma)$ of detaching after exactly σ steps have been completed, regardless of whether they are forward or backward. We begin by computing the probability density of detaching in an interval dt around time t after having completed σ steps, $\phi(t, \sigma)$, which is a variant of (5.33). The idea is the following: let $\mathbf{p}(t; \sigma)$ be the probability vector at time t after σ steps have been completed. We define $\phi(t, \sigma)$ as

$$\phi(t, \sigma)dt = -\mathbf{1} \cdot \hat{K} \cdot \mathbf{p}(t, \sigma)dt. \quad (5.38)$$

Any bound motor, regardless of whether it stepped or not, does not contribute to the count of the motors lost. The probability of detaching after exactly σ steps have been completed is

$$P(\sigma) = \int_0^\infty \phi(t, \sigma)dt = -\mathbf{1} \cdot \hat{K} \cdot \int_0^\infty \mathbf{p}(t, \sigma)dt. \quad (5.39)$$

What is left to do is to determine $\mathbf{p}(t, \sigma)$. When $\sigma = 0$, up to time t there can be no forward or backward steps, that is the time evolution is determined by $\hat{K}^S$ uniquely. It follows that we can write

$$\mathbf{p}(t, \sigma = 0) = e^{\hat{K}^S t} \cdot \mathbf{p}(0), \quad (5.40)$$

and because all the eigenvalues of $\hat{K}^S$ must be negative, we obtain

$$P(0) = \mathbf{1} \cdot \hat{K} \cdot (\hat{K}^S)^{-1} \cdot \mathbf{p}(0). \tag{5.41}$$

In order to compute $P(\sigma = 1)$, we need to find $\mathbf{p}(t, \sigma = 1)$, which means that at a time $t' < t$ a step (either forward or backward) has occurred, whereas before or after the step the system may change conformation but not location. We construct this with the following logic: (i) the probability of not taking any step up to time t' is $\mathbf{p}(t', \sigma = 0) = e^{\hat{K}^S t'} \mathbf{p}(0)$; (ii) the change in probability associated with taking a step around time t' is given by $(\hat{K}^+ + \hat{K}^-) \cdot \mathbf{p}(t', \sigma = 0)dt'$; (iii) the rest of the evolution is obtained using the operator $e^{\hat{K}^S (t-t')} \mathbf{p}(t', \sigma = 1)$. Combining all together,

$$\mathbf{p}(t, \sigma = 1) = \int_0^t dt' e^{\hat{K}^S (t-t')} \cdot (\hat{K}^+ + \hat{K}^-) \cdot e^{\hat{K}^S t'} \cdot \mathbf{p}(0). \tag{5.42}$$

From (5.39), after changing integration variables, we obtain

$$P(\sigma = 1) = -\mathbf{1} \cdot \hat{K} \cdot (\hat{K}^S)^{-1} \cdot (\hat{K}^+ + \hat{K}^-) \cdot (\hat{K}^S)^{-1} \cdot \mathbf{p}(0). \tag{5.43}$$

Inspecting this equation, we understand that the operator that defines the occurrence of a step is

$$\hat{K}_{\text{Step}} = -(\hat{K}^+ + \hat{K}^-) \cdot (\hat{K}^S)^{-1}, \tag{5.44}$$

and it is easy to show that $\hat{K} \cdot (\hat{K}^S)^{-1} = \hat{I} - \hat{K}_{\text{Step}}$ [start with setting $\hat{I} = \hat{K}^S \cdot (\hat{K}^S)^{-1}$ on the r.h.s. and use (5.44)]. It is easy then to convince oneself that

$$P(\sigma) = \mathbf{1} \cdot (\hat{I} - \hat{K}_{\text{Step}}) \cdot (\hat{K}_{\text{Step}})^\sigma \cdot \mathbf{p}(0). \tag{5.45}$$

By extension, if we are interested in the probability of taking exactly m forward steps before detaching, regardless of the number of forward and backward steps, we obtain

$$\begin{aligned} P(m) &= \mathbf{1} \cdot (\hat{I} - \hat{K}_{\text{Fwd}}) \cdot (\hat{K}_{\text{Fwd}})^m \cdot \mathbf{p}(0) \\ \hat{K}_{\text{Fwd}} &= -\hat{K}^+ \cdot (\hat{K}^S + \hat{K}^-)^{-1} \end{aligned} \tag{5.46}$$

and

$$\begin{aligned} P(l) &= \mathbf{1} \cdot (\hat{I} - \hat{K}_{\text{Bwd}}) \cdot (\hat{K}_{\text{Bwd}})^l \cdot \mathbf{p}(0) \\ \hat{K}_{\text{Bwd}} &= -\hat{K}^- \cdot (\hat{K}^S + \hat{K}^+)^{-1} \end{aligned} \tag{5.47}$$

is the probability of completing l backward steps before detachment, regardless of the number of forward steps.

The approach discussed so far is fairly general; however, it has an important limitation. Suppose we are interested in computing the probability distribution of the net displacement by the motor $P(n = m - l)$. We can obtain this quantity from

$$P(n) = \sum_{m=0}^{\infty} \sum_{l=0}^{\infty} P(m, l) \delta_{n, m-l}, \tag{5.48}$$

which requires us to construct $P(m, l)$, as we have done with the two simpler models. Here, we run into a problem, which is best illustrated with an example. Suppose that we want to compute the probability of taking one forward step, one backward step, and then detaching from the track, that is $P(m = 1, l = 1)$. There are two possible pathways to achieve that: first take a forward step, then a backward step, or vice versa. These two pathways provide the following contributions to the probability $P(m = 1, l = 1)$,

$$P(m = 1, l = 1) = \mathbf{1} \cdot (\hat{I} - \hat{K}_{\text{Step}}) \cdot \hat{K}_{\text{Bwd}} \cdot \hat{K}_{\text{Fwd}} \cdot \mathbf{p}(0) + $$
$$+ \mathbf{1} \cdot (\hat{I} - \hat{K}_{\text{Step}}) \cdot \hat{K}_{\text{Fwd}} \cdot \hat{K}_{\text{Bwd}} \cdot \mathbf{p}(0).$$

Unfortunately, $\hat{K}_{\text{Fwd}}$ and $\hat{K}_{\text{Bwd}}$ in general do not commute, that is $\hat{K}_{\text{Fwd}} \cdot \hat{K}_{\text{Bwd}} \neq \hat{K}_{\text{Bwd}} \cdot \hat{K}_{\text{Fwd}}$. This implies that each pathway may have a different weight, and we have to account for each and every one independently. The number of these pathways is of course $(m + l)!/(m!l!)$, as discussed when we derived $f(m, l, t)$ in (5.4) and (5.17). Clearly, the binomial factor $(m + l)!/(m!l!)$ becomes large very quickly, and it is extremely challenging to estimate $P(\sigma > 10)$ numerically. In the same way, computing $P(v)$ is a daunting task.

5.5.3 Average Properties

From (5.45)–(5.47), we can compute the average number of times that the monitored event (e.g. number of forward steps, backward steps, the total number of steps) occurs. In order to do so, we need to compute

$$\bar{\sigma} = \sum_{\sigma=0}^{\infty} \sigma P(\sigma) = \mathbf{1} \cdot (\hat{I} - \hat{K}_{\text{Step}}) \cdot \left[\sum_{\sigma=0}^{\infty} \sigma (\hat{K}_{\text{Step}})^{\sigma} \right] \cdot \mathbf{p}(0).$$

We show in Appendix 5.12 that

$$\bar{\sigma} = \mathbf{1} \cdot \hat{K}_{\text{Step}} \cdot (\hat{I} - \hat{K}_{\text{Step}})^{-1} \cdot \mathbf{p}(0)$$
$$\bar{m} = \mathbf{1} \cdot \hat{K}_{\text{Fwd}} \cdot (\hat{I} - \hat{K}_{\text{Fwd}})^{-1} \cdot \mathbf{p}(0) \tag{5.49}$$
$$\bar{l} = \mathbf{1} \cdot \hat{K}_{\text{Bwd}} \cdot (\hat{I} - \hat{K}_{\text{Bwd}})^{-1} \cdot \mathbf{p}(0).$$

It is also possible to show that $\bar{\sigma} = \bar{m} + \bar{l}$, and therefore, it makes sense to define the probability of forward stepping as

$$P_{\text{FWD}} = \frac{\overline{m}}{\overline{m} + \overline{l}}, \tag{5.50}$$

and the mean run length as

$$\overline{L} = s(\overline{m} - \overline{l}). \tag{5.51}$$

As we did in the previous cases, we can construct a quantity with the units of velocity from the ratio of the average run length and the average run time,

$$v = \frac{\overline{L}}{\overline{t}}, \tag{5.52}$$

where $\overline{L}$ can be obtained from (5.51), and $\overline{t}$ is given by (5.34).

5.5.4   Simple Examples

We illustrate this more general theoretical framework by applying it to the one-state and two-state models described in previous sections.

One-state Model. In the one-state model in Fig. 5.2a, the matrices $\hat{K}^+ = k^+$, $\hat{K}^- = k^-$, $\hat{K}^S = -(k^+ + k^- + \gamma)$, and $\hat{K} = -\gamma$ are all scalar quantities. Similarly, $\mathbf{p}(0) = 1$ and $\mathbf{1} = 1$. It follows that $\hat{K}_{\text{Step}} = (k^+ + k^-)/(k^+ + k^- + \gamma)$ and that

$$P(\sigma) = \frac{\gamma}{k^+ + k^- + \gamma} \left(\frac{k^- + k^-}{k^+ + k^- + \gamma} \right)^\sigma, \tag{5.53}$$

which is an intuitive result and could be obtained from $P(n, m)$ in (5.5) as $P(\sigma) = \sum_{m=0}^{\infty} \sum_{l=0}^{\infty} P(m, l) \delta_{\sigma, m+l}$. We leave this as an exercise—all the ingredients of the derivation are shown in this section. Similarly, we can extract the probability distribution for taking m (l) forward (backward) steps before detaching, regardless of the number of backward (forward) steps completed (we leave these as exercise). From (5.49) to (5.34), we obtain $\overline{\sigma} = \frac{k^+ + k^-}{\gamma}$, $\overline{m} = \frac{k^+}{\gamma}$, $\overline{l} = \frac{k^-}{\gamma}$, $\overline{t} = \frac{1}{\gamma}$, which lead to the following results for the mean run length and average velocity:

$$\overline{L} = s \frac{k^+ - k^-}{\gamma},$$
$$v = s(k^+ - k^-), \tag{5.54}$$

in perfect agreement with (5.7) and (5.10).

Two-state Model. For the two-state model in Fig. 5.2b, the matrices $\hat{K}$, $\hat{K}^S$, $\hat{K}^+$, and $\hat{K}^-$ are

$$\hat{K} = \begin{pmatrix} -k & k^+ + k^- \\ k & -(k^+ + k^- + \gamma) \end{pmatrix}; \hat{K}^+ = \begin{pmatrix} 0 & k^+ \\ 0 & 0 \end{pmatrix};$$

$$\hat{K}^S = \begin{pmatrix} -k & 0 \\ k & -(k^+ + k^- + \gamma) \end{pmatrix}; \hat{K}^- = \begin{pmatrix} 0 & k^- \\ 0 & 0 \end{pmatrix}.$$

$$(5.55)$$

The matrix operations can be straightforwardly carried out and yield $\overline{\sigma} = (k^+ + k^-)/\gamma, \overline{m} = k^+/\gamma, \overline{l} = k^-/\gamma, \overline{t} = (k_T + k)/(k\gamma)$, from which we obtain

$$\overline{L} = s \frac{k^+ - k^-}{\gamma}$$

$$v = s \frac{k(k^+ - k^-)}{k_T + k},$$

$$(5.56)$$

all in perfect agreement with (5.25) and (5.29). In addition, both this matrix formulation and the previous solution of $P(n)$ yield (5.53).

5.6   Experiments Performed Under Constant External Load

So far, we focused our attention on the characterization of the dependence of experimentally measurable characteristics of motility on the motor catalytic cycle and on nucleotide concentration. Another control parameter used in experiments to probe the function of motors is the load (F) applied to the motor by holding a bead via optical tweezers. Experiments have shown that a resistive load increases the likelihood of stepping backward and of detaching. At the so-called stall force, the velocity of the motor averages to zero as the probability of stepping forward and stepping backward is equal.

In the context of stochastic kinetic models, the effect of resistive load may be introduced by modifying the rate of sensitive reactions as $k(F) = k(0)e^{\beta F \delta}$, where $k(0)$ is the rate in the absence of load, $\beta = (k_B T)^{-1}$, and δ is a fitting parameter. For the simple one-state and two-state model, one can assume that k^+ becomes smaller as the load increases, whereas k^- and γ increase with the resistive force. It is interesting to observe the features of the velocity distribution as a function of load. When $k^+ \approx k^-$, $P(v)$ of the one-state and two-state models show a bimodal shape.[5] This is a surprising result if compared with a continuous diffusive motion whose $P(v)$ would be a Gaussian-like shape centered on the mean value. As illustrated in Fig. 5.6, it is the discrete nature of the stepping mechanism which is responsible for the bimodal shape of $P(v)$ [14]. This fascinating and unexpected result is a genuine prediction emerging from the analytical solution of the theoretical model.

---

[5] The bimodality occurs also at $F = 0$, but the probability of recording a negative velocity is extremely low.

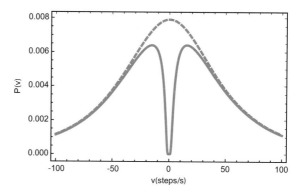

Fig. 5.6 Bimodal velocity distribution plotted from (5.13) (solid line) and its continuous approximation obtained in [14] (dashed line). The parameters used for the plot are $k^+ = 100$ (s^{-1}), $k^- = 100$ (s^{-1}), and $\gamma = 10$ (s^{-1})

5.7 Advantages and Limitations of Stochastic Kinetic Models

Stochastic kinetic models are a convenient framework to model and rationalize the function of molecular motors. They provide valuable tools to fit and interpret the experimental data, and they can be used in order to make predictions that can advance our understanding of the function of molecular motors. In addition, stochastic kinetic models provide a theoretical framework that is ideally suited to study statistical mechanics in non-equilibrium [21]. However, they have two severe shortcomings. First, in the elaborate versions of the models, there are too many fitting parameters; this limits their predictive power. Second, the connection with the structure of the motors is lacking. The only feature related to the structure is δ, a fitting parameter associated with the sensitivity of a particular rate to the applied load. These two problems are not unrelated. By using only rates, we are ignoring the details of structural information which have been made available through a variety of experimental, computational, and theoretical studies. Combining structural and kinetic information (see for instance [22]) is a crucial direction of research in the years to come in order to improve the quality of the models and bring them closer to experiments.

5.8 Appendix A: Mathematical Functions

In this section, we list special functions used to calculate $P(n)$ and $P(v)$. Hypergeometric function:

$$_0F_1(; a; z) = \sum_{k=0}^{\infty} \frac{\Gamma(a)}{\Gamma(a+k)} \frac{z^k}{k!}, \tag{5.57}$$

where Γ is Gamma function.

Gauss hypergeometric function:

$$_2F_1(a, b; c; z) = \frac{\Gamma(c)}{\Gamma(a)\Gamma(b)} \sum_{k=0}^{\infty} \frac{\Gamma(a+k)\Gamma(b+k)}{\Gamma(c+k)} \frac{z^k}{k!}. \tag{5.58}$$

Modified Bessel function of the first kind:

$$I_\sigma(z) = \left(\frac{z}{2}\right)^\sigma \sum_{k=0}^{\infty} \frac{(\frac{z^2}{4})^k}{k!\Gamma(\sigma+k+1)}, \tag{5.59}$$

where σ is a real number.

Modified Bessel function of the second kind:

$$K_{n+\frac{1}{2}}(z) = \sqrt{\frac{\pi}{2z}} \frac{e^{-z}}{\sqrt{z}} \sum_{k=0}^{\infty} \frac{(k+n)!}{k!(-k+n)!}(2z)^{-k}, \tag{5.60}$$

where n is an integer.

5.9 Appendix B: The Distribution of Run Length

The summation for (5.5),

$$P(n) = \sum_{m,l=0}^{\infty} \frac{(m+l)!}{m!l!}\left(\frac{k^+}{k_T}\right)^m \left(\frac{k^-}{k_T}\right)^l \left(\frac{\gamma}{k_T}\right)\delta_{m-l,n}, \tag{5.61}$$

can be carried out for $n > 0$ and $n < 0$, leading to[6]

$$
\begin{aligned}
P(n > 0) &= \left(\frac{k^+}{k_T}\right)^n \left(\frac{\gamma}{k_T}\right) \sum_{l=0}^{\infty} \left(\frac{k^+k^-}{k_T^2}\right)^l \frac{(2l+n)!}{(n+l)!l!} \\
&= \left(\frac{k^+}{k_T}\right)^n \left(\frac{\gamma}{k_T}\right) {}_2F_1\left(\frac{1+n}{2}, \frac{2+n}{2}; 1+n; 4\frac{k^+k^-}{k_T^2}\right), \\
P(n < 0) &= \left(\frac{k^-}{k_T}\right)^{-n} \left(\frac{\gamma}{k_T}\right) \sum_{m=0}^{\infty} \left(\frac{k^+k^-}{k_T^2}\right)^m \frac{(2m-n)!}{(m-n)!m!} \\
&= \left(\frac{k^-}{k_T}\right)^{-n} \left(\frac{\gamma}{k_T}\right) {}_2F_1\left(\frac{1-n}{2}, \frac{2-n}{2}; 1-n; 4\frac{k^+k^-}{k_T^2}\right),
\end{aligned}
\tag{5.62}
$$

---

[6] Hint: use the duplication formula for the Gamma function: $\Gamma(z)\Gamma(z+\frac{1}{2}) = 2^{1-2z}\sqrt{\pi}\Gamma(2z)$ [23].

where $_2F_1$ is Gaussian hypergeometric function (5.58). By using the special case of $_2F_1$ [24]:

$$_2F_1\left(a, \frac{1}{2} + a; 2a; z\right) = 2^{2a-1}(1 - z)^{-\frac{1}{2}}[1 + (1 - z)^{\frac{1}{2}}]^{1-2a}, \qquad (5.63)$$

we obtain the following expression for the run length distribution:

$$P(n \gtrless 0) = \left(\frac{2k^\pm}{k_T + \sqrt{k_T^2 - 4k^+k^-}}\right)^{|n|} \frac{\gamma}{\sqrt{k_T^2 - 4k^+k^-}}. \qquad (5.64)$$

We now want to compute the average of this distribution, that is

$$\bar{n} = g\left[\sum_{n=0}^{\infty} n(a_+)^n + \sum_{n=-\infty}^{0} n(a_-)^{-n}\right] = g\left[\sum_{n=0}^{\infty} n(a_+)^n - \sum_{n=0}^{\infty} n(a_-)^n\right], \quad (5.65)$$

where we defined

$$a_\pm = \frac{2k^\pm}{k_T + \Delta}$$

$$g = \frac{\gamma}{\Delta} \qquad (5.66)$$

$$\Delta = \sqrt{k_T^2 - 4k^+k^-}.$$

With a little bit of algebra, we can prove the following relationships:

$$\frac{1 - a_+a_-}{1 + a_+a_-} = \frac{\Delta}{k_T},$$

$$\frac{a_+ \pm a_-}{1 + a_+a_-} = \frac{k^+ \pm k^-}{k_T}. \qquad (5.67)$$

Now, it is easy to show that

$$\sum_{n=0}^{\infty} na^n = \frac{a}{(1 - a)^2} \qquad (5.68)$$

for $a_\pm$, which are both non-negative and < 1. From this, we obtain that

$$\bar{n} = g\left[\frac{a_+}{(1 - a_+)^2} - \frac{a_-}{(1 - a_-)^2}\right],$$

and after some simple algebra, we can rewrite this expression as

$$\bar{n} = g \left[\frac{a_+ - a_-}{1 + a_+ a_-} \frac{1 - a_+ a_-}{1 + a_+ a_-} \frac{1}{\left(1 - \frac{a_+ + a_-}{1 + a_+ a_-}\right)^2} \right].$$

After plugging in (5.66)–(5.67), we finally get

$$\bar{n} = \frac{k^+ - k^-}{\gamma}. \tag{5.69}$$

5.10   Appendix C: Derivation of the Run Time Distribution

We start from (5.21) and we wish to compute

$$\sum_{m=0}^{\infty} \sum_{l=0}^{\infty} \tilde{f}(m, l, s) = \sum_{m=0}^{\infty} \sum_{l=0}^{\infty} \frac{(m+l)!}{m! l!} [\tilde{f}_+(s)]^m [\tilde{f}_-(s)]^l [\tilde{f}_\gamma(s)].$$

The summations over m and l may be reorganized as $\sum_{\sigma=0}^{\infty} \sum_{m=0}^{\sigma}$, from which we obtain

$$\sum_{m=0}^{\infty} \sum_{l=0}^{\infty} \tilde{f}(m, l, s) = \sum_{\sigma=0}^{\infty} \sum_{m=0}^{\sigma} \frac{\sigma!}{m!(\sigma - m)!} [\tilde{f}_+(s)]^m [\tilde{f}_-(s)]^{\sigma-m} [\tilde{f}_\gamma(s)].$$

Now, the σ-th power of the sum of two numbers a and b is $(a+b)^\sigma = \sum_{m=0}^{\sigma} \frac{\sigma!}{m!(\sigma-m)!} a^m b^{\sigma-m}$; therefore,

$$\sum_{m=0}^{\infty} \sum_{l=0}^{\infty} \tilde{f}(m, l, s) = [\tilde{f}_\gamma(s)] \sum_{\sigma=0}^{\infty} [\tilde{f}_+(s) + \tilde{f}_-(s)]^\sigma,$$

and because $\tilde{f}_+(s) + \tilde{f}_-(s) < 1$, we use the geometric series and obtain

$$\sum_{m=0}^{\infty} \sum_{l=0}^{\infty} \tilde{f}(m, l, s) = \frac{\tilde{f}_\gamma(s)}{1 - \tilde{f}_+(s) - \tilde{f}_-(s)}.$$

By plugging in the definitions of the Laplace transforms of $f_+(t)$, $f_-(t)$, and $f_\gamma(t)$, we obtain

$$\sum_{m=0}^{\infty} \sum_{l=0}^{\infty} \tilde{f}(m, l, s) = \frac{k\gamma}{(s + k)(s + k^+ + k^- + \gamma) - k(k^+ + k^-)}.$$

The two roots of the second-order polynomial at the denominator are

$$\lambda_{\pm} = \frac{-(k_T + k) \pm \sqrt{(k_T + k)^2 - 4k\gamma}}{2}, \tag{5.70}$$

so that we obtain

$$\sum_{m=0}^{\infty} \sum_{l=0}^{\infty} \tilde{f}(m, l, s) = \frac{k\gamma}{(s - \lambda_+)(s - \lambda_-)}.$$

With a little bit of algebraic manipulations, we can write

$$\tilde{P}(s) = \frac{k\gamma}{\lambda_+ - \lambda_-} \left(\frac{1}{s - \lambda_+} - \frac{1}{s - \lambda_-} \right).$$

Computing the inverse Laplace transform is now easy, as it is nothing more than the sum of two exponentials, which leads to our desired result,

$$P(t) = \frac{k\gamma}{\lambda_+ - \lambda_-} \left(e^{\lambda_+ t} - e^{\lambda_- t} \right).$$

Note that $\lambda_{\pm} < 0$.

5.11 Appendix D: Velocity Distribution

5.11.1 One-State Model

Equation 5.3 reduces to

$$f(m, l, t) = \frac{(m + l)!}{m! l!} (k^+)^m (k^-)^l \gamma e^{-k_T t}$$
$$\int_0^t dt_{m+l} \int_0^{t_{m+l}} dt_{m+l-1} \cdots \int_0^{t_3} dt_2 \int_0^{t_2} dt_1. \tag{5.71}$$

The time-ordered integration can be evaluated recurrently, which gives $\frac{t^{m+l}}{(m+l)!}$, thus,

$$f(m, l, t) = \left(\frac{t^{m+l}}{m! l!} \right) (k^+)^m (k^-)^l \gamma e^{-k_T t}. \tag{5.72}$$

We obtain $f(n, t)$ by imposing the condition $m - l = n$,

$$f(n, t) = \sum_{m,l=0}^{\infty} \left(\frac{t^{m+l}}{m!l!} \right) (k^+)^m (k^-)^l \gamma e^{-k_T t} \delta_{m-l,n}$$

$$= \left(\frac{k^+}{k^-} \right)^{\frac{n}{2}} \gamma e^{-k_T t} (t\sqrt{k^+k^-})^n \sum_{l=0}^{\infty} \frac{(t^2 k^+ k^-)^l}{l!(n+l)!} \quad (5.73)$$

$$= \left(\frac{k^+}{k^-} \right)^{\frac{n}{2}} \gamma e^{-k_T t} I_n (2t\sqrt{k^+k^-}).$$

5.11.2   Two-State Model

The inverse Laplace transform $f(m, l, t) = \mathcal{L}^{-1}[\tilde{f}(m, l, s)]$ is defined as

$$\mathcal{L}^{-1}[\tilde{f}(m, l, s)] = \frac{(m+l)!}{m!l!}(k)^{\sigma+1}(k^+)^m(k^-)^l \gamma \frac{1}{2\pi i} \int_{c-i\infty}^{c+i\infty} \frac{e^{st}}{(s+\xi_1)^{\sigma+1}(s+\xi_2)^{\sigma+1}} ds,$$

$$(5.74)$$

where for the sake of conciseness we defined $\sigma = m + l$, $\xi_1 = k$, and $\xi_2 = k^+ + k^- + \gamma$. We evaluate the inverse Laplace transform using the residue theorem (see [25] for details),

$$\int_C f(z)dz = 2\pi i \sum_{i=1}^{n} \text{Res}_{z=-\xi_i} f(z). \quad (5.75)$$

A function with a singular point in $-\xi_1$ of order $\sigma + 1$ with $\sigma \geq 0$ can be written as $f(z) = \chi(z)/(z + \xi_1)^{\sigma+1}$, where χ is analytic and different from zero at the singular point $-\xi_1$. In this case, the residue of the function in ξ_1 is given by

$$\text{Res}_{z=-\xi_1} f(z) = \frac{\frac{d^\sigma}{dz^\sigma} \chi(z)|_{z=-\xi_1}}{\sigma!}. \quad (5.76)$$

In the case of (5.74), under the assumption that $\xi_1 \neq \xi_2$,[7] there are two poles of order $\sigma + 1$ to compute, one around $-\xi_1$, and the other around $-\xi_2$. Therefore,

$$\text{Res}\left[\frac{1}{(s+\xi_1)^{\sigma+1}(s+\xi_2)^{\sigma+1}} e^{st} \right]\Big|_{s=-\xi_1} = \frac{1}{\sigma!} \frac{d^\sigma}{ds^\sigma} \left[\frac{1}{(s+\xi_2)^{\sigma+1}} e^{st} \right]\Big|_{s=-\xi_1}$$

$$= \frac{1}{\sigma!} \sum_{p=0}^{\sigma} (-1)^p \binom{\sigma}{p} \frac{t^{\sigma-p} e^{-\xi_1 t}}{(-\xi_1 + \xi_2)^{\sigma+p+1}} \frac{(\sigma+p)!}{\sigma!},$$

$$(5.77)$$

---

[7] The case for $\xi_1 = \xi_2$ is left as an exercise, and the solution may be found in [15].

$$\mathrm{Res}\left[\frac{1}{(s+\xi_1)^{\sigma+1}(s+\xi_2)^{\sigma+1}}e^{st}\right]\Big|_{s=-\xi_2} = \frac{1}{\sigma!}\frac{d^\sigma}{ds^\sigma}\left[\frac{1}{(s+\xi_1)^{\sigma+1}}e^{st}\right]\Big|_{s=-\xi_2}$$

$$= \frac{1}{\sigma!}\sum_{p=0}^{\sigma}(-1)^p\binom{\sigma}{p}\frac{t^{\sigma-p}e^{-\xi_2 t}}{(-\xi_2+\xi_1)^{\sigma+p+1}}\frac{(\sigma+p)!}{\sigma!}.$$

$$(5.78)$$

Thus, (5.74) ($f(m,l,t) = \mathcal{L}^{-1}[\tilde{f}(m,l,s)]$) can be written as

$$f(m,l,t) = \frac{\gamma(k)^{\sigma+1}(k^+)^m(k^-)^l}{m!l!}\left[\sum_{p=0}^{\sigma}(-1)^p\binom{\sigma}{p}\frac{t^{\sigma-p}e^{-\xi_1 t}}{(-\xi_1+\xi_2)^{\sigma+p+1}}\frac{(\sigma+p)!}{\sigma!}\right.$$

$$\left.+\sum_{p=0}^{\sigma}(-1)^p\binom{\sigma}{p}\frac{t^{\sigma-p}e^{-\xi_2 t}}{(-\xi_2+\xi_1)^{\sigma+p+1}}\frac{(\sigma+p)!}{\sigma!}\right]$$

$$= \frac{\gamma(k)^{\sigma+1}(k^+)^m(k^-)^l}{m!l!}\frac{t^\sigma}{\sqrt{\pi}}e^{-\frac{\xi_1+\xi_2}{2}t}\left[\frac{\sqrt{(\xi_1-\xi_2)t}}{(\xi_2-\xi_1)^{\sigma+1}}K_{\sigma+\frac{1}{2}}(\frac{\xi_1-\xi_2}{2}t)\right.$$

$$\left.+\frac{\sqrt{(\xi_2-\xi_1)t}}{(\xi_1-\xi_2)^{n+1}}K_{\sigma+\frac{1}{2}}(\frac{\xi_2-\xi_1}{2}t)\right],$$

$$(5.79)$$

where K is the modified Bessel function of the second kind (5.60). We then take advantage of an identity that relates K with the modified Bessel function of the first kind, I (see 5.59), which can be found in Eq. 10.34.2 of [26],

$$K_{n+\frac{1}{2}}(-x) = -i\left(\pi I_{n+\frac{1}{2}}(x) + (-1)^n K_{n+\frac{1}{2}}(x)\right) \quad (x > 0). \tag{5.80}$$

Using this identity, we rewrite (5.79) as follows: For $\xi_1 - \xi_2 > 0$,

$$f(m,l,t) = \frac{\gamma}{m!l!}\sqrt{\pi}e^{-\frac{\xi_1+\xi_2}{2}t}t^{m+l}\frac{k^{m+l+1}(k^+)^m(k^-)^l}{(\xi_1-\xi_2)^{m+l+1}}\sqrt{(\xi_1-\xi_2)t}\,I_{m+l+\frac{1}{2}}(\frac{\xi_1-\xi_2}{2}t).$$

$$(5.81)$$

If $\xi_2 - \xi_1 > 0$, then

$$f(m,l,t) = \frac{\gamma}{m!l!}\sqrt{\pi}e^{-\frac{\xi_1+\xi_2}{2}t}t^{m+l}\frac{k^{m+l+1}(k^+)^m(k^-)^l}{(\xi_2-\xi_1)^{m+l+1}}\sqrt{(\xi_2-\xi_1)t}\,I_{m+l+\frac{1}{2}}(\frac{\xi_2-\xi_1}{2}t).$$

$$(5.82)$$

Both cases are written as

$$f(m,l,t) = \frac{\gamma\sqrt{\pi}}{m!l!}e^{-\frac{\xi_1+\xi_2}{2}t}t^{m+l}\frac{k^{m+l+1}(k^+)^m(k^-)^l}{|\xi_2-\xi_1|^{m+l+1}}\sqrt{|\xi_2-\xi_1|t}\,I_{m+l+\frac{1}{2}}(\frac{|\xi_2-\xi_1|}{2}t).$$

$$(5.83)$$

Finally, as explained in the main text, we obtain the velocity distribution by changing variables to $v = (m - l)/t$ with the help of Dirac's delta function. For $m - l > 0$, we obtain

$$P(v > 0) = \sum_{\substack{m,l \\ m>l}} \int_0^\infty f(m, l, t) \delta(v - \frac{m-l}{t}) dt$$

$$= \sum_{\substack{m,l \\ m>l}} \frac{m-l}{v^2} \frac{\gamma \sqrt{\pi}}{m!l!} e^{-\frac{\xi_1+\xi_2}{2} \frac{m-l}{v}} \left(\frac{m-l}{v}\right)^{m+l+\frac{1}{2}} \tag{5.84}$$

$$\frac{k^{m+l+1}(k^+)^m(k^-)^l}{|\xi_2 - \xi_1|^{m+l+\frac{1}{2}}} I_{m+l+\frac{1}{2}}\left(\frac{|\xi_2 - \xi_1|}{2} \frac{m-l}{v}\right).$$

Analogously, when $m - l < 0$, we have

$$P(v < 0) = \sum_{\substack{m,l \\ l>m}} \int_0^\infty f(m, l, t) \delta(v - \frac{m-l}{t}) dt$$

$$= \sum_{\substack{m,l \\ l>m}} \frac{l-m}{v^2} \frac{\gamma \sqrt{\pi}}{m!l!} e^{-\frac{\xi_1+\xi_2}{2} \frac{m-l}{v}} \left(\frac{m-l}{v}\right)^{m+l+\frac{1}{2}} \tag{5.85}$$

$$\frac{k^{m+l+1}(k^+)^m(k^-)^l}{|\xi_2 - \xi_1|^{m+l+\frac{1}{2}}} I_{m+l+\frac{1}{2}}\left(\frac{|\xi_2 - \xi_1|}{2} \frac{m-l}{v}\right).$$

5.12 Appendix E: Averages in the N-state Model

We want to compute

$$\bar{\sigma} = \sum_{\sigma=0}^\infty \sigma P(\sigma) = \mathbf{1} \cdot (\hat{I} - \hat{K}_{\text{Step}}) \cdot \left[\sum_{\sigma=0}^\infty \sigma (\hat{K}_{\text{Step}})^\sigma\right] \cdot \mathbf{p}(0). \tag{5.86}$$

In order to do so, we use the two following results. Given a matrix $\hat{M}$ whose eigenvalues μ_i are such that $|\mu_i| < 1$ we have [19]

$$\sum_{n=0}^\infty \hat{M}^n = (\hat{I} - \hat{M})^{-1}. \tag{5.87}$$

Furthermore, if $\hat{I} - \alpha \hat{M}$ is invertible (α is a number), then we have [19]

$$\frac{d}{d\alpha}(\hat{I} - \alpha\hat{M})^{-1}|_{\alpha=1} = (\hat{I} - \hat{M})^{-1} \cdot \hat{M} \cdot (\hat{I} - \hat{M})^{-1}. \tag{5.88}$$

We can write

$$\sum_{\sigma=0}^{\infty} \sigma(\hat{K}_{Step})^{\sigma} = \lim_{\alpha\to1} \alpha\frac{d}{d\alpha} \sum_{\sigma=0}^{\infty}(\alpha\hat{K}_{Step})^{\sigma},$$

and if the eigenvalues of $\hat{K}_{Step}$ are all less than 1 (which is necessary for convergence), the summation converges as long as the limit is taken from below 1. Using (5.87), we get

$$\sum_{\sigma=0}^{\infty} \sigma(\hat{K}_{Step})^{\sigma} = \lim_{\alpha\to1} \alpha\frac{d}{d\alpha}(\hat{I} - \alpha\hat{K}_{Step})^{-1},$$

and using (5.88), the expression becomes

$$\sum_{\sigma=0}^{\infty} \sigma(\hat{K}_{Step})^{\sigma} = (\hat{I} - \hat{K}_{Step})^{-1} \cdot \hat{K}_{Step} \cdot (\hat{I} - \hat{K}_{Step})^{-1}.$$

Plugging in this expression in (5.86), we derive

$$\overline{\sigma} = \mathbf{1} \cdot \hat{K}_{Step} \cdot (\hat{I} - \hat{K}_{Step})^{-1} \cdot \mathbf{p}(0). \tag{5.89}$$

The other averages in (5.49) are obtained in a similar way.

Finally, we show that $\overline{m} + \overline{l} = \overline{\sigma}$. In order to prove this, consider the following:

$$\hat{K}_{Fwd} \cdot (\hat{I} - \hat{K}_{Fwd})^{-1} + \hat{K}_{Bwd} \cdot (\hat{I} - \hat{K}_{Bwd})^{-1} =$$
$$- \hat{K}^{+} \cdot (\hat{K}^{S} + \hat{K}^{-})^{-1} \cdot [\hat{I} + \hat{K}^{+} \cdot (\hat{K}^{S} + \hat{K}^{-})^{-1}]^{-1}$$
$$- \hat{K}^{-} \cdot (\hat{K}^{S} + \hat{K}^{+})^{-1} \cdot [\hat{I} + \hat{K}^{-} \cdot (\hat{K}^{S} + \hat{K}^{+})^{-1}]^{-1}.$$

We replace the first identity matrix with $(\hat{K}^{S} + \hat{K}^{-}) \cdot (\hat{K}^{S} + \hat{K}^{-})^{-1}$, and the second one with $(\hat{K}^{S} + \hat{K}^{+}) \cdot (\hat{K}^{S} + \hat{K}^{+})^{-1}$, and remembering that $(\hat{M} \cdot \hat{X})^{-1} = \hat{X}^{-1} \cdot \hat{M}^{-1}$, we obtain

$$\hat{K}_{Fwd} \cdot (\hat{I} - \hat{K}_{Fwd})^{-1} + \hat{K}_{Bwd} \cdot (\hat{I} - \hat{K}_{Bwd})^{-1} = \hat{K}_{Step} \cdot (\hat{I} - \hat{K}_{Step})^{-1}. \tag{5.90}$$

With this relationship, it is trivial to show that $\overline{m} + \overline{l} = \overline{\sigma}$ (just multiply by $\mathbf{1}$ on the left and $\mathbf{p}(0)$ on the right).

References

1. B. Alberts, *Molecular Biology of the Cell* (2008)
2. T. Soldati, M. Schliwa, Nat. Rev. Mol. Cell Biol. **7**(12), 897 (2006)
3. D.D. Hackney, Ann. Rev. Physiol. **58**, 731 (1996)
4. M.L. Mugnai, C. Hyeon, M. Hinczewski, D. Thirumalai, Rev. Mod. Phys. **92**(2), 025001 (2020)
5. J. Howard et al., *Mechanics of Motor Proteins and the Cytoskeleton*, vol. 743 (Sinauer Associates Sunderland, MA, 2001)
6. E.L. Holzbaur, Y.E. Goldman, Current Opin. Cell Biol. **22**(1), 4, 025001 (2010)
7. G. Bhabha, G.T. Johnson, C.M. Schroeder, R.D. Vale, Trends Biochem. Sci. **41**(1), 94, 025001 (2016)
8. N. Kodera, D. Yamamoto, R. Ishikawa, T. Ando, Nature **468**(7320), 72 (2010)
9. A.D. Mehta, M. Rief, J.A. Spudich, D.A. Smith, R.M. Simmons, Science **283**(5408), 1689 (1999)
10. A. Yildiz, P.R. Selvin, Acc. Chem. Res. **38**(7), 574, 025001 (2005)
11. J. Andrecka, Y. Takagi, K. Mickolajczyk, L. Lippert, J. Sellers, W. Hancock, Y. Goldman, P. Kukura, Methods Enzymol. **581**, 517, 025001 (2016)
12. M.J. Schnitzer, K. Visscher, S.M. Block, Nat. Cell Biol. **2**(10), 718, 025001 (2000)
13. W.J. Walter, V. Beránek, E. Fischermeier, S. Diez, PloS One **7**(8), e42218 (2012)
14. H.T. Vu, S. Chakrabarti, M. Hinczewski, D. Thirumalai, Phys. Rev. Lett. **117**(7), 078101 (2016)
15. R. Takaki, M.L. Mugnai, Y. Goldtzvik, D. Thirumalai, Proc. National Acad. Sci. **116**(46), 23091, 078101 (2019)
16. M.L. Mugnai, M.A. Caporizzo, Y.E. Goldman, D. Thirumalai, Biophys. J. **118**(7), 1537, 078101 (2020)
17. K.J. Mickolajczyk, N.C. Deffenbaugh, J.O. Arroyo, J. Andrecka, P. Kukura, W.O. Hancock, Proc. Nat. Acad. Sci. **112**(52), E7186, 078101 (2015)
18. H. Isojima, R. Iino, Y. Niitani, H. Noji, M. Tomishige, Nat. Chem. Biol. **12**(4), 290, 078101 (2016)
19. D.S. Bernstein, *Matrix Mathematics: Theory, Facts, and Formulas* (Princeton University Press, 2009)
20. I. Oppenheim, K.E. Shuler, G.H. Weiss, *Stochastic Processes in Chemical Physics: the Master Equation* (1977)
21. T.L. Hill, *Free Energy Transduction and Biochemical Cycle Kinetics* (Dover Publications, Inc., 2005)
22. M. Hinczewski, R. Tehver, D. Thirumalai, Proc. Nat. Acad. Sci. **110**(43), E4059, 078101 (2013)
23. F.W. Olver, A. Olde Daalhuis, D. Lozier, B. Schneider, R. Boisvert, C. Clark, B. Miller, B. Saunders, H. Cohl, M. McClain. Nist digital library of mathematical functions. http://dlmf.nist.gov/, Release1.1.1of2021-03-15. http://dlmf.nist.gov/
24. M. Abramowitz, I.A. Stegun, R.H. Romer, *Handbook of Mathematical Functions with Formulas, Graphs, and Mathematical Tables* (Dover, 1964)
25. J.W. Brown, R.V. Churchill, *Complex Variables and Applications Eighth Edition* (McGraw-Hill Book Company, 2009)
26. F.W. Olver, D.W. Lozier, R.F. Boisvert, C.W. Clark, *NIST Handbook of Mathematical Functions Hardback and CD-ROM* (Cambridge University Press, 2010)

Chapter 6
Physics of the Cell Membrane

Ben Ovryn, Terrance T. Bishop, and Diego Krapf

Abstract The plasma membrane is a phospholipid bilayer that links the cell to the outside world. The surface of eukaryotic cells consists of both hydrophilic and lipophilic lipid molecules and essentially an equal amount of protein. Some of these proteins are easily removed, while others are deeply integrated with the membrane. The amphipathic lipid molecules self-assemble in aqueous environments to spontaneously form bilayers with their hydrophilic head toward the water and hydrophobic face hiding toward the interior. In this chapter, we first discuss the variety of membrane proteins and evaluate a few examples. Then, we deal with membrane fusion, a process that is ubiquitous to cellular function. Membrane fusion is a process that requires concerted action of proteins and lipids. We dissect the features that make possible this complex process and its tightly regulated temporal evolution. At last, we examine how the material properties of the membrane give rise to its fluid mosaic elastic behavior and present a rigorous mathematical exploration of the energy associated with membrane bending. We introduce the Helfrich Hamiltonian and apply it to the shape and free energy of a membrane as it de-adheres from a flat substrate.

6.1 The Phospholipid Bilayer

Robert Hooke made the first documented microscopic observations of cells more than 350 years ago [1], but these early visualizations were not able to reveal organelles or nuclei. More than 200 years would elapse between Hooke's coining of the term "cells" and an understanding that cells had independent walls [2]. It would take at least

B. Ovryn
New York Institute of Technology, New York, NY, USA
e-mail: bovryn@nyit.edu

T. T. Bishop
Southern Illinois University, Carbondale, IL, USA
e-mail: Terrance.bishop@siu.edu

D. Krapf (✉)
Georgetown University, Washington, DC, USA
e-mail: diego.krapf@colostate.edu

© Springer Nature Switzerland AG 2022
K. B. Blagoev and H. Levine (eds.), *Physics of Molecular and Cellular Processes*,
Graduate Texts in Physics, https://doi.org/10.1007/978-3-030-98606-3_6

Fig. 6.1 2D structure of a glycerophospholipid. The glycerophospholipid phosphatidylcholine, one of the four main lipids in eukaryotic membranes, has glycerol esterified to two hydrophobic (nonpolar) fatty acid tails and a phosphate group with a hydrophilic choline head. When aligned in a bilayer, the choline head group, phosphate and glycerol have a high probability of being close to the aqueous interface, while the lipid tails, beginning with the C2 carbons, (i.e., the carbonyl group) form the hydrocarbon core. (Image from ChemIDplus TOXNET database (chem.nlm.nih.gov/chemidplus), Dipalmitoylphosphatidylcholine, registry number 2644-64-6)

another 100 years until the fluid mosaic model was articulated to describe the plasma membrane of eukaryotic cells. It is now understood that the phospholipid bilayer of eukaryotic cells consists of both hydrophilic and hydrophobic lipid molecules and essentially an equal amount of protein. These amphipathic lipid molecules self-assemble in aqueous environments to form bilayers with their hydrophilic head toward the water and hydrophobic face hiding toward the interior via an energetically favorable process (in the latter part of the chapter, the more complicated role of membrane fusion is introduced). Although the term lipid bilayer underscores the protein content of the plasma membrane, this name more aptly conveys the significant amount of molecules like glycolipids and cholesterol in the membrane. These molecules combine to provide both the fluid and the mechanical properties of eukaryotic plasma membranes (the contributions of cytoskeletal filaments to the properties of the plasma membrane are not emphasized in this chapter). Often the plasma membrane is depicted as a highly symmetric arrangement of amphipathic lipid molecules, however, there is actually a significant difference in the lipid distribution between the inner and outer monolayers.

The phospholipids in the eukaryotic plasma membrane are predominately glycerophospholipids, consisting of two hydrophobic fatty acid tails (hydrocarbon chains) and a hydrophilic head group. Phosphatidylcholine, a glycerophospholipid that has a terminal choline head group (Fig. 6.1), constitutes nearly a fifth of the lipid content of the red blood cell plasma membrane and nearly 40% of the endoplasmic reticulum membrane and the inner and outer mitochondrial membranes. By contrast, the cell membrane of Escherichia coli, considered the quintessential bacterial membrane [3], essentially consists of a single lipid (phosphatidylethanolamine).

The extended structure of lipid molecules can be approximated as either cylindrical, when the cross-sectional area of the head group is essentially equivalent to the tails, or conical, when there is an asymmetry between these two regions. When cylindrical lipids are arrayed in a bilayer, the relaxed state produces a planar shape. Conversely, if lipid molecules with a large head group are arrayed together, the bilayer has an intrinsic positive (convex) curvature, where each monolayer has a complementary curvature. In order to form a planar bilayer, bending a positively curved bilayer requires an energy $E = \kappa c_0^2/2$ to reorient the molecules, where κ and c_0 represent the bending modulus and the intrinsic or spontaneous curvature, respectively. Using typical values cited in published literature, $\kappa \approx 30\ kT$ [4, 5] and $c_0 \approx 0.26$ nm^{-1} [6], yields a bending energy $E \approx 1\ kT$ /nm^2 (these concepts will be revisited in greater detail in a subsequent part of the chapter).

6.2 Membrane Proteins

The phospholipids in the bilayer provide both a barrier that separates the cell from the outside world and a solvent for proteins that enable selective transmembrane communication [6]. The fluid mosaic model developed by S.J. Singer and Garth Nicolson, in 1972, encapsulates the idea that hydrophobic proteins can be a stable and integral part of the hydrophobic lipid bilayer [7]. These membrane proteins are often termed "integral" proteins so as to distinguish them from proteins that are peripherally associated with the membrane and easily removed [8]. In this chapter, we use the terms membrane protein and integral membrane protein interchangeably. It is rather remarkable that hydrophobic proteins can reside inside of the hydrophobic core of the lipid bilayer. In fact, because of an intricate and choreographed set of processes, membrane proteins are inserted and folded in the membrane so as to minimize free energy. Membrane proteins and soluble proteins have structures that are governed by thermodynamics with interiors populated by hydrophobic residues and polar residues toward the exterior. Membrane proteins, however, face an additional constraint imposed by the bilayer orientation, defined by the transmembrane axis (perpendicular to the bilayer), such that all spatially extended membrane proteins reflect the preference of some amino acids for the cytoplasmic face (the "positive-inside" rule) [9]. In order for membrane proteins to traverse the thickness of the bilayer, the α-helices and β-strands are generally longer than soluble proteins. This is illustrated in Fig. 6.2 which shows the seven membrane spanning α-helices (shown as coils) and β strands (shown as ribbons) of rhodopsin in the membrane with the top (red) surface representing the exterior of the bilayer (obtained from the RCSB Protein Data Bank [10]).

The hydrophobic match between the membrane protein and the bilayer drives the fluid mosaic so as to minimize the energy penalty caused by having either longer or shorter protein hydrophobic length than the thickness of the relaxed bilayer hydrophobic core. When the hydrocarbons of the bilayer do not adequately cover the hydrophobic regions of the membrane protein, the tails of the lipids can either extend or the protein can be tilted within the bilayer. Because membrane proteins are two to three orders of magnitude, more rigid than the bilayer, the thickness of the bilayer changes so as to reduce the hydrophobic mismatch which results in an energetic cost as the membrane reconfigures. A quantitative analysis of the energy associated with membrane bending will be examined in considerable detail in the subsequent part of the chapter.

Fig. 6.2 Bovine rhodopsin shown in the plasma membrane. The α-helices and β strands of the seven transmembrane rhodopsin are shown. The exterior portion of the bilayer is on the top (red); the N-terminus of the protein is on the top and the C-terminus is on the bottom right. The hydrodynamic thickness is 3.12 ± 0.15 nm (image from the RCSB Protein Data Bank (www.rcsb.org) of PDB ID 1U19 [10])

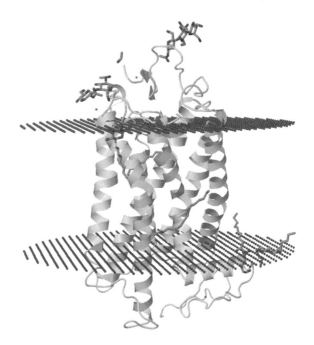

6.2.1 Integral Proteins

Integral proteins are also known as transmembrane proteins because they have one or more segments which span the membrane. These proteins have three segments: exoplasmic, cytosolic, and a membrane spanning region. The exoplasmic and cytosolic domains typically have hydrophilic exteriors to interact with an aqueous environment, where the membrane spanning region is typically hydrophobic to interact with the hydrocarbon core of the phospholipid bilayer. Here, the exoplasmic space is regarded as one of three options: the lumen of the ER or Golgi or the cell exterior. In order to span the membrane, integral membranes typically contain one or more α helices or multiple β barrels. Membrane proteins can be further subdivided into classes (I–IV) that depend upon the orientation of the protein with respect to the cytosolic and exoplasmic space (the transmembrane axis), as well as the number of passes. Each of these four types of integral proteins is synthesized at the ER along with certain GPI-linked proteins as well. The synthesis of each of these five ER membrane proteins is fine-tuned to achieve specific protein orientation [11] (Fig. 6.3).

6.2.1.1 Type I Embedded Proteins

Type I proteins have an N-terminal signal sequence that directs them to the ER as well as a membrane spanning α helix. The signal-recognition particle (SRP) binds to its signal sequence on a nascent peptide chain as well as the large ribosomal sub-

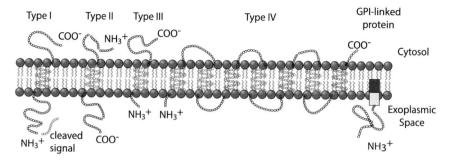

Fig. 6.3 Five potential orientations of integral membrane proteins. The classification of these proteins depends upon their orientation with respect to the membrane, as well as their number of passes through the membrane. The exoplasmic space consists of either the ER or Golgi lumen and the extracellular space in the plasma membrane

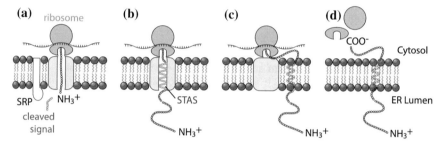

Fig. 6.4 Insertion mechanism for Type I proteins into the ER membrane. a Nascent protein with its N-terminus in the exoplasmic space, i.e., the ER lumen. **b** Protein synthesis pauses at the stop-transfer sequence. **c** The protein is transferred from the translocon to the lipid bilayer. **d** finished Type 1 protein with its C-terminus in the cytoplasm

unit. This nascent protein–ribosome complex is targeted to the SRP receptor on the ER membrane. Once at the membrane, GTP binds the SRP and SRP receptor to strengthen their interaction. The nascent protein is transferred from the SRP/large ribosomal subunit to the translocon, which requires opening the translocation channel. Once in the translocation channel, the protein continues to be synthesized until an internal stop-transfer anchor sequence (STAS) is reached which halts the synthesis. The Stop-transfer sequence is transferred laterally from the translocation channel to the phospholipid bilayer. Upon completion of their synthesis, the ribosomal subunits are released into the cytosol, and the newly synthesized Type I integral proteins are able to freely diffuse in the membrane (Fig. 6.4).

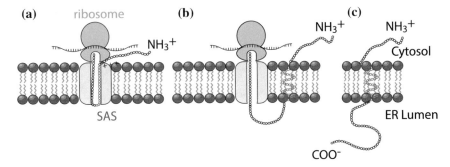

Fig. 6.5 Insertion mechanism for Type II proteins into the ER membrane. a The nascent chain is oriented so that the N-terminus is localized in the cytoplasm. **b** The signal-anchor sequence (SAS) is transferred laterally to the bilayer. **c** A completed Type II protein

6.2.1.2 Type II and Type III Proteins

Type II and Type III proteins do not contain the cleavable signal sequence of a Type I protein, but rather they contain an internal signal-anchor sequence that serves as both a signal sequence to the ER and a membrane anchor sequence. As the Type II and Type III proteins are being translated, their internal signal-anchor sequence (SAS) is synthesized and bound by an SRP which directs the ribosome nascent chain complex to the ER membrane. The orientation of Type III is similar to that of Type I proteins, in that the N-terminus is in the exoplasmic space and the C-terminus is in the cytosol. The orientation of Type II is opposite of Type III orientation. This difference in orientation between Type II and Type III is accomplished by differences in the positioning of the internal signal-anchor sequence. For Type II proteins, the nascent chain is oriented within the translocon with its N-terminal portion toward the cytosol. Next, as the protein chain continues to grow, the Signal-anchor sequence is transferred laterally anchoring the sequence to the phospholipid bilayer. Once protein synthesis is done, the C-terminus of the polypeptide is released into the lumen and the ribosomal subunits are released into the cytosol. Type III synthesis is very similar except the positively charged residues are closer to the C-terminus than the N-terminus resulting in an opposite orientation within the translocon compared to Type II. This results in Type III having its N-terminus in the ER lumen, and its C-terminus in the cytosol (Figs. 6.5 and 6.6).

6.2.1.3 Multiple Topogenic Sequences

Types I, II, and III all utilize a single topogenic sequence. For Type I proteins, there is a single internal stop-transfer anchor sequence, and for Types II and II, there is one internal signal-anchor sequence. Type IV proteins (multipass proteins) have multiple internal topogenic sequences. These membrane spanning α helixes can act to direct the protein to the ER, to anchor the protein, or to stop the transfer of the protein. Type

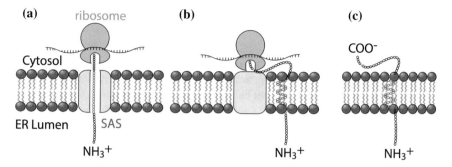

Fig. 6.6 Insertion mechanism for Type III proteins into the ER membrane. Note the presence of a stop-transfer sequence instead of a signal-anchor sequence instead of a stop-transfer anchor sequence. **a** The nascent chain is oriented so that the N-terminus is in the ER lumen. **b** The protein is transferred laterally from the translocon when the SAS is reached. **c** A completed Type III protein

IV proteins are further categorized by the orientation of their N-terminus with respect to the membrane. This N-terminus arrangement is determined by the charge of the sequences that follows the hydrophobic sequence closest to the N-terminus. Type IV proteins are also classified by the number of transmembrane α helices. If there is an even number of membrane spanning regions, the N-terminus and the C-terminus will be facing toward the same side of the membrane, where an odd number of membrane spanning regions would result in opposite orientations of the two termini.

6.2.1.4 Type IV Proteins

Type IV proteins can be classified according to having their N-terminus within the cytoplasmic or extracellular side. Those with N-terminus in the cytosol include proteins like glucose transporters and most ion channel proteins. The hydrophobic section neighboring the N-terminus is responsible for commencing the insertion of the nascent protein sequence into the ER membrane. The first N-terminal α helix and the following odd numbered ones serve as signal-anchor sequences, while the even sequences act as stop-transfer anchor sequences.

Type IV proteins with the N-Terminus in exoplasmic space include the family of G-protein-coupled receptors. The hydrophobic α-helix closest to the N-terminus contains a bunch of positively charged amino acids. These positively charged amino acids act to assist with the insertion of the growing peptide into the translocon with the N-terminus lengthening into the lumen. As the chain continues to grow, it is implanted into the membrane of the ER by alternating type II signal-anchor and stop-transfer sequences.

6.2.1.5 GPI-Linked Proteins

GPI-linked proteins are covalently attached to an amphipathic molecule known as Glycosylphosphatidylinositol (GPI). These GPI-linked proteins are initially synthesized in a similar fashion to other transmembrane proteins, however, a dense sequence of amino acids serves as a signal to be recognized by transamidase. Transamidase simultaneously cleaves the original anchor sequence while transferring the remaining protein to a GPI anchor. One benefit of a GPI anchor is the ability to easily diffuse on the surface of the phospholipid bilayer.

6.2.2 Peripheral Proteins

One method of peripheral protein association is accomplished by disruption of the electrostatic interaction between basic groups of peripheral protein and anionic lipids. The inner leaflet of the plasma membrane of animal cells is composed of 20% anionic lipids that provide negative charges. A peripheral protein can be phosphorylated in the membrane spanning region. This addition of a phosphate is referred to as an electrostatic switch because the phosphate can be cleaved by phosphatases altering the peripheral proteins' membrane binding potential.

Another method of membrane association involves post-translational modification of a peripheral protein to add an acyl chain. When this mechanism is used, a lipid anchor is permanently added in the post-translational modification. This anchor is too weak to associate a peripheral protein to the membrane, so a second acyl chain (or prenyl group) is added. The second chain doubles the strength of the interaction providing sufficient strength to associate the peripheral protein and the membrane. Acyltransferases can be used to modify the affinity of the proteins for the membrane.

The third method of regulation involves the binding of ligands such as nucleotides or Ca^{2+}. Binding of GTP to the G protein ADP- ribosylation factor causes a conformational change in the protein that exposes an amphipathic helix which can insert itself into the bilayer. Ca^{2+} can trigger similar effects. Recoverin is a neuronal calcium binding protein in photoreceptor cells of the eye. When Ca^{2+} binds recoverin, it induces a conformational change where a bound myristoyl group extends out to become a membrane. This "Myristoyl switch" can be seen in other peripheral proteins involved in signal transduction [12]. Finally, peripheral proteins can bind with a specific lipid component of the membrane like DAG or polyphosphorylated inositol.

6.2.3 Receptors

Cell surface receptors are integral membrane proteins that are responsible for communicating external signals to the cell. For surface cell receptors, there are two properties that characterize their interaction with a given ligand, specificity and affinity. Specificity refers to the ability of a protein to refer one molecule preferentially to another, whereas the affinity refers to the strength of the binding interaction. Molecular complementarity is the idea that the shape and the chemical surface of a binding site must be complementary to its ligand. When a ligand binds to its receptor, a conformational change occurs in that receptor which is communicated across the membrane spanning domain resulting in the subsequent activation of protein in the cytosol. The process of taking an extracellular signal and converting it into an intracellular process is referred to as signal transduction.

6.2.3.1 G Protein-Coupled Receptors—GPCR

GPCRs are the largest family of receptors in cells. They typically contain seven transmembrane helices and have an N-terminus in the extracellular space and their C-terminus in the cytoplasmic space. GPCR activation method involves conformational changes of GPCR, especially in the regions facing the cytosol.

Olfaction is a process which utilizes a G protein-coupled receptor. Olfaction neurons express one type of odor receptor, and in order to detect a wide variety of scents, 400 distinct receptors are used in concert to distinguish 10^{12} different scents. The olfaction signaling cascade begins when an odorant binds the 7-transmembrane odorant receptor. The direct effector is adenylyl cyclase which leads to a cascade that results in an influx of calcium ions and sodium ions. This influx leads to an action potential that is communicated to the brain. For olfactory processes, it is important to have a mechanism for desensitization after prolonged activation. This mechanism is accomplished by GPCR kinase (GRK) that phosphorylates the GPCR on multiple sites, and by arrestin that binds to the phosphorylated GPCR, effectively desensitizing it.

6.2.3.2 Receptor Tyrosine Kinases

Receptor tyrosine kinases are integral membrane proteins with a single transmembrane domain. Their extracellular domain is variable and the cytoplasmic part contains a tyrosine kinase domain. The activation mechanism for receptor tyrosine kinases (RTK) is dependent on kinase domain dimerization and transphosphorylation. RTKs accomplish this activation along with the help of dimeric signal ligands, juxtapose positioning, and release of steric hindrance.

Activation of RTK by epidermal growth factor leads to dimerization and autophosphorylation within the C-terminal tail of the receptor. This phosphorylation recruits in a GTP exchange factor (GEF) and various adaptor containing proteins leading to the activation of Ras and downstream MAP kinase pathways. Activated RTKs can also stimulate phospholipase C activity to generate lipid-derived second messengers.

6.2.3.3 Cytokine Receptors

Cytokine receptors are integral membrane proteins with a single transmembrane domain. Cytokine receptors have no kinase domain, instead they have a kinase tightly associated with their intracellular domains. The associated kinase is Janus kinase (JAK, also known as "just another kinase"), which binds to the cytoplasmic tail. The activation of cytokine receptors is dependent on kinase domain dimerization through interaction with dimeric signal ligands. The binding of a dimeric signal ligand leads to conformational changes in pre-formed dimers.

When a ligand binds, it leads to a conformational change that places the dimers of the cytokine receptor closer together. These dimers JAK region trans-phosphorylates one another. These phosphorylated sites act as binding regions for signal transducer and activator of transcription (STAT). STAT is phosphorylated and released to the cytoplasm where it forms a homodimer. The phosphorylated STAT dimer enters the nucleus and leads to the expression of various genes. One Gene that is activated is a suppressor of cytokine signaling which exhibits negative feedback on the system.

6.2.3.4 Receptor Serine/Threonine Kinases

Serine/threonine kinase receptors are integral membrane proteins with a single transmembrane domain. Serine/threonine kinase receptors feature a Type I receptor and a Type II receptor positioned just next to each other within the membrane. The type I receptor has an inactive kinase domain, whereas the type II receptor has an active kinase domain. The human genome encodes for seven type I receptors and five Type II receptors. Activation of serine/threonine occurs through kinase domain dimerization via trans-activation of type I kinase domain by Type II kinase domain.

Transforming growth factor β (TGFβ) brings together the type I and type II receptors. The type II active kinase phosphorylates the Type I's kinase. R-Smad binds the type I receptor and forms a complex with Smad anchor for receptor activation (SARA). The type I kinase phosphorylates R-Smad which promotes its dissociation from the receptor and SARA. Now in the cytoplasm, R-Smad binds Co-Smad, and the Smad hetero-oligomer enters the nucleus. Smad associates with DNA-binding proteins to activate or inhibit the transcription of specific genes.

6.2.3.5 Integrins and the Glycocalyx

Integrin, an important cell surface receptor, is a glycoprotein with an alpha and beta subunit [13]. As a superfamily, integrins are significantly N-linked glycosylated [14]. The hierarchical processes that lead to integrin binding to protein ligands on the extracellular matrix and the formation of an integrin adhesion requires integrin activation [15–17]. Upon activation, the affinity for a ligand increases as ligand-binding site epitopes in integrin's extracellular domains become exposed. Subsequent to integrin activation, the stalks of the conformationally activated heterodimer have relatively short projections (tens of nanometers) from the plasma membrane into the extracellular space. These short integrins molecules diffuse among a sea of molecules, called the pericellular matrix or glycocalyx, which also project toward the extracellular matrix. The glycocalyx consists of proteoglycans that have a core protein and many covalently linked glycosaminoglycan (GAG) chains. Heparan sulfate (HS), chondroitin sulfate, and hyaluronan (HA) are prominent GAGs present on most cells (HA is a GAG without a core protein). The longer diffusing molecules of the glycocalyx suggest that the ligand-binding domains of a homogenous population of activated, short integrins are prevented from reaching the extracellular matrix [5]. An analysis of the free energy of these diffusing molecules and the binding energy associated with the plasma membrane, so as to displace integrins toward the extracellular matrix, predicts the magnitude of the energy barrier which prevents the formation or nucleation of an integrin adhesion patch (a cluster of bound integrins). It is conjectured that a stochastic ensemble of a few actin polymerizing filaments is required to nucleate an adhesion [5]. As the adhesion patch grows in size, it becomes a nascent adhesions or a focal complex [18, 19].

6.2.3.6 Surface Cell Receptor Examples

Endocrine Signaling
Endocrine signaling can carry information in the form of hormones at great distances. These hormones are synthesized by endocrine organs and act on various target cells distant from their site of synthesis. The movement of hormones is typically accomplished by blood or other extracellular fluids.

Hormones fall into four general classes: small water-soluble molecules, peptide hormones, lipophilic molecules, which are detected by extracellular receptors, and lipophobic molecules which are detected by intracellular receptors. Hormones have temporal and structural specificity, only being released when needed, and detected only by the cell which they are meant to target. Some examples of the different hormonal classes are the following:

- Small water-soluble molecules: histamine, epinephrine
- Peptide hormones: insulin, luteinizing hormone
- Lipophobic molecules with extracellular receptors: prostaglandins
- Lipophilic molecules with intracellular receptors: cortisol, progesterone

Paracrine Signaling
Paracrine signaling occurs when signaling molecules released by a cell are destined to affect target cells in close proximity to the signal source. The release of neurotransmitters from synaptic vesicles is a prime example of paracrine signaling. The neurotransmitters diffuse across the synaptic cleft where they bind to dendritic receptors

Autocrine

In autocrine signaling, cells respond to signals that they released themselves. Some growth factors act in this fashion. Autocrine stimulation occurs when a cell produces growth factors that act on itself. Many tumors lack balance in the autocrine signaling processes and over-release growth factors which lead to additional tumor proliferation.

6.3 Membrane Fusion

Membrane fusion is a process shared by a variety of cellular processes ranging from fertilization, intracellular trafficking, viral entry, tissue formation, carcinogenesis, and more. And while the goal of membrane fusion is different for each of these cases, they all share similarities within the mechanism they employ to accomplish their function. This section delves into the mechanisms and driving forces for membrane fusion.

6.3.1 Intermediate Structures

The basic structures leading to the fusion of two lipid bilayers are outlined in Fig. 6.7 and summarized below. Mechanistically, it involves bending of the membranes, forming a hemifusion stalk where the proximal leaflets are allowed to mix their components, expanding this stalk into a diaphragm and eventually forming a pore.

Hemifusion is an intermediate stage of membrane fusion that results in lipid mixing between the proximal leaflets of two membranes, but not mixing of the contents of the distal membranes. Hemifusion leads to a distinct membrane structure known as a fusion stalk (Fig. 6.7c). The formation of this fusion stalk depends on a combination of elastic deformations within the membrane leaflets. Necessary deformations include bending, splaying of the hydrocarbon tails of lipids molecules, and tilting of these lipids with respect to the membrane [20]. A more detailed discussion of the physics of membrane deformations is presented later in this section.

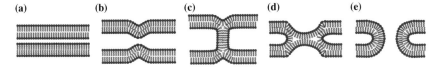

Fig. 6.7 Sketch of membrane fusion leading to the formation of a fusion pore [22]. (**a**) Two membranes prior to fusion initiation. (**b**) A membrane protrusion forms reducing the hydration repulsion energy between the membrane leaflets coming into contact. (**c**) The formation of the hemifusion stalk results in mixing of proximal leaflets but does not involve mixing of the distal leaflets. (**d**) Expansion of the hemifusion stalk results in hemifusion diaphragm. (**e**) A fusion pore forms, either from the hemifusion diaphragm or directly from the fusion stalk

For biological membranes, lipid species can modulate hemifusion formation but generally the physiologically available lipids in the plasma membrane do not provide sufficient driving forces for hemifusion to take place. For example, the hemifusion diaphragm forms spontaneously in fully hydrated membrane systems only when the membrane lipid composition is almost entirely phosphatidylethanolamine (PE) [21]. This condition is not realistic in cell membranes, so it becomes apparent that additional forces via proteins integrated into the membrane are necessary to drive the reaction to completion. Within the cell fusion processes, the hemifusion pore is almost universally formed as an intermediate step to successful fusion events. In these successful events, the hemifusion diaphragm must transition into an expanding fusion pore.

A key difference between the hemifusion diaphragm and the fusion pore is the mixing of lipids from both the proximal and the distal layers of the cell membrane as well as the mixing of the contents initially separated by the apposed membranes. A key similarity between the hemifusion diaphragm and the fusion pore is that in both the cases, their formation depends on specific lipid compositions. Fusion pores in viral fusion and exocytosis have been seen to open on the order of microseconds with a diameter near 2 nm. For the next few milliseconds to seconds, the fusion pore can exhibit an irregular opening and closing pattern known as flickering. However, for cases where there is a successful fusion event, the fusion pore exhibits a gradual expansion that is irreversible. It has been shown that fusion pores are essentially lipidic, that is, they can form in exclusively lipid environment without the help of additional proteins [23]. However, additional fusion proteins can have a strong influence on their properties.

6.3.2 Membrane Tension as a Driving Force

One of the most energetically unfavorable stages of the fusion process is the formation and the expansion of a fusion pore. During this process, the cell drives the expansion of fusion pores by actively modulating local membrane surface tension. Furthermore, membrane tension plays key roles in prefusion and early fusion stages [24].

Due to the fluid nature of lipid membranes, besides the traditional membrane mechanical forces, membrane tension is also affected by lipid flow share rates. These dynamic contributions can be highly anisotropic. On the other hand, static tension can be considered in terms of an analogue to hydrostatic pressure in the 3D liquid case. Therefore, one would expect static tension to be homogeneous across the entire membrane. However, anisotropies can easily develop in the plasma membrane tension. For example, membrane domains, which are widespread in the plasma membrane have different tensions on both sides of the domain boundary. In the same way, that two bulk phases intersect at a surface, which is associated with a surface tension γ_s; two membrane phases intersect at a line, associated with an interfacial line tension γ_l. The tension difference $\Delta \tau$ within a domain of radius R can be computed from the interfacial line tension: $\Delta \tau = \gamma_l / R$. Interfacial line tensions in lipid bilayer domains have been measured from the thermodynamics of domain nucleation rate, given that in thermodynamic equilibrium, the nucleation rate J follows an Arrhenius law

$$J = J_0 \exp(-\Delta G/k_B T), \qquad (6.1)$$

where $\Delta G \propto \gamma_l^2$ is the activation barrier to nucleation and $k_B T$ is thermal energy and J_0 is a constant [25]. Additionally, membrane line tensions are also estimated by shape analysis [26].

Typical tensions of the plasma membrane are in the range of 0.01–0.04 mN/m but, in specialized cells, it can reach much higher values. As mentioned above, the cells modulate membrane tension as a mechanism to drive specific processes such as fusion pore expansion. The forces that modulate membrane tension are related to four general processes:

1. Hydrostatic pressure is controlled by osmotic imbalance between the cytosol and the extracellular medium, i.e., differences in the total ionic concentrations. Due to the nature of hydrostatic pressure, its contributions are homogenous across the cell membrane.
2. The plasma membrane interacts with the cortical cytoskeleton adjacent to the cell membrane. These forces can introduce local heterogeneities and specific local membrane curvatures. Interactions with the cytoskeleton can further promote or inhibit the formation of membrane domains.
3. Adhesion processes provide strong forces that tether the membrane at specific locations.
4. The consumption or abscission of membrane material effectively increases or decreases the area of the membrane, which in turn, alters membrane tension.

6.3.3 Fusion Proteins

Successful fusion events rely on the help of a variety of proteins expressed on the surface of each cell participating in the fusion event. These proteins are necessary to bring the membranes within close contact with one another, catalyze the formation of the initial intermediates, and drive the process to a successful completion. Accomplishing these tasks relies on the coordinated effort of numerous families of proteins [27]. A handful of the key players are highlighted below.

6.3.3.1 Soluble N-Ethylmaleimide-Sensitive-Factor Attachment Protein Receptors - SNAREs

Soluble SNAREs are a superfamily of small proteins that have been termed the workhorses for fusion. There is great variability in sizes and structures of SNAREs, the only truly identifiable feature between them being a SNARE motif. The Snare motif contains 60–70 amino acids that includes eight heptad repeats for coiled coil domains. Most SNARES contain a C-terminal domain that serves as a membrane anchor. SNARE proteins from opposing membranes associate with core complexes during fusion, and dissociate after, giving rise to a SNARE cycle.

Combining the correct SNARE proteins will lead to spontaneous assemblage into four helical bundles in which the SNARE motifs are arranged in a parallel fashion (Fig. 6.8). When this parallel complex is assembled, SNARE proteins complete their function with an elegant mechanism. SNAREs in opposite membranes intertwine with one another and pull their opposing membranes into closer proximity. There exist four subfamilies of SNARE proteins: Qa-SNAREs, Qb-SNAREs, Qc-SNAREs, and R-SNAREs. All functional SNARE complexes rely on a SNARE from each subfamily. There exists ample evidence that SNARE proteins are responsible for driving membranes into close proximity, but it is still unclear exactly how SNAREs accomplish this mechanism. SNARE proteins remain the best candidate for inducing

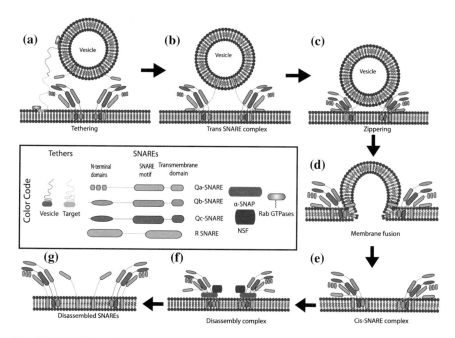

Fig. 6.8 SNARE-mediated vesicle fusion. **a** Tethering process whereby surface molecules on both the membrane target and the vesicle initiate the fusion process at distances up to 50 nm. **b** Formation of the trans-SNARE complex from proteins on the vesicle and the target membrane. **c** Zippering that occurs as the SNAREs on apposed membranes coil together in a spontaneous process. As the SNAREs coil, they form a four-helix complex and the two membranes are brought together. **d** The two membranes merge and their contents are combined. **e** The cis-SNARE complex refers to the energetically favorable complex that is formed over the course of the fusion reaction. **f** Disassembly of cis complex requires recruitment of the cytosolic protein α-SNAP, which, in turn, recruits the ATPase NSF to disassemble the cis-SNARE complex in an energy-dependent manner. **g** Disassembled SNAREs can be used again in additional membrane fusion events

successful fusion by inducing and stabilizing the transition state that leads to the opening of a fusion pore.

6.3.3.2 Sec1/Munc18-Like Proteins—SM

SM proteins are hydrophilic and typically contain 650–700 residues. When SM proteins are deleted, the respective fusion event fails, demonstrating the necessity of SM proteins for fusion. In the human genome, there exists seven varieties of SM proteins which suggests that they are versatile fusion agents that are capable of functioning in multiple fusion reactions. All SM proteins are arch-shaped molecules consisting of three domains and a major v-shaped cleft in the middle. There are four distinct ways that SM proteins interact with SNAREs. The most common type of interaction occurs when the SM binds directly to the N-terminus of a corresponding Qa-SNARE. This binding is sequence specific and has been observed in all fusion reactions within the Golgi complex and the endoplasmic reticulum which suggests its importance in secretory processes. SM proteins like Munc18a, 18b, and 18c can also bind directly to their corresponding Qa-SNARE. This process requires a distinct closed conformation of syntaxins.

SM mediated fusion in yeast cells has been shown not to bind to Qa-SNAREs but rather to fully assembled SNARE complexes. The exact mechanism of this process is yet to be determined. The final mechanism occurs in endosomal/vacuolar fusion via SM proteins interacting with SNAREs indirectly via SNARE binding proteins. Current evidence suggests that SM proteins regulate SNARE assembly in a fashion that is coupled to membrane attachment. It has also been suggested that SM proteins may act as syntaxin chaperones. SM proteins can interact with individual SNAREs, or SNARE complexes to help drive fusion to completion.

6.3.3.3 Rab Proteins

Rab proteins are members of the GTP-binding family. Similar to other proteins in this family, when in a GTP bound state, Rab proteins' initiate downstream effector proteins. The activation and deactivation of Rab proteins is modulated by both GTP exchange factors and GTPase activating proteins. Rab proteins range between 21 and 25 kDa and contain many regions that are highly conserved across the Ras superfamily. Most Rab proteins are tightly associated with the membrane via post-translation addition of two geranylgeranyl (20-carbon polyisoprenoid) groups. The Rab family contains approximately 70 members who act as the master regulators of intracellular vesicle transport. A small fraction of Rab proteins are found within the cytosol, where majority of Rab proteins are localized to membranes of transport vesicles and their target compartments.

6.3.4 Electrostatic Forces

From a physical point of view, the fusion of lipid bilayers in an aqueous environment takes place in two steps [28]. In the first step, the membranes are brought together, so that the process must overcome the repulsive electrostatic forces between the two outer leaflets. The second step involves destabilization of the boundary between the hydrophilic and hydrophobic portion of the bilayer. These transition states rely on forces that minimize the exposure of nonpolar surfaces to water. The successful culmination of these two steps involves the formation of an aqueous fusion pore. The time scale for a successful fusion reaction has been found to depend exponentially on membrane tension, with increased membrane tension leading to more rapid fusion processes.

Example 6.1 Example of Membrane Fusion—Viral Entry Viruses with an envelope require membrane fusion to infect the cell. To achieve this, first and foremost, the virus expresses many fusion proteins locked into a prefusion conformation [29]. Two transitions must occur for the virus cell to become fusogenic. The first process is known as priming and commonly occurs via proteolytic cleavage. The second process is known as triggering and occurs as a result of ligand binding. Examples of ligands that trigger this transition are protons and co-receptor expressed on a cell surface. Viral fusion proteins have been termed "suicide enzymes" because they undergo a irreversible priming step and only function once. This is vastly different when compared to the mechanism of SNARE proteins which exhibit a ATP dependent cycle.

  Viral fusion proteins fall into three categories. The first class of fusion proteins are trimers of a single-chain precursor, which requires proteolytic cleavage to become active. The second class achieves priming by cleavage of a heterodimeric partner protein, and the final class of fusion protein has no priming mechanism. When the virus fuses its envelope to its target membrane, its internal contents are released into the cell to wreak havoc.

6.4 Energy Required to Bend a Membrane

6.4.1 Fluid Properties of the Plasma Membrane

Because the membrane has both fluid-like and elastic properties, it is useful to begin with an examination of a curved fluid surface under tension and subsequently look at the effect of elasticity. Consider a transition between two fluids (e.g., water and air). If the transition between the two media occurs over a very thin layer, it can be treated as a distinct boundary which is commonly called a surface. Ultimately, the free energy of the plasma membrane will be analyzed, but it is often more common to analyze the forces and the shape of the surface. Figure 6.9 illustrates a pressure

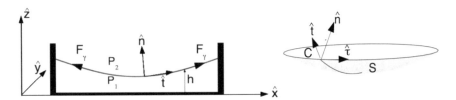

Fig. 6.9 Forces on a fluid surface. Two forces are shown acting on the fluid interface: (1) a pressure difference $\Delta P = P_2 - P_1$, acting in the direction $\hat{\mathbf{n}}$ normal to the surface S, and (2) a surface tension γ, which acts along the boundary C, in the direction $\hat{\mathbf{t}}$ tangent to S and perpendicular to C. The unit vector $\hat{\boldsymbol{\tau}}$ points along the boundary C

difference $\Delta P = P_2 - P_1$, across a surface. Because small creatures can walk on the surface of a fluid, there must be enough force generated to balance their weight. This force arises from the surface tension of the fluid. Surface tension, γ, has units of force/length (i.e. N/m) which is effectively an energy density (J/m^2); water at 25 °C, has a surface tension $\gamma \approx 0.072$ N/m [30].

The shape of the surface S, shown in Fig. 6.9, may be characterized by a height $z = h(x, y)$, above the xy plane; a surface with this parameterization is known as a Monge patch and the Monge parameterization will be invoked throughout this chapter [31, 32]. At any point, $\mathbf{r}(x, y, h(x, y))$, on this smooth surface there is a unit normal vector $\hat{\mathbf{n}}$, which points from medium 1 (with pressure P_1) to medium 2. If the surface is in equilibrium, the pressure difference in the normal direction $\hat{\mathbf{n}}$ must be balanced by the surface tension $\mathbf{F}_\gamma$. The surface tension along the boundary C acts in the direction $\hat{\mathbf{t}}$, where $\hat{\mathbf{t}}$ is a unit vector tangent to the surface. Force balance implies the total force $\mathbf{F}_T$ must cancel [33],

$$\mathbf{F}_T = \iint_S (P_2 - P_1)\, \hat{\mathbf{n}}\, dS + \oint_C \gamma \hat{\mathbf{t}}\, d\lambda = 0, \qquad (6.2)$$

where the integrals are over the surface and the line perimeter C, respectively. In order to show how these two forces balance, the line integral can be re-stated as a surface integral using Stokes' theorem [34]. From the geometry of Fig. 6.9, Stokes' theorem states $\oint_C \mathbf{f} \cdot \hat{\boldsymbol{\tau}}\, d\lambda = \iint_S (\nabla \times \mathbf{f}) \cdot \hat{\mathbf{n}}\, dS$ where $\mathbf{f}$ is any vector and $\hat{\boldsymbol{\tau}}$ points along C. Letting $\mathbf{f} = \gamma \hat{\mathbf{n}} \times \boldsymbol{\mu}$, where $\boldsymbol{\mu}$ is a constant vector, Stokes' theorem yields

$$\oint_C (\gamma \hat{\mathbf{n}} \times \boldsymbol{\mu}) \cdot \hat{\boldsymbol{\tau}}\, d\lambda = \iint_S \left[\nabla \times (\gamma \hat{\mathbf{n}} \times \boldsymbol{\mu}) \right] \cdot \hat{\mathbf{n}}\, dS. \qquad (6.3)$$

The line integral in Eq. 6.3 now has a similarity to the second integral in Eq. 6.2, but it is important to note that Eq. 6.3 is an integral along $\hat{\boldsymbol{\tau}}$, whereas the second integral in Eq. 6.2 is an integral along $\hat{\mathbf{t}}$. Using the vector identity $(\mathbf{a} \times \mathbf{b}) \cdot \mathbf{c} = \mathbf{b} \cdot (\mathbf{c} \times \mathbf{a})$ yields $(\gamma \hat{\mathbf{n}} \times \boldsymbol{\mu}) \cdot \hat{\boldsymbol{\tau}} = \boldsymbol{\mu} \cdot (\gamma \hat{\boldsymbol{\tau}} \times \hat{\mathbf{n}})$ and from Fig. 6.9, $\hat{\mathbf{t}} = \hat{\boldsymbol{\tau}} \times \hat{\mathbf{n}}$, so the integrand on the left-hand side of Eq. 6.3 becomes $(\gamma \hat{\mathbf{n}} \times \boldsymbol{\mu}) \cdot \hat{\boldsymbol{\tau}} = \boldsymbol{\mu} \cdot \gamma \hat{\mathbf{t}}$. The integrand on the right hand side of Eq. 6.3 can be simplified using the vector identity $\nabla \times$

$(\mathbf{A} \times \mathbf{B}) = \mathbf{A}(\nabla \cdot \mathbf{B}) - \mathbf{B}(\nabla \cdot A) + \mathbf{B} \cdot \nabla \mathbf{A} - \mathbf{A} \cdot \nabla \mathbf{B}$, such that $\left[\nabla \times (\gamma \hat{\mathbf{n}} \times \boldsymbol{\mu})\right] \cdot \hat{\mathbf{n}} = \left\{\boldsymbol{\mu} \cdot \nabla(\gamma \hat{\mathbf{n}}) - \boldsymbol{\mu} \left[\nabla \cdot (\gamma \hat{\mathbf{n}})\right]\right\} \cdot \hat{\mathbf{n}} = \boldsymbol{\mu} \cdot \left\{\hat{\mathbf{n}} \cdot \nabla(\gamma \hat{\mathbf{n}}) - \hat{\mathbf{n}} \left[\nabla \cdot (\gamma \hat{\mathbf{n}})\right]\right\}$. Therefore, Eq. 6.3 becomes

$$\oint_C \boldsymbol{\mu} \cdot \gamma \hat{\mathbf{t}} \, d\lambda = \iint_S \boldsymbol{\mu} \cdot \left\{\hat{\mathbf{n}} \cdot \nabla(\gamma \hat{\mathbf{n}}) - \hat{\mathbf{n}} \left[\nabla \cdot (\gamma \hat{\mathbf{n}})\right]\right\} \, dS. \tag{6.4}$$

For a constant surface tension γ, Eq. 6.4 simplifies considerably,

$$\oint_C \gamma \hat{\mathbf{t}} \, d\lambda = -\iint_S \gamma \hat{\mathbf{n}}(\nabla \cdot \hat{\mathbf{n}}) \, dS. \tag{6.5}$$

Therefore, Eq. 6.2 can be written as $\mathbf{F}_T = \iint_S \left[\Delta P - \gamma \left(\nabla \cdot \hat{\mathbf{n}}\right)\right] \hat{\mathbf{n}} \, dS = 0$, yielding

$$\Delta P = \gamma \left(\nabla \cdot \hat{\mathbf{n}}\right), \tag{6.6}$$

which relates the pressure difference to the curvature of the surface, $\nabla \cdot \hat{\mathbf{n}}$. If the pressure difference is not zero, then the surface must be curved. Conversely, for a flat surface in equilibrium $\Delta P = 0$. Introducing $H = -(\nabla \cdot \hat{\mathbf{n}})/2$ into Eq. 6.6, the Young–Laplace equation (6.7) is obtained [33]

$$\Delta P = P_{inside} - P_{outside} = -2H\gamma. \tag{6.7}$$

With the height of the surface characterized as $z = h(x, y)$, a straightforward approach to obtain the unit normal vector $\hat{\mathbf{n}}$ is to consider the surface defined by the equation $f(x, y, z) = z - h(x, y) = 0$ and use Eq. 6.8 [31],

$$\hat{\mathbf{n}} = \frac{\nabla f}{\sqrt{\left(\frac{\partial f}{\partial x}\right)^2 + \left(\frac{\partial f}{\partial y}\right)^2 + \left(\frac{\partial f}{\partial z}\right)^2}} = \frac{\nabla(z - h(x, y))}{|\nabla(z - h(x, y))|}. \tag{6.8}$$

Therefore

$$\hat{\mathbf{n}} = \frac{\left\{-\frac{\partial h(x,y)}{\partial x}, -\frac{\partial h(x,y)}{\partial y}, 1\right\}}{\sqrt{\left[\frac{\partial h(x,y)}{\partial x}\right]^2 + \left[\frac{\partial h(x,y)}{\partial y}\right]^2 + 1}}. \tag{6.9}$$

In order to simplify the notation, it is easier to let $h = h(x, y)$ and introduce a shorthand for the partial derivatives $h_i = \frac{\partial h}{\partial i}$, $h_{ii} = \frac{\partial^2 h}{\partial i^2}$, and $h_{ij} = \frac{\partial^2 h}{\partial i \partial j}$ with i and j being x or y. With $H = -(\nabla \cdot \hat{\mathbf{n}})/2$ and Eq. 6.9,

$$H = \frac{h_{xx}\left(h_y^2 + 1\right) - 2h_x h_y h_{xy} + h_{yy}\left(h_x^2 + 1\right)}{2\left(h_x^2 + h_y^2 + 1\right)^{3/2}}. \tag{6.10}$$

For a 2D surface with h independent of y, i.e., $h = h(x)$, Eq. 6.10 simplifies to

Fig. 6.10 A surface characterized by two radii of curvature. The surface is characterized in the Monge parameterization with $h(x, y)$. Two orthogonal planes show the two different normals and their osculating circles

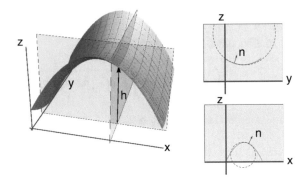

$$H = \frac{h''(x)}{2\left[1 + h'(x)^2\right]^{3/2}}. \tag{6.11}$$

For a sphere of constant radius R, the height becomes $h(x, y) = \sqrt{R^2 - x^2 - y^2}$ and $H = -1/R$; the minus sign is consistent with a normal vector which points outward. Therefore, for a spherical bubble, $\Delta P = 2\gamma/R$.

Figure 6.10 shows a somewhat more complicated surface and the curvatures (with their osculating circles) in two orthogonal planes chosen along the xz and yz planes. Although there is nothing necessarily special about these two orthogonal planes, there are two orthogonal planes, called the principal planes, for which the curvatures will assume extreme values (a maximum and minimum). Once these two curvatures (k_1 and k_2) are determined, the curvature in other planes may be obtained from their linear combination. Additionally, the curvature H is the mean of these curvatures, $H = \frac{1}{2}(k_1 + k_2)$ [31, 32]. In order to completely characterize the bending of the surface, another curvature, called the Gaussian curvature, is required. The Gaussian curvature is $K = k_1 k_2$. For a surface characterized as $h = h(x, y)$, the Gaussian curvature is given as Eq. 6.12 and the principal curvatures can be determined.

$$K = \frac{h_{xx}h_{yy} - h_{xy}^2}{\left(h_x^2 + h_y^2 + 1\right)^2}. \tag{6.12}$$

$$k_1 = H + \sqrt{H^2 - K}, \tag{6.13}$$
$$k_2 = H - \sqrt{H^2 - K}.$$

The mean curvature takes on a special significance because surfaces characterized by $k_1 = -k_2$ ($H = 0$ and $K < 0$) are called minimal surfaces because they have minimal surface area; a necessary condition for a minimal surface is that it has vanishing mean curvature. Any perturbation in the curvature of a minimal surface will increase the surface area. Clearly, a sphere is not a minimal surface, but it does

Fig. 6.11 The catenoid is a minimal surface. When $h(x, y) = c \cosh^{-1}\left(\frac{\sqrt{x^2+y^2}}{c}\right)$, we obtain $H = 0$

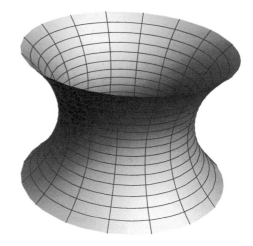

have the important property of minimizing the ratio of its surface area to volume. Certain surfaces of revolution (e.g., plane, cylinder, catenoid), have been known since the mid 1800s to be minimal surfaces. Although it is straightforward (as shown below) to solve the forward problem and test if a surface is minimal, the inverse problem of searching for minimal surfaces is much more difficult and is of historical importance in the development of the calculus of variations; in the mid-1700s, Lagrange applied variational methods to search for minimal surfaces; in Sect. 6.4.3, an Euler–Lagrange equation will be used to find the surface of minimal energy associated with bending of the plasma membrane.

The problem of finding a surface of smallest area that is bounded by a closed contour is referred to as Plateau's Problem because Joseph Plateau performed many cool experiments with soap films (mixtures of water, glycerin and an emulsifier, sodium oleate) between brass wire frames [35]. Figure 6.11 shows a catenoid which is the shape a thin film adopts when two soapy rings of fixed radius are slowly pulled apart; before the film ruptures, the surface forms catenoid shape so as to minimize its total energy. With $h(x, y) = c \cosh^{-1}\left(\frac{\sqrt{x^2+y^2}}{c}\right)$, it can be readily verified using Eqs. 6.10 and 6.13, that $H = 0$.

6.4.2 Bending Energies and the Helfrich Hamiltonian

Unlike a Newtonian fluid, the phospholipid bilayer has elastic properties, but like a fluid, the membrane cannot support in-plane shear. Energy is required, however, for other types of deformations, such as bending. From the theory of elasticity, the energy density associated with bending a thin plate so that it obtains a curvature, C, is $E_\kappa = \frac{1}{2} \int_{membrane} \kappa\, C^2\, dA$, where the bending modulus $\kappa = Y I$ is the product of the Young's modulus, Y, and the principal moment of inertia [36, 37]. For phospholipid membranes, the bending modulus $\kappa \approx 30\,kT$ [4, 5]. Therefore, the free energy associated with bending the plasma membrane should have a dependence upon the bending modulus and the square of the curvature.

The free energy of membrane stretching and bending is a scalar and can only be a function of curvature terms which are also invariant with respect to a change in coordinate systems. Both the mean curvature, H, and Gaussian curvature, K, are invariant with respect to a change of coordinates and depend only upon the direction of the surface normal, $\hat{n}$ [32]. For a closed membrane, such as a vesicle, it makes sense to distinguish between an inward and outward pointing normal vector, however, this does not hold for the symmetry associated with a flat membrane. Accordingly, the free energy associated with membrane deformation is expected to have a quadratic dependence upon the mean curvature, but not a linear dependence. By contrast, the Gaussian curvature is an intrinsic property of the surface and does not depend upon the sign of $\hat{n}$ [32]. Equation 6.14, which shows the dependence of the free energy associated with membrane deformation upon the curvatures, is often referred to as the Helfrich Hamiltonian because it is analogous to Helfrich's 1973 derivation of the energy required to stretch and bend the membrane [31, 38–40]. Helfrich referred to the two elastic moduli (κ and κ_G in Eq. 6.14), as the splay and saddle splay modulus to emphasize that his derivation followed closely from work on the curvature elasticity of liquid crystals [31, 32]. Equation 6.14, is called the Helfrich equation and often the Canham–Helfrich free energy to afford credit to Canham who formulated an energy functional in 1970: [41]

$$E = \iint_S \left(2\kappa\,(H - c_o)^2 + \gamma + \kappa_G K\right) dA + \iiint_V \Delta P\,dV \qquad (6.14)$$

Equation 6.14 demonstrates that the energy is proportional to the square of the mean curvature and contains a term, c_o, which represents the spontaneous curvature of the membrane. When $H = c_o$, the free energy is minimal. It is often appropriate to assume that the spontaneous curvature equals zero because the membrane is symmetric. If the molecules have a preferential bending direction or in the presence of proteins which bend the membrane, than the spontaneous curvature will not be zero. In many instances, the topology of the surface does not change and the dependence of E upon the Gaussian curvature K can be ignored because the integral of the Gaussian curvature is a constant. For example, for a sphere, $K = 1/R^2$ and $E(K) = \kappa_G \iint K\,dA = \frac{\pi}{2}\kappa_G$. Therefore, assuming that both c_o and ΔP equal zero and ignoring the Gaussian term, the Helfrich free energy can be written as

$$E = \iint \left(2\kappa H^2 + \gamma\right) dA. \qquad (6.15)$$

The free energy required to bend a flat membrane may be obtained from the Helfrich integral (Eq. 6.15) which integrates the energy density, $e_m = 2\kappa H^2 + \gamma$, over the entire membrane. Equation 6.15 is a complicated, non-linear differential equation because both the mean curvature H and differential area dA depend upon gradients in $h(x, y)$. With $\mathbf{r}(x, y, z) = \{x, y, h(x, y)\}$, the area element dA can be determined from the tangent vectors $\bar{\mathbf{e}}_x = \{1, 0, h_x\}$ and $\bar{\mathbf{e}}_y = \{0, 1, h_y\}$ as $dA = |\mathbf{e}_x(x, y) \times \mathbf{e}_y(x, y)| \, dx \, dy$, which yields $dA = dx \, dy \sqrt{1 + h_x^2 + h_y^2}$. Fortunately, it may generally be assumed that the bending is weak such that the gradients in $h(x, y)$ can be approximated. Under this simplification, called the small gradient approximation, terms of $\mathcal{O}\left[(\nabla h)^2\right]$ and higher are ignored. The small gradient approximation applied to Eq. 6.10 yields: $H \approx \frac{1}{2}(h_{xx} + h_{yy})$ and $dA \approx \left[1 + \frac{1}{2}(h_x^2 + h_y^2)\right] dxdy$. Inserting these approximations into Eq. 6.15 (and disregarding constant surface integrals), yields the simplified form of the Helfrich equation

$$E = \frac{1}{2} \iint \left[\kappa(h_{xx}^2 + h_{yy}^2) + \gamma(h_x^2 + h_y^2)\right] dx \, dy. \tag{6.16}$$

6.4.3 Free Energy and Shape of a Bent Membrane

As an example, applying the Helfrich Hamiltonian to the geometry of the membrane in Fig. 6.12, the membrane shape is given as $h = h(x)$ and the free energy in Eq. 6.16 becomes

$$E = \frac{1}{2} \iint (\kappa h_{xx}^2 + \gamma h_x^2) dx \, dy. \tag{6.17}$$

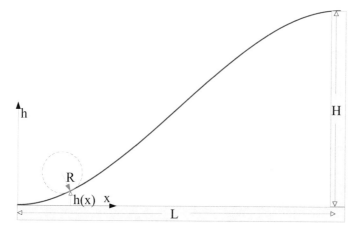

Fig. 6.12 Bending of the plasma membrane to form a projection of height H at a distance L from a point adhesion. Assuming an initially flat membrane which remains adhered to the ECM at the origin, energy bends the membrane to a maximum distance H at length L from the origin

The equilibrium shape, $h(x)$, will minimize the energy subject to the membrane's boundary conditions. It is easy to test if a given shape, $h(x)$, satisfies Eq. 6.17, however, not every shape will minimize the total energy. Therefore, this forward method does not prove that the shape, $h(x)$ minimizes the energy. Fortunately, the inverse problem of predicting the shape that will produce a stationary value of the energy, E, can be solved using the calculus of variations developed by Euler and Lagrange [36, 42]. Their variational approach states that the solution, $h(x)$, satisfies the Euler–Lagrange equation. In 1D, the Euler–Lagrange equation is particularly simple to formulate: given the functional $F = \int_a^b H\left[x, f(x), f'(x)\right] dx$, the Euler–Lagrange equation becomes $\frac{\partial F}{\partial f} - \frac{d}{dx}\frac{\partial F}{\partial f'} = 0$. The Euler–Lagrange equation with the Helfrich energy (Eq. 6.17) as the functional yields

$$\kappa\, h''''(x) - \gamma h''(x) = 0, \tag{6.18}$$

for the shape of the membrane $h(x)$. This differential equation is sometimes called the shape equation. It can be verified by direct substitution that the solution is

$$h(x) = \frac{\kappa e^{-\frac{\sqrt{\bar{\gamma}}x}{\sqrt{\kappa}}}\left(c_1 e^{\frac{2\sqrt{\bar{\gamma}}x}{\sqrt{\kappa}}} + c_2\right)}{\gamma} + c_3 + c_4 x. \tag{6.19}$$

We can solve for the four unknowns in Eq. 6.19 by applying the appropriate boundary conditions. As shown in Fig. 6.12, the membrane is bound to the ECM at the origin and at $x = L$. Therefore, we have $h(0) = 0$; $h'(0) = 0$; $h(L) = H$ and $h'(L) = 0$. The solution to the four equations readily yields four constants. Introducing the variable $\lambda = \sqrt{\kappa/\gamma}$, which has units of length, the constants are given by

$$c_1 = -\frac{\gamma H}{\sqrt{\bar{\gamma}}\sqrt{\kappa}L\left(e^{\frac{\sqrt{\bar{\gamma}}L}{\sqrt{\kappa}}} + 1\right) - 2\kappa\left(e^{\frac{\sqrt{\bar{\gamma}}L}{\sqrt{\kappa}}} - 1\right)}, \tag{6.20}$$

$$c_2 = -\frac{\gamma H e^{\frac{\sqrt{\bar{\gamma}}L}{\sqrt{\kappa}}}}{\sqrt{\bar{\gamma}}\sqrt{\kappa}L\left(e^{\frac{\sqrt{\bar{\gamma}}L}{\sqrt{\kappa}}} + 1\right) - 2\kappa\left(e^{\frac{\sqrt{\bar{\gamma}}L}{\sqrt{\kappa}}} - 1\right)}, \tag{6.21}$$

$$c_3 = \frac{H\sqrt{\kappa}\left(e^{\frac{\sqrt{\bar{\gamma}}L}{\sqrt{\kappa}}} - 1\right)}{2\sqrt{\kappa}\left(e^{\frac{\sqrt{\bar{\gamma}}L}{\sqrt{\kappa}}} - 1\right) - \sqrt{\bar{\gamma}}L\left(e^{\frac{\sqrt{\bar{\gamma}}L}{\sqrt{\kappa}}} + 1\right)}, \tag{6.22}$$

$$c_4 = \frac{\sqrt{\bar{\gamma}}H\left(e^{\frac{\sqrt{\bar{\gamma}}L}{\sqrt{\kappa}}} + 1\right)}{\sqrt{\bar{\gamma}}L\left(e^{\frac{\sqrt{\bar{\gamma}}L}{\sqrt{\kappa}}} + 1\right) - 2\sqrt{\kappa}\left(e^{\frac{\sqrt{\bar{\gamma}}L}{\sqrt{\kappa}}} - 1\right)}. \tag{6.23}$$

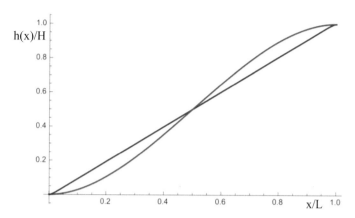

Fig. 6.13 Shape of the membrane $h(x)$ depends upon ratio of the length scale λ. For very small λ, the large tension prevents bending of the membrane ($\lambda/L = 0.01$, blue). By contrast, when the tension is very low, the membrane bends considerably ($\lambda/L = 1$, red)

Inserting these four constants into Eq. 6.20, the shape $h(x)$ is given by

$$h(x) = \frac{H\left(\lambda - \lambda e^{L/\lambda} + x e^{L/\lambda} + \lambda e^{\frac{L-x}{\lambda}} - \lambda e^{x/\lambda} + x\right)}{2\lambda + L e^{L/\lambda} - 2\lambda e^{L/\lambda} + L}. \tag{6.24}$$

Thus, the energy required to bend the membrane into this shape can be obtained by inserting Eq. 6.24 into Eq. 6.17

$$
\begin{aligned}
E &= \int_0^L \kappa \left(h''(x)^2 + \frac{h'(x)^2}{\lambda^2}\right) dx \\
&= \frac{H^2 \kappa \left(e^{L/\lambda} + 1\right)}{\lambda^2 \left(2\lambda + L e^{L/\lambda} - 2\lambda e^{L/\lambda} + L\right)} \\
&= \frac{H^2 \kappa}{\lambda^2 \left(L - 2\lambda \tanh\left(\frac{L}{2\lambda}\right)\right)}. \tag{6.25}
\end{aligned}
$$

The shape of the membrane, which depends upon the length scale λ, is shown in Fig. 6.13. In the limiting case where surface tension goes to zero, $\gamma \to 0$, the shape, $h(x)$, is given by

$$\lim_{\lambda \to \infty} h(x) = \frac{3Hx^2}{L^2} - \frac{2Hx^3}{L^3}. \tag{6.26}$$

References

1. R.. Hooke, *Micrographia: Or Some Physiological Descriptions of Minute Bodies Made by Magnifying Glasses, with Observations and Inquiries Thereupon* (Dover Publications Reprint Edition, 2003)
2. J.L. Robertson, J. Gen. Physiol. **150**(11), 1472 (2018)
3. C. Sohlenkamp, O. Geiger, FEMS Microbiol. Rev. **40**(1), 133 (2016)
4. E. Evans, W. Rawicz, Phys. Rev. Lett. **64**(17), 2094 (1990)
5. E. Atilgan, B. Ovryn, Biophys. J . **96**(9), 3555 (2009)
6. O.S. Andersen, R.E. Koeppe, Ann. Rev. Biophys. Biomol. Struct. **36**, 107 (2007)
7. S.J. Singer, G.L. Nicolson, Science **175**(4023), 720 (1972)
8. B. Alberts, A. Johnson, J. Lewis, M. Raff, K. Roberts, P. Walter, ISBN **1174808063**, 1392 (2007)
9. S.H. White, W.C. Wimley, Ann. Rev. Biophys. Biomol. Struct. **28**(1), 319 (1999)
10. T. Okada, M. Sugihara, A.N. Bondar, M. Elstner, P. Entel, V. Buss, J. Mol. Biol. **342**(2), 571 (2004)
11. H. Lodish, A. Berk, C.A. Kaiser, C. Kaiser, M. Krieger, M.P. Scott, A. Bretscher, H. Ploegh, P. Matsudaira, et al., *Molecular Cell Biology* (Macmillan, 2008)
12. M. Luckey, *Membrane Structural Biology: with Biochemical and Biophysical Foundations* (Cambridge University Press, 2014)
13. R.O. Hynes, Science **300**(5620), 755 (2003)
14. J. Gu, Regulation of integrin functions by N-glycans. Glycoconj. J. **21**, 9 (2004)
15. R. Zaidel-Bar, C. Ballestrem, Z. Kam, B. Geiger, J. Cell Sci. **116**(22), 4605 (2003)
16. M. Cohen, D. Joester, B. Geiger, L. Addadi, ChemBioChem **5**(10), 1393 (2004)
17. C.G. Galbraith, K.M. Yamada, J.A. Galbraith, Science **315**(5814), 992 (2007)
18. M.A. Schwartz, Cold Spring Harb. Perspect. Biol. **2**(12), a005066 (2010)
19. B. Geiger, K.M. Yamada, Cold Spring Harb. Perspect. Biol. **3**(5), a005033 (2011)
20. Y. Kozlovsky, L.V. Chernomordik, M.M. Kozlov, Biophys. J . **83**(5), 2634 (2002)
21. L.V. Chernomordik, M.M. Kozlov, Cell **123**(3), 375 (2005)
22. H.R. Marsden, I. Tomatsu, A. Kros, Chem. Soc. Rev. **40**(3), 1572 (2011)
23. L.V. Chernomordik, M.M. Kozlov, Nat. Struct. Mol. Biol. **15**(7), 675 (2008)
24. M.M. Kozlov, L.V. Chernomordik, Curr. Opin. Struct. Biol. **33**, 61 (2015)
25. C.D. Blanchette, W.C. Lin, C.A. Orme, T.V. Ratto, M.L. Longo, Langmuir **23**(11), 5875 (2007)
26. T. Baumgart, S.T. Hess, W.W. Webb, Nature **425**(6960), 821 (2003)
27. D. Ungar, F.M. Hughson, Ann. Rev. Cell Dev. Biol. **19**(1), 493 (2003)
28. R. Jahn, T. Lang, T.C. Südhof, Cell **112**(4), 519 (2003)
29. S.C. Harrison, Nat. Struct. Mol. Biol. **15**(7), 690 (2008)
30. J.C. Berg, *An Introduction to Interfaces & Colloids: The Bridge to Nanoscience* (World Scientific, 2010)
31. S. Safran, *Statistical Thermodynamics Of Surfaces And Membranes* (CRC Press, Interfaces, 2018)
32. R.D. Kamien, Rev. Mod. Phys. **74**(4), 953 (2002)
33. L. Landau, E. Lifshitz. *Fluid Mechanics: Course of Theoretical Physics*(butterworth, 1987)
34. G.B. Arfken, H.J. Weber, *Mathematical Methods for Physicists* (Academic, 2001)
35. R. Courant, H. Robbins **2**, 901 (1956)
36. A.S. Saada, *Elasticity: Theory and Applications* (Krieger Publishing Company, 1983)
37. L. Landau, E. Lifshitz, *Theory of Elasticity* (Pergamon Press, Oxford, 1959)
38. W. Helfrich, Zeitschrift für Naturforschung c **28**(11–12), 693 (1973)
39. H. Deuling, W. Helfrich, Journal de Physique **37**(11), 1335 (1976)
40. H. Deuling, W. Helfrich, Biophys. J. **16**(8), 861 (1976)
41. P.B. Canham, J. Theor. Biol. **26**(1), 61 (1970)
42. G.A. Maugin, in *Continuum Mechanics Through the Eighteenth and Nineteenth Centuries* (Springer, 2014), pp. 7–32

Chapter 7
Introduction to Models of Cell Motility

Youyuan Deng and Herbert Levine

Abstract Biological cells are quintessentially active objects, using their stored energy to power motion. The field of cell motility is quite broad, ranging from bacteria to mammalian cells, from swimming to crawling to more exotic self-propulsion methods, to moving on surfaces or through liquid or via traversing fibrous gels, and to moving individually versus moving in a coordinated collective manner. Also, cell motion can be guided by external information in the form of chemical gradients, stiffness variations, and adhesion to oriented fiber networks. This chapter reviews the type of models that have been used to address these questions, focusing on mechanistic approaches grounded in the field of non-equilibrium statistical physics.

7.1 Introduction

Biological cells are motile and use their motility in a variety of functional contexts [1]. Mechanisms underlying this capability can differ, ranging from the flagellar rotation for swimming bacteria to surface crawling based on interacting polymers. In all of these cases, energy sources are ultimately generated by metabolic reactions which enable the non-equilibrium processes essential for active motility. In this regard, moving cells are quintessential active particles.

Because of their importance, much theoretical effort has been devoted to understanding quantitative features of moving cells [2–4]. As a general rule, these efforts need to be carefully chosen to reflect the questions being asked. These questions can range from the tracked behavior of individual cells, the connection between cell dynamics and cell morphology, and the correlated motion of cells when they move collectively. At the individual cell level, we will focus first on descriptive Langevin equations that govern the center-of-mass motion and subsequently consider much

Y. Deng
Rice University, Houston, TX, USA
e-mail: youyuan.deng@rice.edu

H. Levine (✉)
Northeastern University, Boston, MA, USA
e-mail: h.levine@northeastern.edu

© Springer Nature Switzerland AG 2022
K. B. Blagoev and H. Levine (eds.), *Physics of Molecular and Cellular Processes*,
Graduate Texts in Physics, https://doi.org/10.1007/978-3-030-98606-3_7

more detailed models that focus on cell morphology. Afterward we will turn to similar model dichotomies for the case of collective motion.

There is no sense in which this chapter is a complete review; doing that would require a book all by itself. Instead, we hope to reveal to the reader the various aspects of this vast research field through the lens of the research as it impacted the interests over several decades of one of the authors (HL). The reader is of course encouraged to construct for his/her/themselves a different parsing of all this material and thereby contribute innovative new ideas to the ongoing challenges of cell motility.

7.2 Random-Walk Models

Let us start with the simplest model. Cells often can be thought of as persistent random walkers; this means that they move at roughly constant speed and change direction randomly with some typical persistence time. For example, Dictyostelium amoebae were observed to exhibit this type of crawling motility in the absence of any directional cues [5] (Fig. 7.1). A particle of this kind can be described by a Langevin equation for the velocity direction, just a single angle in two dimensions:

$$\frac{d\theta}{dt} = \eta(t) \ ; \ \frac{d\mathbf{x}}{dt} = (v\cos\theta, v\sin\theta) \tag{7.1}$$

Here v is a fixed speed and η stands for white noise with correlation

$$< \eta(t)\eta(t') > = \sigma^2\delta(t - t')$$

This stochastic ODE is equivalent to a Fokker–Planck equation [6] corresponding to angular diffusion,

$$\frac{\partial P(\theta, t)}{\partial t} = \frac{\sigma^2}{2}\frac{\partial^2 P(\theta, t)}{\partial \theta^2} \tag{7.2}$$

Here P is the probability of finding the cell with velocity in the θ direction at time t.

One could of course try to connect this simple phenomenological approach with explicit dynamics observed for a specific cell. An example of this arises in the study of E. coli run-and-tumble dynamics. For this simple bacterium, cells are observed to move in straight lines for a period of time; this is called a run and results from the coherent bundling and rotation of the multiple flagella attached to the cell body [7]. Then, a reversal in rotation direction leads to a tumble in which the flagella unbundle and the cell randomly reorients. In a uniform medium this leads on the macroscopic scale to the same type of diffusional processes as given above; See [8] for an early and more detailed discussion of how to go from a microscopic description to a Fokker–Planck representation.

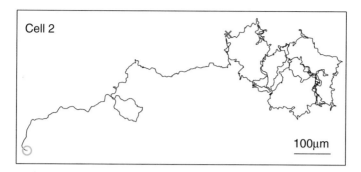

Fig. 7.1 Typical trajectory of a Dictyostelium cell, produced by time-lapse recording of an amoeba every 5 second. This motion can be modeled as a persistent random walk. Taken from [5]

Often, cells do not just move isotropically. Instead, cells can use directional cues from their environment, or as we shall see later from neighboring cells, to bias their motion. This general notion is known as taxis and comes in a number of varieties: chemotaxis denotes bias in motility direction due to a chemical gradient [9], durotaxis refers to moving directionally based on medium stiffness, electrotaxis involves response to electric fields [10], etc. The simplest possibility is that each individual cell can sense the external gradient and align its velocity direction to match. Again we can use a phenomenological approach [11] and assume

$$\frac{d\theta}{dt} = -\alpha \sin (\theta - \theta_0) + \eta(t) \tag{7.3}$$

where θ_0 is the gradient direction and α the strength of the guidance. Note that the guidance term must be a periodic function of the motion direction and sin is just the simplest possible choice. As before, this Langevin dynamics can be reformulated as a drift-diffusion equation

$$\frac{\partial P(\theta, t)}{\partial t} = \frac{\partial}{\partial \theta} \left(\alpha \sin (\theta - \theta_0) P(\theta, t) \right) + \frac{\sigma^2}{2} \frac{\partial^2 P(\theta, t)}{\partial \theta^2} \tag{7.4}$$

In steady state, this predicts the angular distribution

$$\ln P/N = \frac{2\alpha}{\sigma^2} \cos (\theta - \theta_0) \tag{7.5}$$

with a peak in the gradient direction and where N is a normalization constant. Sometimes, one sees an equation of this general form in the context of a PDE for a cell population, If we assume that α varies linearly with gradient strength and ignore the persistence effect in favor of a simplified Brownian motion model, we can write for chemotaxis due to a spatially varying concentration c [12]

$$\frac{\partial \rho(\mathbf{x}, t)}{\partial t} = -\alpha_0 \mathbf{\nabla} \cdot (\rho(\mathbf{x}, t) \mathbf{\nabla} c) + D \mathbf{\nabla}^2 \rho(\mathbf{x}, t) \qquad (7.6)$$

Here ρ is the cell density and we have ignored fluctuations due to small particle numbers in deriving a deterministic equation.

It is also interesting to consider in this context guidance by cells that stick to extracellular fibers in a 3D matrix. This is usually called contact guidance [13]. A related system might be cells moving on surface containing topographic ridges which bias their motility [14]. In both of these cases, the directional information provided by the external medium is *nematic*, that is along a line but with no preferred direction either forward or backward. This leads to an obvious modification of the previous equation

$$\frac{d\theta}{dt} = -\alpha \sin 2(\theta - \theta_0) + \eta(t) \qquad (7.7)$$

leading to a bimodal steady-state distribution

$$\ln P/N = \frac{\alpha}{\sigma^2} \cos 2(\theta - \theta_0) \qquad (7.8)$$

Of course, this type of treatment does not take into account the molecular basis for the gradient sensing capability. Because of this shortcoming, we do not get any sense of how the parameters vary with important experimental conditions. For example, the ability to detect chemical gradients should depend on the mean value of the imposed concentration; it is unrealistic to expect a cell to detect a 10 nM difference in concentration between front and back if the mean value across the cell is 100 mM, From the perspective of biology, it is also interesting to understand how external signals actually lead to the biased motion, i.e., how the motors are driven by the sensors to accomplish this navigational feat. We will turn to these questions in the next section.

Recent work has highlighted the possibility of another type of taxis, one that operates only at the population level [11]. To understand the population taxis idea, we imagine that cells have a speed which depends on an experimental parameter such as a chemical concentration. This dependence does not impart any direct directional information to an individual cell but it does lead to a population bias in the presence of a chemical gradient. Specifically, our general approach would now lead to the stochastic process

$$\frac{d\theta}{dt} = \eta(t) \; ; \; \frac{d\mathbf{x}}{dt} = v(\mathbf{x})(\cos \theta, \sin \theta) \equiv \mathbf{u}(\mathbf{x}) \qquad (7.9)$$

giving rise to a description in terms of a multi-dimensional probability distribution equation [15]

$$\frac{\partial P(\mathbf{x}, \theta, t)}{\partial t} = -\mathbf{\nabla} \cdot (\mathbf{u}(\mathbf{x}) P(\mathbf{x}, \theta, t)) + \frac{\sigma^2}{2} \frac{\partial^2 P(\mathbf{x}, \theta, t)}{\partial \theta^2} \qquad (7.10)$$

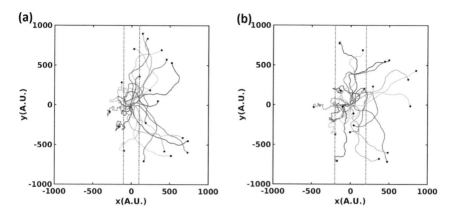

Fig. 7.2 Direct simulation of a model where persistence varies with stiffness. Here there is a graded stiffness profile, with a soft substrate in the left region and hard substrate in the right region. The central region has a constant stiffness gradient and varying width L = **a** 100, **b** 200; Adapted from [15]

and σ^2 is again just the magnitude of the η correlation function. This can easily be extended to the case where the persistence is spatially dependent; σ^2 just becomes a function of position and the same equation is valid if we only keep terms up to the usual second derivative—see [15] for a derivation. A direct numerical computation depicted in Fig. 7.2 easily shows that an initial set of cells will preferentially congregate in the high persistence regions. This is because once cells reach the high persistence region they keep going further inward, whereas cells in the low persistence area rapidly turn around. This creates a trapping effect which can also be found directly by simulating individual cell motion. It is important to note that population taxis does not require any interaction between the individual cells. Thus, it is distinct from mechanisms that have been proposed to account for collective chemotaxis [16, 17] or collective durotaxis [18] when cells cooperate with each other to determine gradients that they individually cannot determine. These cases fall under the general topic of collective cell motion which will be tackled later on.

Two important systems seem to exhibit modulation of persistence as an underlying mechanism driving taxis. The more recent suggestion involves the possibility that mammalian cells moving on hard substrates have cytoskeletal organization that tends to be more polarized and more rigid then ones on softer substrates [19]. This difference can then lead to greater persistence of motion, i.e., turning is harder, on hard substrates. A much more well-known example brings us back to E. coli. It has been well established that the rate of transition from a run to a tumble is directly modulated by changes in the concentration of various attractants. On an intermediate time scale, this seems to map directly to our previous discussion. However, as we shall discuss in the next section, E. coli exhibits an extra wrinkle; that is, at longer timescales, persistence returns back to a set point as the system adapts to new concentration values. This means that cells are actually responding to temporal changes in concentration and using this measurement while moving as a surrogate for spatial gradients. This wrinkle will require an extension of our existing framework that takes into account the adaptation dynamics. This extension will be taken up in the next section, after discussing the simpler case of direct gradient measurement.

7.3   Looking Under the Hood

What is the molecular machinery that supports a cell's capability to effectively chemotax? We will investigate this question in two popular model organisms, Dictyostelium and E. coli. They are quite different. A Dicty cell measures gradients across its own length by comparing the occupancy of its receptors on different parts of its body. This information is then used to polarize the cell, i.e., to create different chemical states (front versus back) inside the cell—see Fig. 7.3. E. coli presumably is too small to work this way and in any case has all of its receptors clustered in one location. Instead it adopts the idea that we already mentioned, the modulation of persistence by concentration.

7.3.1   Dicty

Let us start with Dictyostelium and its sensory apparatus. A Dicty cell has a uniform distribution of receptors for the cAMP molecule, which acts as a chemoattractant during the aggregation phase of its lifecycle [20]. One can formulate the cell's gradi-

Fig. 7.3 A Dictyostelium cell moves left to right under the influence of a gradient of cAMP. The parts of the cell membrane in red indicate regions with active actin polymerization giving rise to protrusions, whereas green labels myosin, responsible for cell contraction at the rear. This is a clear example of a cell that is polarized by means of external chemical information

ent detection task as one of maximum likelihood estimation [21]. To be specific, the probability of a given receptor to be bound is determined by two-state equilibrium

$$p_n = \frac{c(x_n)}{c(x_n) + K_b}$$

where $c(x_n)$ is the concentration at the physical location of receptor n on the cell membrane. For illustrative purposes, we will assume an exponential profile

$$c(x_n) = c_0 e^{\frac{p}{2}\cos(\phi_n - \phi)} \tag{7.11}$$

where the cell is assumed circular with radius $\ell/2$, the receptors are placed at the equally spaced point $\phi_n = \frac{2\pi n}{N}$ and $p = \ln\frac{c_{max}}{c_{min}}$ denotes the dimensionless gradient strength. The gradient direction is determined by ϕ and it is the object of primary interest. The joint probability distribution for all the receptors given these parameters is

$$L(p, \phi; \{y_n\}) = \prod_n^N p_n^{y_n}(1 - p_n)^{1-y_n} \tag{7.12}$$

where the y values are either 0 (unoccupied) or 1 (occupied). This probability becomes the likelihood when we plug in the observed state of all the y_n at a particular time.

The estimates of ϕ and p are then just the values of these parameters that maximize this likelihood. One can then determine the signal-to-noise ratio, equivalently the accuracy of the estimate, by determining how sharply peaked the likelihood functions are in the vicinity of these estimates. Mathematically, this is related to the Cramer–Rao bound on the accuracy of an estimator [22]. The details of this calculation are presented in [21] with the following results; define

$$z_1 = p\cos\phi \ , \quad z_2 = p\sin\phi$$

The estimates of these quantities from the data are

$$\hat{z}_1 = \frac{1}{\mu}\sum_{n=1}^N y_n \cos\phi_n \ , \quad \hat{z}_2 = \frac{1}{\mu}\sum_{n=1}^N y_n \sin\phi_n$$

where

$$\mu = \frac{Nc_0 K_b}{4(c_0 + K_b)^2}$$

This then gives $\hat{\phi} = \tan^{-1}(\hat{z}_1/\hat{z}_2)$. We also obtain the width of the distribution at the estimator locations

$$\frac{\partial^2 \ln L}{\partial z_{1,2}^2} = \frac{\mu}{2}$$

Putting it all together gives the uncertainty in determining the gradient direction as

$$\sigma_\phi^2 = \frac{8(c_0 + K_b)^2}{Np^2 c_0 K_b} \tag{7.13}$$

which has a minimum at $c_0 = K_b$.

A few things are worth noting about this final expression. Various aspects of the formula are reasonable; the uncertainty depends inversely on the gradient strength p and the number of receptors N and is also proportional to the variance of the typical individual measurement at mean concentration c_0, which is just $q(1 - q)$ with $q = \frac{c_0}{c_0 + K_b}$. The latter is why the best accuracy is attained at half-filling ($q = 1/2$), if we maintain a fixed value of the dimensionless gradient strength p (but not the actual absolute gradient) and vary c_0.

It is important to understand that the above derivation gives the optimal answer and requires cells to know the position of all their receptors and properly interpret the binding results; how well cells actually match this optimum depends on exactly how this binding data interrogation is accomplished and with what processing error. Finally, this accuracy is the one obtained from utilizing a single snapshot of the receptor-based observations. One can get better accuracy by time averaging, losing of course the ability to rapidly track changing environments. Comparison of this and related results regarding sensing with experimental measurements are presented in [23, 24].

Of course, we must also study how receptor occupancy measurements are actually used to bias motion and how close this gets to the just calculated information-theoretic bound on possible performance. This brings us to the study of the cell's internal circuitry. Implementing directed motion must mean that the cell polarizes, i.e., establishes a different macroscopic chemical state at the leading edge versus the trailing one. For example, actin should only be polymerizing at the leading edge, as is clearly shown in Fig. 7.3. But, as pointed out initially by Devreotes and Parent [25], this cannot be accomplished by purely local chemical reactions. A simple thought experiment explains why. Imagine that a cell is exposed to a concentration at its right edge of 100 nM. The cell decision-making chemical network can not know, based solely on this local information, whether it should become a front region, say if the left edge is at 90 nM. or whether becoming a back region is the right decision, say if the left edge is at 110 nM. In other words there needs to be cross-cell comparison in order to properly respond to a gradient.

To date the most compelling resolution of this issue is provided by the LEGI— local excitation, global inhibition—concept [26, 27]. Essentially, the response to the detection of ligand-receptor binding is twofold, as both excitatory and inhibitory mechanisms are activated. The excitation remains localized whereas the inhibitory one is assumed to diffuse throughout the cell interior. The simplest version of this class of models is given by the equations along the membrane

$$\frac{\partial A}{\partial t} = \lambda_A R - d_a A$$

$$-D\hat{n} \cdot \nabla I|_m = \lambda_I R - d_I I_m$$

$$\frac{\partial E}{\partial t} = k_+ A(1 - E) - k_- I_m E \tag{7.14}$$

Here A is the local activation, I is the diffusing inhibitor, and E the downstream effector, defined to lie between 0 and 1, that will eventually drive the motility machinery toward being front-like. The fields A and E are defined along the membrane whereas I is a field defined everywhere inside the cell with I_m the value it takes right at the membrane; in the cell interior I satisfies the diffusion equation $\frac{\partial I}{\partial t} = D\nabla^2 I$. The second of the equations above serves as a boundary condition for this diffusion equation and $\hat{n}$ is the inward pointing unit normal. The input signal is denoted by R and has been taken to be the deterministic result

$$R(x) = \frac{c(x)}{c(x) + K_b}$$

If necessary, this can be replaced by the stochastic signal [28] including the fluctuations that arise from the binary distribution, as discussed above.

To illustrate how the model works and how it can be extended to include signal amplification, we retreat to one spatial dimension where the interior of the cell lies between $x = 0$ and $x = L$ and the membrane consists of just two points. Then one can easily solve the model for the steady-state concentrations given the values of the signal at the right and left edges, defined as $R_0 \pm R_1/2$. For the activator, we obtain

$$A_R = \frac{\lambda_A (R_0 + R_1/2)}{d_A}$$

$$A_L = \frac{\lambda_A (R_0 - R_1/2)}{d_A} \tag{7.15}$$

For the inhibitor, we have $I(x) = I_L + \frac{x}{L}(I_R - I_L)$. The boundary conditions in the above set of equations become

$$-\frac{D}{L}(I_R - I_L) = \lambda_I (R_0 - R_1/2) - d_I I_L$$

$$+\frac{D}{L}(I_R - I_L) = \lambda_I (R_0 + R_1/2) - d_I I_R \tag{7.16}$$

giving

$$\frac{I_R + I_L}{2} = \frac{\lambda_I R_0}{d_I} \quad , \quad I_R - I_L = \frac{\lambda_I R_1}{d_I + 2D/L} \tag{7.17}$$

This shows that the difference in inhibitor concentrations depends only on the gradient of R and becomes small for large diffusion constant. We can now immediately determine that

$$E_{R,L} = \frac{\frac{k_+\lambda_A}{d_A}(R_0 \pm R_1/2)}{\frac{k_+\lambda_A}{d_A}(R_0 \pm R_1/2) + \frac{k_-\lambda_I}{d_I}(R_0 \pm \frac{d_I}{d_I+2D/L}R_1/2)} \qquad (7.18)$$

As a sanity check, we note that at zero diffusion the external gradient does not affect the effector levels; the LEGI concept requires communication through the cell interior.

The important feature to notice is the perfect adaptation; if $R_1 = 0$, the effector level is independent of the stimulus strength R_0. This feature was tested experimentally in [27] and found to be quite accurate, except perhaps at very high ligand concentration. In the presence of a positive gradient, the deviation from this fixed baseline on the right-hand side value is always positive whereas the left side is always decreased. Our previous problem with the direct gradient measurement having a floating mean has disappeared. This result allows us to imagine tuning an amplifier to a stimulus value slightly above the baseline; this thereby guarantees that the front of the cell is pushed into a front mode whereas the rest of the cell is not affected and thereby becomes the back. This amplification is necessary as the system clearly exhibits a nonlinear polarization response even with a very small imposed gradient [29].

We can go one step further and exhibit one possible amplification procedure [30]. Suppose we replace the downstream effector equation with the nonlinear alternative

$$\frac{\partial E}{\partial t} = \frac{k_+ A(1 - E)}{(1 - E) + K_A} - \frac{k_- I E}{E + K_I}. \qquad (7.19)$$

Note that this version still perfectly adapts as the steady-state value of $E = E_0$ depends only on the ratio of A to I for which R_0 drops out. We now insist that at E_0 the forward and backward rates are roughly equal and imagine that the equilibrium constants K_A and K_I are small. This assumption is usually referred to as ultrasensitivity [31]. This ultrasensitive model will give rise to a response that swings from E around 0 on the smaller concentration side to E around 1 on the high side. This can be seen as follows. First, the aforementioned condition of equality means that

$$V_a \equiv k_+\lambda_A/d_A \simeq V_b \equiv k_-\lambda_I/d_I$$

With this condition the level of E_0 is just $K_I/(K_A + K_I)$. Next, to make analytic progress we assume that the diffusion constant is large and hence I is uniform across the cell. The gradient response is then obtained by comparing the effector level at $R = R_0 \pm R_1/2$ corresponding to the front and back of the cell, assuming that I remains uniform. By directly solving the steady-state equation for E, we find that [27, 30]

$$E_\pm = \frac{\alpha + \sqrt{\alpha^2 \pm 4V_a^2(1 \pm \frac{R_1}{2R_0})K_I R_1/(2R_0)}}{\pm V_a R_1/R_0} \qquad (7.20)$$

with

$$\alpha \equiv V_a \left((1 - K_I)(1 \pm R_1/2R_0) - 1 - K_A \right)$$

If $R_1 \rightarrow 0$ at fixed values of the K's, α is always negative. Since α^2 is large compared to the second term in the square root, we can expand. We recover a linear result with the edges always being respectively higher or lower than the fixed baseline; this is analogous to the one obtained in the previous linear effector model in the small R_1 limit. In detail, we have

$$E_\pm = \frac{K_I}{K_A + K_I} \left(1 \pm \frac{R_1}{2R_0} (1 + \frac{1 - K_I}{K_I + K_A}) \right) \tag{7.21}$$

If on the other hand $R_1/R_0 > K_I, K_A$, α becomes positive at the front and the square root is approximately equal to α. Hence for a large enough gradient we find that $E_R \simeq \frac{2\alpha}{V_a R_1/R_0} \simeq 1$; for the back we still get a negative α, an almost complete cancelation, and as a consequence $E_L \simeq 0$. In this ultrasensitive LEGI model, the equal V condition is an explicit realization of the idea that we must insist on a non-trivial parameter constraint to make sure that the amplification threshold is set at the baseline value obtained for a uniform stimulus (R_0 here). How the cell maintains this condition in the presence of inevitable rate fluctuations is still uncertain.

7.3.2   E. Coli

We have just seen how a cell can utilize its size to resolve concentration differences and measure gradient direction. But E. coli cannot do this and instead couples persistence to the adapted concentration; the adaptation means that in the end, the time derivative of the concentration as measured by the moving cell is what drives chemotaxis. Here we know quite a bit about the molecular players involved in this mechanism. The receptor binding leads to the activation of the CheY kinase via phosphorylation and this then determines the switching properties of the flagellar motor; increased binding leading to increased CheY activation and delays the transition to tumbling. The adaptation is dependent on the enzymes CheR and CheB which respectively add and remove methyl groups from the receptor. This modification changes the relationship between occupancy and CheY activation; see [32] for a physicist-friendly review.

We can then go through the same type of analyses as presented for Dictyostelium, determining the sensitivity with which the cell can measure absolute concentrations and then introducing a simple phenomenological model that allows for the exploration of the response to various imposed gradients. The sensitivity problem is quite interesting, as it makes contact with the classic work of Berg and Purcell [33, 34] in estimating the role of ligand diffusion in limiting concentration sensing. Following [35], we can break up the problem into the results that can be obtained from one snapshot (as before) of the occupancy of N receptors and the results that can be obtained after time averaging. For a single snapshot, we define Z as the percentage

Fig. 7.4 The time-averaged variance as a function of the measurement time T for different number of receptors. The collection of points where the measurement time equals to the correlation time for different N is drawn as a dashed line. Below this line, the variance approaches the instantaneous variance σ_Z^2 while above this line the variance scales as $1/T$

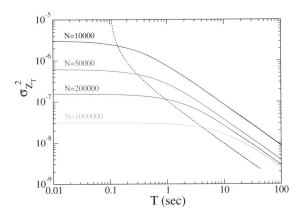

of bound receptors, which clearly is an estimate of $\frac{\bar{c}}{\bar{c}+K_b}$. The variance of Z is just $\sigma_Z^2 = A_0/N$ with the coefficient A_0 equal to

$$A_0 = \frac{\bar{c}K_d}{(\bar{c}+K_d)^2} \tag{7.22}$$

where $\bar{c}$ is the mean concentration, i.e., the time-average of the fluctuating concentration being sensed; A_0 is just the single receptor variance. This simple finding arises from the fact that instantaneous measurements at separate receptors are uncorrelated and the mean has an uncertainty that scales as $1/N$. Thus, utilizing a one-time measurement, the cell can attain arbitrary accuracy in its evaluation of a signal concentration.

This result appears to disagree with the Berg–Purcell conclusion that accuracy is limited by diffusive noise, but that is not the case. The Berg–Purcell formula refers to the accuracy achieved after time averaging by time T and predicts

$$\left(\frac{\delta c}{\bar{c}}\right)^2 \simeq \frac{1}{DT\bar{c}R} \tag{7.23}$$

Here D is the diffusivity of the molecule with mean concentration $\bar{c}$, R is the cell radius and T is the measurement time. In [35], it is shown that a snapshot de-correlate with time constant

$$\tau_c = \frac{1}{k_-} + \frac{\rho\bar{c}/K_d}{J_{diff}} = \frac{1}{k_-} + \frac{N}{4\pi DRK_d} \tag{7.24}$$

For times in excess of this constant, we need to multiply the previous estimate by τ/T. This recovers the classical result of (7.23) for large N, with the caveat that it is only valid for long enough times. A plot of the error as a function of both N and T is shown in Fig. 7.4.

Finally, we turn to the behavioral response, assuming for simplicity a deterministic signal. Let the average methylation level be labeled M and the CheY activation a. The response can be modeled via the dynamics [36]

$$a = G(c, M)$$
$$\frac{dM}{dt} = F(a, c, M) \tag{7.25}$$

Here G is the receptor cluster response to the incoming signal $c(t)$ and F relates to the adaptation. The key idea is that there exists an activity value a_0 for which $F = 0$. The methylation M will automatically relax to this value if given enough time. Thus in steady state the value of a will not care about the signal, i.e., the system will exhibit perfect adaptation. Interestingly, the circuitry responsible for adaptation here, termed integral feedback, is different than what we saw in the LEGI model where the activator and inhibitor legs canceled out in a mechanism relying on incoherent feedforward topology. Interested readers can look at [37] for the variety of small circuits able to a accomplish prefect adaptation. Returning to this case, a time-dependent c varying faster than the methylation relaxation rate will give rise to a non-zero activity change, roughly proportional to $\frac{dc}{dt}$. This is worked out in [36] quantitatively for specific realizations of the F and G functions.

7.4 Shapes

So far, our treatment has dealt with point-like objects or with rigid objects whose motion could still be described just by the center of mass. But, crawling cells are flexible and their morphology is intimately connected to their motion. This connection can be a static one whereby cells such as keratocytes adopt fixed shapes that will vary according to external conditions [38]. Alternatively, cell shape can be a dynamic variable with motion coupled to cycles of membrane protrusions and cell body contractions, as in the motion of amoeboid cells [39]. Treating the shape degrees of freedom as explicit variables will lead us to the consideration of two specific approaches, the cellular Potts model CPM [40] and the phase-field formulation [41, 42]. First, however, we discuss a simpler approach that puts in by hand a caricature of shape dynamics in the form of a contraction–protrusion cycle.

Let us assume that a canonical shape such as an ellipse is a good lowest order approximation to the shape of a cell moving on a 2d substrate. We assume, following the work of del Alamo [43] that cells undergo a dynamical morphology cycle. A cell starts off with its full length. Then, it contracts and stochastically breaks and re-forms cell-substrate bonds; this continues until it reaches the minimum length. It then "protrudes" by placing its back at the position of the rear-most adhesion, and reinitializes all adhesions. Mechanical equilibrium is maintained at each step via potential energy minimization. Details can be found in [44, 45].

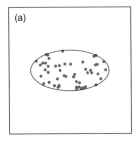 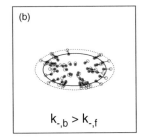 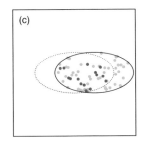

Fig. 7.5 a Cells have elastic bonds with the substrate, identified with focal adhesions. **b** Cells have a constant contraction rate during the contraction phase of the cycle, thereby stretching the bonds. The formation of the attachments is homogeneous but detachment occurs at a spatially varying, force-dependent rate. Throughout the contraction cycle, force balance is maintained by moving the cell. **c** After contraction is complete, the rear of the cell relocates to the site of the left-most adhesion

To complete this picture, we must describe the dynamics of the adhesive bonds between the cell and the substrate (see Fig. 7.5). The substrate can be assumed rigid or can be modeled as an elastic medium. A bond starts with zero stretching, but during cell contraction, the cell versus substrate attached ends will have differing displacements, and the stretching tension builds up. This tension leads the bonds to stochastically break, obeying the off-rate

$$k_{off} = k_{off}^{(0)} \exp(\text{bond tension}) \tag{7.26}$$

Now, a moving cell is polarized and specifically has front-back asymmetry of cell-substrate adhesions. To take this into account, we set the baseline (zero stretching) off-rate $k_{off}^{(0)}$ of adhesion bonds to vary linearly according to their relative position along the cell:

$$k_{off}^{(0)}(x) = k_{off}^{(0)}(x_b) + \frac{k_{off}^{(0)}(x_f) - k_{off}^{(0)}(x_b)}{x_f - x_b}(x - x_b),$$
$$k_{off}^{(0)}(x_f) < k_{off}^{(0)}(x_b) \tag{7.27}$$

where x denotes the coordinate of the cell end of a bond projected along the cellular long axis, and $x_f > x_b$ are that of the front, back ends of the cell.

This model can do a reasonable job of explaining the pattern of forces exerted on the substrate as well as how the cell speed varies with substrate properties. But clearly, the shape of the cell should not be postulated but calculated. We therefore turn to methods to carry out that calculation.

7.4.1 Cellular Potts Model

CPM is the simpler of the two approaches. Here we work on a lattice and define the cell configuration by a set of points with binary spin variables S equal to 1; all other points have $S = 0$ [40]. The initial version of the CPM uses an equilibrium free energy to describe the dynamics of this configuration. Specifically, the free energy function contains a volume term and a surface energy,

$$E = \lambda \sum_i (S_i - V_0)^2 + J \sum_{i,j} (1 - \delta(S_i, S_j)) \tag{7.28}$$

with the second sum being taken over nearest neighbors and where $\delta(S_i, S_j) = 1$ if $S_i = S_j$, 0 otherwise. The first term imposes a soft constraint on the total volume of the cell whereas the second is proportional to its perimeter. This overall free energy is used in a standard Metropolis update scheme; a site is chosen and a spin flip is attempted. The change in E is then computed; if the energy is lowered the change is automatically accepted and if it is higher it is accepted with probability $e^{-\Delta E/T}$ where T is a parameter characterizing the strength of membrane fluctuations. Later we will see that this formalism offers a natural extension to the case of interacting cells undergoing collective motion.

The basic problem with the cellular Potts approach is that cells are active particles and hence do not have dynamics that just serve to minimize some free energy. One way to begin to include this idea in the CPM is via the addition of a self-propulsion force. To accomplish this we assign to a cell a polarization direction $\hat{p}$, assumed to behave just like the velocity direction for a persistent random-walk discussed earlier. We then add a new term to the free energy [4, 28]

$$E_{sp} = -\alpha \sum_i S_i \, \hat{p} \cdot \mathbf{x_i} \tag{7.29}$$

where $\mathbf{x}_i$ is the location of each lattice point. This new term will favor update moves that push the cell in the direction of its polarization. Unless there is some assumed directional bias, the orientation of $\hat{p}$ will be assumed to diffuse and the center-of-mass motion will again be that of an unbiased persistent random walk; if there is directional bias the equation of $\hat{p}$ can be modified accordingly.

We can mention one recent application of the CPM that extends the model just considered to the case of a more complex environment. Imagine the cell interacting with a set of fibers of polymers such as collagen or fibronectin lying atop a standard homogeneous substrate. We expect that adhesion with these fibers will be much stronger than adhesion to the rest of the substrate. One can take this into account [46] by adding a term to the energy of the form

$$\sum J_f \xi_i S_i \tag{7.30}$$

where $\xi_i = 1$ for a site occupied by fiber and zero otherwise. The fibers themselves can be laid down randomly or with some level of nematic order, depending on how the surface is prepared experimentally. This model has been used to explain data regarding the connection between cell aspect ratio and fiber ordering; see [46] for a full discussion.

7.4.2 Phase Field Model

An alternate approach to cell morphodynamics uses a continuum formulation and tries to make more direct contact with the actual driving force for crawling cell motility, namely, the actin cytoskeleton and its interaction with a variety of other proteins, especially including myosin motors [47]. The idea for this model arises from the literature on modeling the dynamics of physical systems undergoing phase transitions, hence the word "phase". The original formulation was developed to study solidification of a pure material [48]. Let us assume that the material in question has its thermodynamics modeled by a Landau free energy [49, 50]

$$F = \gamma \int d^2x \left[\frac{\epsilon}{2} |\nabla\phi|^2 + \frac{(\phi^2 - 1)^2}{2\epsilon} + \frac{T - T_m}{T_m} \phi(1 - \phi^2/3) \right] \qquad (7.31)$$

Here T_m is the melting temperature and at this point there is a double-well potential ensuring that the two thermodynamic phases $\phi = 1$ (solid) and $\phi = -1$ (liquid) have equal free energy. As the temperature is increased past T_m, the liquid becomes the lowest free energy, i.e., the system undergoes a first-order phase transition. The gradient term penalizes the existence of solid-liquid interfaces, as would occur during the process of solidification from an initial seed. We then take as the equation of motion for ϕ the purely relaxational kinetics

$$\beta\dot{\phi} = -\frac{\delta F}{\delta\phi} = \gamma \left[\epsilon\nabla^2\phi - \frac{2\phi(\phi^2 - 1)}{\epsilon} - \frac{T - T_m}{T_m}(1 - \phi^2) \right] \qquad (7.32)$$

The phase-field model then couples the ϕ to the temperature dynamics by invoking the release of latent heat L at the interface as the solidification progresses. This process gives rise to the final equation

$$D\nabla^2 T = -\frac{L}{c_p} \frac{\dot{\phi}}{2} \qquad (7.33)$$

where c_p is the specific heat. For the case of a growing solid, these equations are augmented with the boundary conditions that there is solid at the origin, liquid at far distances and the temperature at infinity approaches $T_\infty = T_M - \frac{L\Delta}{c_p}$ where Δ is the dimensionless undercooling.

We are interested in the case of the interface width (related, as we shall soon see, to ϵ) taken to be very small compared to any other length in the system. On this scale, the interface is almost straight and we can look at leading order in ϵ for a propagating state obeying

$$0 = \epsilon^2 \phi'' - 2\phi(\phi^2 - 1) \tag{7.34}$$

The solution of this equation takes the from $\phi_0 = -\tanh(\frac{x-vt}{\epsilon})$, where x is the coordinate normal to the interface. To this order the propagation velocity v is undetermined. Going to next order, the velocity term, the correction to the Laplacian operator due to curvature κ of the interface, and the temperature term must all add together to allow for a perturbative solution even in the presence of a translational zero mode of the lowest order solution. This leads to the requirement

$$\int_{-\infty}^{\infty} dx \, \phi_0' \left[(\beta v + \gamma \kappa)\phi_0' - \gamma \frac{T - T_m}{T_m}(1 - \phi_0^2) \right] = 0 \tag{7.35}$$

Continuing to assume that the temperature does not change significantly over the interface width leads to the effective condition applied locally

$$\gamma \frac{T_m - T_I}{T_m} = \gamma \kappa + \beta v_n \tag{7.36}$$

which relates the undercooling at the interface $T_m - T_I$ to the surface energy contribution and to the normal velocity v_n. Finally, the temperature equation can also be simplified in this limit to be just a pure diffusion equation for the temperature together with interface-localized source given as the discontinuity of the gradient

$$- disc \, \hat{n} \cdot \nabla T = \frac{L}{c_p} v_n \tag{7.37}$$

These last two results recapitulate standard equations in the "free surface" limit [51] and show how they arise naturally from a computationally tractable PDE formulation.

We are now ready to adapt this general idea to the cell morphology system [41, 42]. Following Shao et al. [52], we will use the phase field to depict the difference between the inside of the cell ($\phi = 1$) and the outside ($\phi = 0$). For cells, the membrane velocity will be taken to be that of the actin cytoskeleton in the vicinity of the membrane. This leads to the membrane evolution equation

$$\frac{\partial \phi}{\partial t} = -\mathbf{u} \cdot \nabla \phi + \Gamma \left(\epsilon \nabla^2 \phi - \frac{G'}{\epsilon} + \kappa \epsilon |\nabla \phi| \right) \tag{7.38}$$

where the advection term couples the phase field to the local actin flow velocity, $\mathbf{u}$, and as before ϵ is the parameter controlling the width of the cell boundary, G is again a double-well potential, now with minima at $\phi = 1$ and $\phi = 0$. Note that the curvature term is added to explicitly cancel the curvature term coming from the

Laplacian. The actin velocity is then determined by solving a force balance equation in the interior of the cell. These forces arise both from the intrinsic hydrodynamics of the actomyosin system as well as from forces exerted by the membrane and by the substrate. Using the simplest form of a compressible active fluid for the actin, we have

$$\nu_0 \nabla \cdot [\phi(\nabla \mathbf{u} + \nabla \mathbf{u}^T)] + \nabla \cdot \sigma_{\text{myo}} + \nabla \cdot \sigma_{\text{poly}} + \mathbf{F}_{\text{mem}} + \mathbf{F}_{\text{adh}} = \mathbf{0} \qquad (7.39)$$

Note that ϕ is coupled to the differential operators in a manner which guarantees that the actin fluid cannot leak out of the cell. One can create more realistic models of the actin system by taking into account nematic polarization degrees of freedom [53] as well as treating the interior of the cell as a two-fluid system of actin embedded in cytosol. One can even imagine coupling a mesoscopic simulation taking into account actual actin fibers (for example using a program such as Cytosim [54] or MEDYAN [55]) to the phase-field formulation, although this has not yet been attempted. The adhesion forces can either be treated as a continuous source of friction or via discrete models of actual focal adhesions. Finally, the membrane forces arising from stretching or bending can be written directly in terms of the phase field. Interested readers should consult the aforementioned references which present all the details of these methods.

One hidden feature of the above equations concerns the active stress terms σ_{myo} and σ_{poly} corresponding to contraction generated by myosin motors and polymerization occurring at the positive ends of uncapped actin filaments. Aside from the specific form of these terms, one needs to indicate how the cell decides where these events are taking place. Connecting to our previous discussion of chemotaxis, the cell needs to decide which part of the membrane is in a "front" state and hence has actively polymerizing actin and which is in a back state dominated by myosin-based contraction. For a single cell, this polarization process can either be spontaneously determined by having the solution break the rotational symmetry [56] or can be controlled by external condition such as the chemical gradient. The spontaneous symmetry breaking will typically lead to the type of persistent random walks described phenomenologically above or to noisy guided motion; the persistence or noise level will depend on the details of the cell, exhibiting behavior ranging from keratocytes with very persistent paths coupled with steady shapes [38] to Dictyostelium with rapidly fluctuating shapes even in the presence of a strong gradient. Results from a phase-field computation of how keratocyte shapes vary with external parameters are shown in Fig. 7.6.

There are many directions in which the phase-field approach could be extended. We have already mentioned coupling to more realistic representations of the actin system and coupling to external gradient signals. These efforts would enable the explanation of the variety of shapes taken on by different cell types under different conditions. Several examples of instabilities of the basic phase-field equations have been used to try to explain observed transitions from steady-state behavior to more complex dynamics [57]. We also should mention the extension to 3d cells moving

Fig. 7.6 Keratocyte shape gallery via phase-field calculations, as the adhesiveness of the substrate and contractility of the cell are varied. Taken from [52]

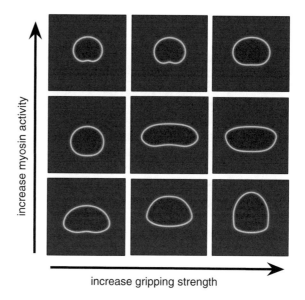

increase myosin activity

increase gripping strength

through more complex environments [58, 59]. As we shall see below, there are also extensions of the phase-field concept to the multicellular milieu.

This ends our discussion of single cell motility. We now turn to collective motility and the approaches that have been used to understand the behavior of collections of cells.

7.5   Models of Collective Motility

Collective cell motility is a widespread phenomenon in biology [60]. It has roles in developmental systems [61], in wound healing [62], and unfortunately in cancer metastasis [63]. We will define collective motility as any situation in which interactions between cells play an important role in determining their individual motions. This most strikingly occurs when adhesive cells move as a coordinated sheet, strand, or cluster, but adhesion is not strictly necessary as cell motions can become correlated just by collisions as they move en masse through confined spaces [64].

There are a variety of modeling frameworks that have been used to help understand the observed phenomenology. As was the case of single cell motility, models must be chosen so as to make contact with particular experimental observables. We will start our discussion with the simplest "agent-based" approaches and then proceed to more complete descriptions addressing more detailed aspects of the data.

7.5.1 Agent-Based Approaches

The simplest approach to collective cell motility starts with the Ben-Jacob/Vicsek (BJV) model [65] motivated by the literature on bird flocking and fish schooling [66]. This model considers cells to be point-like objects characterized only by their physical location and by their polarization, i.e., the direction in which they are currently moving. Each cell is assumed to move with a constant speed and crucially, its direction is chosen to be the average of the directions of all its neighbors within a finite interaction radius. A simple simulation proceeds by computing the velocity vectors of all the cells given their current configuration and then moving the cells using a finite time update with these velocities. This process then repeats with the new cell positions. This model bears some similarity to a classical statistical mechanics system called the XY model in which 2D unit vector spins interact with the nearest neighbors, perhaps on a square lattice. The Hamiltonian H for such a system equals

$$H = -J \sum_{i,j \ nn} \mathbf{S}_i \cdot \mathbf{S}_j \tag{7.40}$$

where "nn" means nearest-neighbor. But there is a critical catch. Because of the cell motion, the BJV model cannot be described by an equilibrium free energy minimization. In fact the BJV model is arguably the work that started the entire field of what is now called active matter [67]. The reason for this is that the model gives rise to a result which epitomizes how non-equilibrium effects can dramatically alter behavior expected for a set of interacting objects. Specifically, a classical finding in the equilibrium physics of the XY model is that there can be no long-range ordering at finite temperature. This is actually a special case of the Mermin–Wagner theorem [68, 69] and arises due to large unbounded fluctuations around any purported broken symmetry state. Assume that we are in a broken symmetry state analogous to what would occur for a low temperature magnet in three dimensions. Because the broken symmetry was continuous, there are fluctuations perpendicular to the assumed order parameter which at long wavelengths have no restoring force; instead they simply rotate the state to a symmetry-related one. These so-called Goldstone modes have dramatic consequences in two spatial dimensions.

To see this feature of the equilibrium XY model quantitatively, we will use a continuum field version of the problem and write a Langevin equation for the vector order parameter field $\boldsymbol{\phi}$ as

$$\Gamma \frac{\partial \boldsymbol{\phi}}{\partial t} = D\nabla^2 \boldsymbol{\phi} - a\boldsymbol{\phi} - b(\boldsymbol{\phi}^2)\boldsymbol{\phi} + \boldsymbol{\eta} \tag{7.41}$$

where ϕ is a two component vector and η is a white noise term accounting for finite temperature. For $a < 0$, the proposed broken symmetry state would take the form $\boldsymbol{\phi} = \sqrt{-a/b}\,\hat{n} + \hat{z} \times \hat{n}\,\delta\phi$ where we only are including fluctuations transverse to the symmetry-breaking direction $\hat{n}$ in the xy-plane. Note that $\boldsymbol{\phi}^2$ is unchanged to leading order by this perturbation. Expanding to linear order in the fluctuation amplitude immediately yields the driven diffusion equation

$$\Gamma \frac{\partial \delta\phi}{\partial t} = D\nabla^2 \delta\phi + \eta_T \tag{7.42}$$

where only the transverse component of the noise matters. We can then calculate the variance of these fluctuations as follows. The solution of the above equation is

$$\delta\phi(\mathbf{x}, t) = \int dt' d^2x' \frac{d^2k d\omega}{(2\pi)^3} \frac{e^{i\mathbf{k}\cdot(\mathbf{x}-\mathbf{x}')}e^{-i\omega(t-t')}}{-\Gamma\omega + Dk^2} \eta_T(\mathbf{x}', t') \tag{7.43}$$

We can compute the variance $< \delta\phi(x, t)^2 >$ by employing the above formula and then using $< \eta_T(\mathbf{x}, t)\eta_T(\mathbf{x}', t') >\sim \delta^2(\mathbf{x} - \mathbf{x}')\delta(t - t')$ to do the integral over one of the sets of space-time integration variables. This leads to

$$< \delta\phi(x, t)^2 >\sim \int dt' d^2x' \frac{d^2k d\omega}{(2\pi)^3} \frac{d^2k' d\omega'}{(2\pi)^3} \frac{e^{i(\mathbf{k}+\mathbf{k}')\cdot(\mathbf{x}-\mathbf{x}')}e^{-i(\omega+\omega')(t-t')}}{(-\Gamma\omega + Dk^2)(-\Gamma\omega' + Dk'^2)} \tag{7.44}$$

which clearly equals

$$\int \frac{d^2k d\omega}{(2\pi)^3} \frac{1}{(-\Gamma\omega + Dk^2)(\Gamma\omega + Dk^2)} \tag{7.45}$$

Doing the ω integral leaves us with simple expression

$$< \delta\phi(x, t)^2 >\sim \int \frac{d^2k}{k^2} \tag{7.46}$$

which diverges at long wavelengths; the divergence at short wavelength does not matter since there is anyway an underlying lattice limiting the maximal magnitude of k. In other words, the proposed broken symmetry state cannot be sustained in the presence of long wavelength thermal fluctuations. This does not happen in three spatial dimensions because the integration measure is now d^3k.

But, the BJV model does not exhibit the same feature because, as we have already mentioned, this is no free energy whose relaxation governs the dynamics. Instead, at low enough temperature the model has a phase in which there is long-range order. This can be seen by analogous calculation in the continuum approach to this model, to be discussed later. One can see this in Fig. 7.7, taken from [70]. There has been a vast literature exploring the details of the various phases and how they depend on specific features of the model; we refer the interested reader to several review articles on this subject [71, 72].

The discussion so far deals with bulk systems and does not include any mechanisms whereby the active "self-propelled" particles can stick together to form a

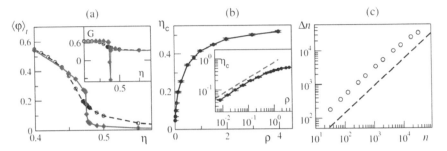

Fig. 7.7 Properties of the BJV model ($d = 2$, $\rho = 2$, $v_0 = 0.5$, periodic boundary conditions). **a** Time-averaged order parameter versus noise strength η for L $= 64$ (black circles) and L $= 256$ (red diamonds), time averages computed over 2×10^7 time-steps; Inset: corresponding Binder cumulants curves; a sharp minimum toward negative values is a signature of a first-order-like transition. **b** Asymptotic phase diagram for the transition to collective motion; Inset: Log–log plot to compare the low density behavior with the mean field predicted behavior fluctuations; **c** Δn scales approximately like $n^{0.8}$. Taken from [70], see there for details

localized cluster. In fact, there are several extensions of the original model which add additional interaction terms mimicking features such as repulsion if particles get too close to each other or adhesion if they are at some more intermediate separation length scale. A simple version of this idea first appeared in [73] where active particles were simulated with the dynamics

$$m\frac{d\mathbf{v}_i}{dt} = \alpha \hat{f}_i - \beta \mathbf{v}_i - \nabla U \tag{7.47}$$

with the inter-particle potential

$$U = -\sum_{i,j} C_a e^{-|\mathbf{x}_i - \mathbf{x}_j|/\ell_a} + \sum_{i,j} C_r e^{-|\mathbf{x}_i - \mathbf{x}_j|/\ell_r}$$

representing a short-range soft-core repulsion and a longer range ($\ell_a > \ell_r$) attraction. The term proportional to α is the self-propulsive force in the direction set by the unit vector $\hat{f}$. In the BJV model, this is set to the average of other particles within some neighborhood. One might also consider other possibilities such as independent persistent random walks (random updates to $\hat{f}_i$ for each particle), or other mechanisms such as one used in [74] when particles tend to align their self-propulsion with the direction that they are actually moving in—this is the cell version of going with the flow. We will soon consider yet another possibility, that the force direction is determined by a mechanism known as contact inhibition of locomotion, essentially that cells tend to move away from directions with the highest cell density.

A simulation of the aforementioned model with the BJV rule leads to a surprising finding. Not only can the cells form moving flocks, but they can also form rotating structures. An example of such a structure is shown in Fig. 7.8. In this state, all particles are moving at their "free" speed α/β in the $\hat{\phi}$ direction and the force arising

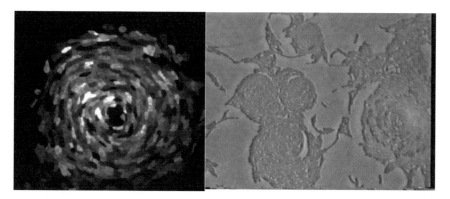

Fig. 7.8 Microorganisms can organize themselves into rotating vortices as a balance between self-propulsion and inter-particle forces. Here we show two examples; Dictyostelium cells which have aggregated into a two-dimensional "puddle" of cells (left) and bacterial cells on hard agar surfaces [75]; Picture courtesy of Eshel Ben-Jacob

from the particle interaction provides the centripetal acceleration required for the circular paths. If we use a continuum description of this active fluid, in steady state we would need to balance

$$(\mathbf{v} \cdot \nabla)\mathbf{v} = -\nabla \int d^2x' U(\mathbf{x} - \mathbf{x}')\rho(\mathbf{x}') \tag{7.48}$$

where ρ is the density. This leads to the vortex shape equation (see [73])

$$D - (\alpha/\beta)^2 \ln r = \int d^2x' U(\mathbf{x} - \mathbf{x}')\rho(\mathbf{x}') \tag{7.49}$$

where D is a constant which determines the total number of cells in the vortex. Predictions from solving this steady-state equation agree with direct simulations of the particle motion. We should note that this type of rotating state has been seen in many systems [76], ranging from fish schools to microorganism colonies to mammalian epithelial cells during glandular development [77].

Another use of this type of approach arises in the context of collective chemotaxis. Recall that single cells can directly bias their motion due to gradients, as long as the difference in receptor occupancy across the cell is sufficiently large. In recent years, several instances have emerged in which gradients that are not large enough at the single cell level can nonetheless be resolved by cell clusters [78, 79]. One possible explanation could in principle be just noise averaging, namely, that coupling together cells that each make a very noisy estimate of the gradient direction can lead to a much larger signal-to-noise ratio for the aggregate. In at least one experimental system, this explanation has been ruled out. Here, increasing the gradient leads to single cell motion that is in the opposite direction than that of clusters; hence the latter cannot just be based on averaging the former. This has been studied [16, 17] using a slight variation of the previous model with a new term reflecting CIL, the contact inhibition of locomotion.

$$\partial_t \mathbf{x}_i = \mathbf{p}_i + \sum_{j \neq i} \mathbf{F}^{ij} \tag{7.50}$$

$$\partial_t \mathbf{p}_i = -\frac{1}{\tau}\mathbf{p}_i + \sigma \boldsymbol{\xi}_i(t) + \beta_i \sum_{j \sim i} \hat{x}_{ij} \tag{7.51}$$

Here $\hat{x}_{ij}$ is a unit vector pointing from the position of cell j to cell i and ξ is a white noise term. In line with the BJV idea, the sum over j is taken only over cells within some typical interaction distance from i. CIL is a well-established phenomenon describing the tendency for cells to polarize away from other cells [80, 81]. One version of CIL arises when signals from the front region of a cell suppresses the corresponding front region of a neighboring cell if the two cells are on a collision course [82]. In the model used here, this is argued to give rise to a modulation of the cell's propulsive force (here determined by the polarization vector $\mathbf{p}$) so that it points away from nearby cells; note that cells in the interior of a cluster for which there is no sparsely populated direction are "stalled", i.e., have their propulsion turned off.

Now, as stated above, this model was used to study collective chemotaxis. Following [16], we assume that the CIL coefficient β_i is proportional to an external chemical concentration which varies in space. Therefore for cell j, $\beta_i = \bar{\beta} + \mathbf{g} \cdot \mathbf{x}_i$ where $\mathbf{g}$ is the chemical gradient. This assumption causes the cells at the high concentration side of the cluster to feel a stronger propulsive force than cells at the low side, and of course the force at the edge always points outward. The simplest way to analyze this system is to take the inter-particle forces to be sufficiently large that cells settle into a cluster configuration which does not depend on the fluctuating polarization. Then one can derive the cluster velocity

$$\langle \mathbf{V} \rangle_c \approx \bar{\beta}\tau \mathcal{M} \cdot \mathbf{g} \tag{7.52}$$

where the approximation is true for shallow gradients. And $\langle \cdots \rangle$ indicates an average over the fluctuating polarization (due to the noise) but with a fixed configuration of cells. The matrix $\mathcal{M}$ only depends on the cells' configuration,

$$\mathcal{M}_{\mu\nu} = \frac{1}{N}\sum_i q_{i,\mu}x_{i,\nu} \tag{7.53}$$

where $\mathbf{q}_i = \sum_{j \sim i} \hat{x}_{ij}$; i, j denotes a cell index and μ, ν Cartesian component indices. Interestingly, one can also find parameter values for which the clusters rotate while they chemotax; this in fact has been observed in the bacterial colony context [83].

Finally, we should mention that much use has been made of agent-based models in the context of tumor growth [84, 85]. In this context one tries to include a variety of biological mechanisms such as cell division and cell death, nutrient and oxygen limitations and possible angiogenesis (formation of new blood vessels) as a response, and of course cell motility and its dependence on chemical gradients and the properties of the medium through which the cells are moving. This is a whole subject matter unto itself and interested readers are referred to the above references for introductions to this subject matter.

7.5.2 Subcellular Elements

A natural extension of agent-based models considers cells to be composed of several internal particles, sometimes refer to as sub-cellular elements [86]. These degrees of freedom of a single cell interact with each other and with those degrees of freedom in nearby cells. What is the advantage of this formulation over the simplest approach? Essentially, one can see the answer by observing the force pattern of a single cell moving on a substrate; the technique for measuring this force pattern is called traction force microscopy [87]. The substrate is chosen to be a relatively soft gel with embedded fluorescent spheres. As the cell passes over a given region, the spheres are displaced by traction forces creating elastic deformations in the gel and the measured displacements can be inverted to determine those tractions. Now the net force applied by the cell to the substrate (and hence to the cell, by Newton's third law) is close to 0; it only has to match the relatively small viscous drag of the liquid medium. Instead, the force pattern tends to be a dipole, either of constant magnitude for a stable moving cell [88] or with some oscillating pattern for amoeboid cells such as Dicty [43]. This leading order behavior can be captured by two-particle sub-cellular element models, but not by point-like agent-based approaches.

To discuss what one of these models looks like in practice, we follow the discussion in [89]. A cell is represented by two particles whose position is updated using simple over-damped dynamics. Each sub-cellular element obeys the equation

$$\mathbf{v} = \frac{1}{\xi} \left(\mathbf{M} + \mathbf{F}_{int} + \mathbf{F}_{cont} \right) \qquad (7.54)$$

Here, ξ is the friction from the substrate; F_{int} is the force between sub-cellular elements of neighboring cells. This force consists of a potential with short-range repulsion and longer range attraction; this is already familiar from our previous discussion of agent-based models. The intra-cell contractile force term is new and represents myosin-based active contraction responsible for the aforementioned dipole traction pattern. In the simplest approach, this is just treated as a time-independent nonlinear spring, again encompassing repulsion and attraction. In more complex models one might impose a cycle in which contraction alternates with actin-based protrusion representing propulsion; this would then lead to an extension of the already discussed single cell of [45] to the collective case. In our simpler case, propulsion will go on simultaneously with contraction and be described separately as an additional motility force $\mathbf{M}$.

For a motile cell, we assume that cell has some polarization (which may or may not be allowed to change dynamically) and hence we can identify a front particle and a rear one. Each particle feels a force pushing it outward ($\mathbf{M}_f$ and $\mathbf{M}_b$ respectively) and balancing the contractile force; as already mentioned, this balance may represent a time-averaged version of the dynamics of more complex model that explicitly contains a contraction–protrusion cycle. For an isolated cell, it clearly would have net cellular motion with $|\mathbf{v}| = (M_f - M_b)/\xi$ and directed along the polarization. The remaining piece of the model is the inclusion of the effect of CIL. Again, this is accomplished following the approach used in the point-like models (recall (7.51)); in detail, for each particle i, we calculate the sum $\mathbf{R_i}$ of the unit vectors $\hat{r}_{ij}$ pointing from particle i to nearby particles j (including the partner particle of the same cell) within a distance R_{int}, specifically

$$\mathbf{R_i} = \sum_{j, r_{ij} < R_{int}} \hat{r}_{ij} \quad \text{and} \quad n_i = \sum_{j, r_{ij} < R_{int}} 1 \qquad (7.55)$$

$$\text{then} \quad \mathbf{M_i} = -M_{f/b}\, \mathbf{R_i}/n_i \qquad (7.56)$$

We can give two examples of the use of this class of models. First. imagine a tissue of moving cells that encounter a "dead zone", i.e., a circular region of substrate in which the cells cannot adhere. The cells flow around this region, giving rise to a steady-state pattern of velocities and traction forces. In Fig. 7.9 we show a simulation of this process where these objects are computed. These results are in good agreement with an experimental study by Trepat and co-workers [90].

A second example is perhaps a bit more physiologically relevant. Now, we imagine making a small hole inside an otherwise confluent tissue. This system was studied in a rather comprehensive set of experiments as reported in [91]. Now, there is one important biological detail which must be added to the two-particle model in order to make contact with the data. That is, cells at the wound edge mechanically connect their actin-based cortices so as to form a supra-cellular cable that can apply forces to the cells via contracting; this cable is clearly visible in for example bright-field microscopy images of expanding cellular monolayers [92]. So, there are two possible mechanisms of wound closure, "purse-string" contraction versus active crawling. These two mechanisms are not mutually exclusive, instead, they often coexist and interact. Experiments have tried to deconvolve these two effects by a variety of ideas; making the substrate under the hole non-adhesive (thereby preventing crawling) versus altering the geometry of the wound which changes the cable force. This then presents an excellent opportunity for computational studies to help clarify underlying mechanisms. To include the supra-cellular actomyosin cable around the wound, we need to distinguish boundary cells around the wound boundary; these boundary cells are mechanically linked with their neighbors by a contraction force $\mathbf{F}_{cable}$ behaving like the intracellular contraction $\mathbf{F}_{contr}$. These connections of neighboring cells form a whole purse-string contraction ring, In simulations, the cable locations can be altered by changing the connecting rules. This model was then used to investigate both the closure kinetics and the traction forces seen for various experimental protocols; see

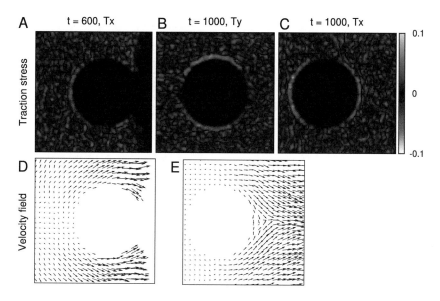

Fig. 7.9 Simulation of cells moving around an obstacle. Inability of cells to adhere on the circular obstacle is modeled by setting the propulsion force to zero when particles enter. **a** X-component of average traction stresses at $t = 600$. **b** Y-component of average traction stresses at $t = 1000$. **c** X-component of average traction stresses at $t = 1000$. **d** Average velocity field at $t = 600$. **e** Average velocity field at $t = 1000$. Averages are taken over five different simulations, and for velocities over a time $t_{relax} = 50$ each. The range for averaging traction stresses is $R_{trac} = 1.0$. The obstacle diameter is 30 in simulation units $\sim$30 typical cell sizes (in the experiment: 1 mm); see [89] for more details

[93]. Note that a more complete treatment would explicitly consider the intracellular cytoskeletal dynamics that leads to cable formation and couple these dynamics to the collective motion. As of yet this has not been attempted.

7.5.3 Vertex/Voronoi Models

The collective motility models we have discussed so far do not consider cell morphology in any serious way. To better take into consideration the cell morphology for the specific case of epithelial cell layers, researchers have introduced a different strategy based on the vertex model [94]. These models assume that cells can be approximated with polygonal shapes. In the vertex model, a cell is parameterized by a set of vertices that mark the common point of three or more neighboring cells. A piece of a vertex model configuration is shown in Fig. 7.10a and for comparison a picture of an epithelial layer is presented in Fig. 7.10b. A variant of this approach is based on the Voronoi construction [96]. In the Voronoi model, the common borders of neighboring cells are determined by this construction, i.e., a cell is defined by its

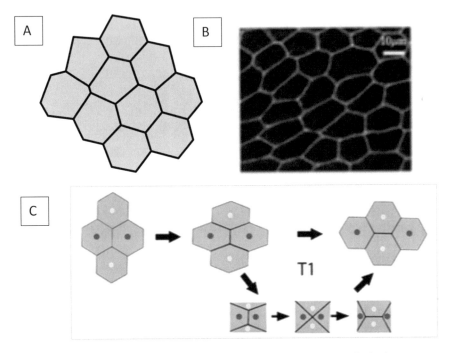

Fig. 7.10 **a** A basic configuration of the vertex model, showing polygonal cells that have no gaps between them. **b** A photograph of a confluent epithelial layer of cells, from [95]. **c** A schematic illustration of a T1 transition, giving rise to local re-arrangements and enabling liquid-like unjammed behavior

center position and consists of all points that are closer to this cell's center than to any other cells' center. In these models, the cell shape itself, the adhesion between cells and many other aspects related to the cell shape and boundary can be treated in a more natural way.

As with all such models, once one has identified the degrees of freedom encoding the geometry, we need to decide on the dynamics. Just as in the previous discussion of the cellular Potts model, the initial approach to this question assumed that the cell could be treated by minimizing a free energy defined with the help of some effective temperature arising from active fluctuations occurring in the cell. We can write down the total energy of the system based on the area and perimeters of each cell,

$$E = \sum_{i=1}^{N} \left[K_A (A_i - A_0)^2 + K_P (P_i - P_0)^2 \right] \tag{7.57}$$

Here A_i and P_i are respectively the area and perimeter for cell i and A_0, P_0 are the preferred values of those objects. In this expression, we have assumed these parameters are constant throughout the layer but one can easily extend this framework to encompass imposed variations due to environmental gradients and features such as the aforementioned cable. K_A is the area modulus arising from the resistance to changes in projected cell area, K_P is an equivalent perimeter modulus due to cells attempting to limit their surface area, which when projected to 2D is just the perimeter. The preferred value for the perimeter arises when one also takes into account the negative energy of adhesion between neighboring cells; see [97, 98] for a more complete discussion. Then, a more general expression of the energy takes the form

$$E = \sum_i K_{A,i} \left(A_i - A_{0,i} \right)^2 + \sum_i K_{P,i} P_i^2 + \sum_{<i,j>} \Lambda_{i,j} P_{i,j}, \qquad (7.58)$$

where the last term is a sum of nearest-neighbor links. This reduces to the previous form aside from an irrelevant constant, if all the parameters are taken to be uniform among cells.

Perhaps the most important result to come out of detailed investigations of the vertex model concerns the notion of unjamming. Jamming is an idea originally developed [99] in systems such as granular material to describe how such a system transitions from liquid-like behavior at low density to a solid-state replete with shear modulus as the density passes a threshold value. The reverse of this might be most familiar in the context of an avalanche flowing down a hillside when the top layer of snow becomes fluidized. In the vertex model, however, the cell layer is always confluent (there is no bare substrate) and hence density defined in terms of coverage fraction does not vary. Nonetheless, there is still a transition that occurs at a critical value of $P_0/\sqrt{A_0} = 3.81$. One estimate of this point can be found by considering the perimeter to area ratio of a perfect pentagon, a common cell motif in an epithelial layer; this leads to

$$P_0/\sqrt{A_0} \simeq q_5 \qquad (7.59)$$

with $q_n = \sqrt{4n \tan \left(\frac{\pi}{n} \right)}$. Below this point, in other words when the desired perimeter is small, cortical tension dominates over cell-cell adhesion and produces finite energy barriers to cell re-arrangement.

This idea was given a tremendous boost when measurements on lung epithelial cells [100] confirmed the basic unjamming picture and the predicted role of the critical shape factor in determining the transition point. This then leads to the question, how do the cells in a confluent layer actually move? The key is what is known as a T1 re-arrangement [101], depicted in Fig. 7.10c. Essentially, the cell pattern undergoes a local topology change in which nearest-neighbor relationships are altered. In the unjammed state, the excess available perimeter allows for these transitions to occur without an energy barrier; in the jammed state the tension along the edges creates such a barrier.

As we have already discussed for the cellular Potts model, the basic equations of motion driven by the free energy may not fully reflect the active nature of cell motility. Bi and co-workers introduced an idea akin to that of (26) into the Voronoi model [98]. Since the degrees of freedom in this model are just the center-of-mass positions, we can use the same idea as described above, namely, to introduce a new term in the energy

$$E_{sp} = -\alpha \sum_i \hat{p}_i \cdot \mathbf{x_i} \tag{7.60}$$

This then leads to a generalized equation for the cell centers

$$\frac{d\mathbf{x}_i}{dt} = -\mu \nabla_i E + v_0 \hat{p}_i \tag{7.61}$$

with $v_0 = \mu\alpha$. Again, this leaves us with a choice for the dynamics of the polarization of cell i, namely, $\hat{p}_i$. These choices can be cell-independent, essentially assuming that the persistent random-walk idea continues to apply even in the presence of other particles. Conversely, we could assume some way in which neighboring cell polarizations influence a given cell, using ideas directly from the BJV neighbor-averaging approach or by introducing correction more indirectly by assuming that a cell will rotate its polarization to accommodate its actual velocity. For the simplest persistent walker idea, Bi and co-workers showed that one could find a generalized jamming/unjamming phase transition, as shown in the phase diagram depicted in Fig. 7.11.

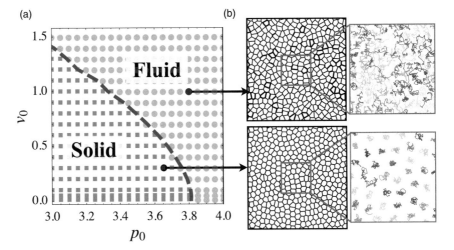

Fig. 7.11 **a** Phase diagram for the Voronoi model with active motility; p_0 is the dimensionless shape parameter **b** Instantaneous tissue snapshots show the difference in cell shape across the transition. Cell tracks also show dynamical arrest due to caging in the solid phase and diffusion in the fluid phase

There have been a wide range of papers dealing with extensions of the basic vertex model approach as well as detailed applications to different experimental situations; it is beyond our scope to present a detailed review of all of these works. In one recent example, Koride et al. [102] combined the vertex model with biochemical regulation of contractility to model collective cellular migration in confluent epithelia. This allows more direct predictions of traction force pattern, akin to the benefit of the inclusion of contractility in the sub-cellular element approach. We also mention the treatment [103] of the wound closure experiment discussed above. Perhaps a more non-trivial modification allows the Voronoi approach to account for the physical separation of cells, i.e., going beyond confluent tissues. There are a few works which have allowed for non-confluent tissues [104, 105] in Voronoi-based models. The basic idea is to introduce a length scale, requiring that each cell lie entirely within a distance ℓ from the dynamical point degree of freedom that describes its location. Thus, the cell boundaries may consist not only of polygonal segments but also circular arcs, and there can be intercellular regions between the cells. Teomy et al. [106] worked out the consequences of this assumption for the basic phase diagram (in the absence of active propulsion). Their results, shown in Fig. 7.12, indicate that the jamming–

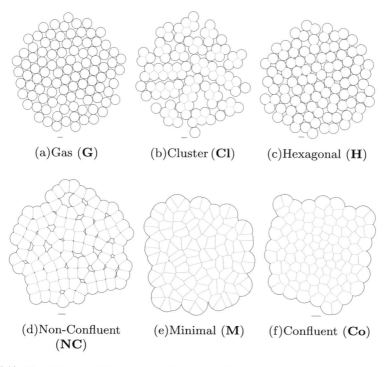

(a)Gas (**G**)   (b)Cluster (**Cl**)   (c)Hexagonal (**H**)

(d)Non-Confluent   (e)Minimal (**M**)   (f)Confluent (**Co**)
(**NC**)

Fig. 7.12 Simulation results for the state of the non-confluent Voronoi model system, exemplifying each of the six phases, starting from the same initial condition with $N = 100$ cells and zero temperature $T = 0$. Green lines are boundaries between adjacent cells, and red lines are the outer boundaries of the cells. See [106] for full details

unjamming transition now becomes embedded in a more general set of phases that includes a gas with no important particle interactions and a clustered structure which agrees in general terms with some experimental observation of sub-confluent layers [107].

7.5.4 Shapes, Revisited

Once we move away from the confluent limit, the vertex model loses its ability to account for the morphological degrees of freedom of the individual cells. For example, the extension to non-confluent models described above leads to cell shapes that are only first approximations to the real shape as determined, as we have seen, by the balance of a number of differing aspects of cell mechanics. Because of this, it is often convenient to start with spatially extended models of individual cells and then add interactions so as to develop a collective model. This is quite easy in the Cellular Potts Model and in fact the collective version of this model was the first motivation for this entire approach. To do this, all we need to do is to introduce a different spin index for each cell [40]. In other words, lattice plaquettes are now assigned a spin index ranging from 0 to N, where N is the total number of cells. This then defines a multicellular configuration, as seen for example in Fig. 7.13. The other advantage offered by CPM is the existence of open source software and user community under the name Compucell3D [108]. Anyone interested in the wide variety of problems that have been tackled using the basic CPM framework should definitely check out this resource.

To highlight one specific use of the CPM model augmented by the active motility force already discussed above, we focus on rotating states seen in many active

Fig. 7.13 Example of a configuration of the CPM; individual cells are labeled by a positive integer number and underlying substrate by a "0". The dynamics of the model are given as a set of rules for how individual square plaquettes change their assigned number

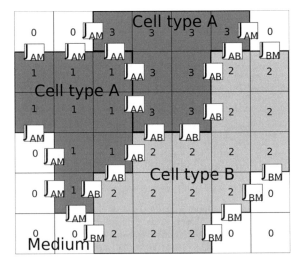

Fig. 7.14 A snapshot from the simulation of a rotating locally confluent vortex of Dictyostelium cells, using the active version of the CPM. See [28] for details

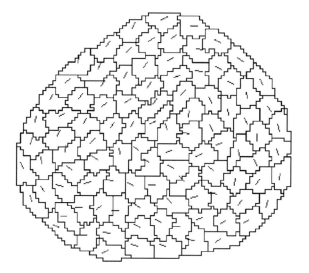

systems, including during the aggregation of Dictyostelium amoebae. It was long been known that starved Dictyostelium cells will aggregate under the influence of chemotactic guidance, forming a dense cell mass that eventually will lead to fruiting body formation [20]. If the aggregation process is limited to two dimensions (by having it constrained to lie in a finite size gap in the vertical direction), one can arrest the full morphogenetic process and allow for the creation of a long-lived rotating vortex state; we have already seen this state as one of the examples shown in Fig. 7.8. As mentioned above, this state can be understood in general by agent-based models, but these may not be quantitatively comparable to experimental data for confluent arrangements of cells. In Fig. 7.14, we show a snapshot from the simulation study of [28] using the CPM approach methodology and an active motility rule that relaxes a cell's polarization direction to its actual velocity direction. One can then compare the cell velocity versus radius with measured data; see this reference for more details of this application.

One can adopt a similar strategy to construct a multicellular phase-field model. Each cell will be assigned its own phase-field ϕ_i which aside from a small interfacial region will be non-zero only in the interior of cell i. We then must add the free energy an additional term reflecting cell-cell interactions, again taken as short-range repulsion and longer range adhesion. Within this framework, Camley and co-workers [109] studied the effects of different assumptions regarding the coupling cell polarizations, ranging from independent persistent random walks to the BJV assumption to the inclusion of CIL. Interestingly, the prevalence of the vortical state we have been seeing is rather sensitive to the form of this cell–cell orientational coupling. These patterns could be compared to experiments on mammalian cells confined to a circular box [110].

7.6 Continuum Models

Our chapter has focused on many classes of models, but all of them treated cells as important "individuals" whose motion should form the basis for comparing models to experiments. This of course makes complete sense for individual cells and for collective motion involving modest numbers of individual cells. At some point, however, it makes sense to consider a continuum hydrodynamics treatment, where the dynamical variables become cell density and cell velocity. The decision to go down this route is trickier in living systems than, for example, in the treatment of water molecules via continuum fluid mechanics. In living systems, individual cells can have physical characteristics (aka phenotypes) that are distinct from the cells around them. One clear example of this is seen in the process of angiogenesis, the creation of new blood vessels. These vessels consist of cells that sprout out from the side of existing vessels, led forward by specialized "tip" cells that organize the entire motion [111]. In a different example, stresses imposed on specific cells in an epithelial layer can cause the layer to buckle into the third dimension and form functional 3D structures [112]. The decision of which cells become phenotypically specialized is made by cell–cell communication both via chemical and mechanical channels. These can be treated by embedding cellular signaling networks into the models we have been discussing, as each cell can be equipped with additional internal degrees of freedom which are then modulated by external signals impinging on the cell. Studying these examples is beyond the scope of this chapter and indeed beyond the scope of most physics-based treatments of cell motility.

We will therefore focus on hydrodynamic models that treat all the cells as identical. We will start out assuming that there is no cell growth or death. This is the case for some experimental systems (the Dicty vortex, for example) but not for others; this of course depends on the duration of the experiment as compared to a typical cell division time, the latter being modulated both by a cell's phenotype and by a cell's mechanical state in the moving tissue. The basic equation for cell density is then the obvious conservation law

$$\frac{\partial \rho}{\partial t} + \nabla \cdot (\rho \mathbf{v}) = 0 \qquad (7.62)$$

Note that the density here refers to the number of individuals per area (for motion on a substrate) or volume; in this language the density of a confluent layer of cells can vary if cells become compressed or extended. This has to be coupled to an equation for the velocity itself. There are two key considerations here. First, cells are exerting active forces on some external medium, either a solid substrate for cells crawling in 2D, a fibrous extracellular matrix for cells moving in 3D materials, or possibly a background fluid for swimming cells.

The existence of this static background means that the hydrodynamic equations no longer have to satisfy Galilean invariance, namely, that the equations look the same in moving reference frames. The details of all possible terms that arise because of the lifting of this symmetry, as compared to the terms in the normal compressible Navier–Stokes equation, are quite complicated. Interested students should read the original papers by Toner and Tu [113, 114] for a complete discussion. Here we will give the approximate form

$$\rho\frac{\partial \mathbf{v}}{\partial t} + \lambda_1(\mathbf{v}\cdot\nabla)\mathbf{v} + \lambda_2(\nabla\cdot\mathbf{v})\mathbf{v} = \alpha\mathbf{v} - \beta|\mathbf{v}|^2\mathbf{v} - \nabla\tilde{P} + D_T\nabla^2\mathbf{v} \qquad (7.63)$$

where we have ignored some higher order terms and have absorbed a term of the form $\nabla|\mathbf{v}|^2$ and $\nabla(\nabla\cdot\mathbf{v})$ into an effective pressure term. Of interest are several features. The velocity magnitude relaxes on a fast timescale to $\sqrt{\alpha/\beta}$, leaving however the direction of the velocity as a slow variable. This mode is familiar from our discussion above of the equilibrium XY model but as shown by Toner and Tu, this system of equations gives rise to an ordered flocking state which does not suffer from arbitrarily large fluctuations which destroy the assumed order. Also notice that the coefficient of the advective term $(\mathbf{v}\cdot\nabla)\mathbf{v}$ is not forced to equal 1 and other terms containing two velocity factors together with one derivative operator are allowed. Form the point of view of active matter physics, it is important to study the behavior of this equation and also its derivation from a more microscopic picture. Form the point of view of collective cell motility, it may be the case that this type of continuum approach is somewhat limited in its range of applicability; it is perhaps a more useful description of the flocking behavior of independently acting individuals rather than the highly coordinated motions of epithelial sheets or cellular clusters. Other researchers may of course have a different opinion on this matter; see, e.g., [115].

One of the interesting directions of research in this vein has considered the role of cell growth using a continuum. Of course, cell growth means that the density equation no longer takes the form of a conservation law, but instead becomes

$$\frac{\partial \rho}{\partial t} + \nabla\cdot(\rho\mathbf{v}) = b(P)\rho \qquad (7.64)$$

where we have assumed that the net birth rate (i.e., birth–death) depends on the mechanical state of the cell through a pressure dependence. Basan and co-workers [116] considered a thought experiment of a malignant tumor lying next to normal tissue, as indicated in Fig. 7.15. They assumed that the net growth rate could be taken to be of the form

$$b(P) = k\,(P_h - P) \qquad (7.65)$$

where the value of the transition point, referred to as the homeostatic pressure, takes into account the fact that imposing pressure tends to increase the cell death rate. As shown in the figure, different types of tissues will have different homeostatic pressures and hence in general one will invade into the other. One can also use the complete continuum equations to study conditions under which the interface between two tissues becomes unstable to transverse fluctuations; details are presented in [117].

While on the subject of interfacial instabilities, we close this chapter by briefly mentioning attempts to understand the formation of finger-like structures in the spreading of epithelial sheets. This example is actually a useful instance of the diver-

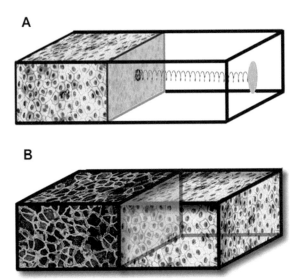

Fig. 7.15 Depiction of the idea of homeostatic pressure. In this conceptual experiment as the tissue grows, the pressure exerted by the piston on the tissue increases until we reach a net cellular birth rate of zero; this defines the homeostatic pressure. In **b** the tissue with the lower homeostatic pressure compressed to a cell density above its homeostatic point can be invaded by the other tissue

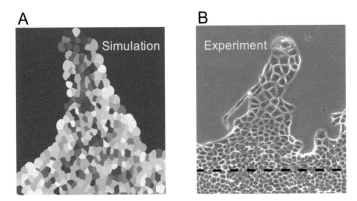

Fig. 7.16 An example of a finger, theory compared to experiment **a** Voronoi cell representation of a finger from a sub-cellular element simulation; details can be found in [118]. The coloring is just for visualization purposes. **b** Finger image from experiment [119]

gence between the interests of physicists and biologists and the related divergence in methodology, i.e., continuum approaches versus models resolving the varying behavior of individual cells. As shown in Fig. 7.16, fingers can be seen emerging during the in vitro spreading of tissue into free space. These fingers can be studied by creating a sub-cellular element model [118]. This model, however, is rather complicated and has a number of rules as to how cells in different parts of the finger communicate with each other and modulate their biophysical properties accordingly. From a physics perspective, it is important to understand the extent to which all this complexity, even if it reflects actual biological reality, is necessary to establish the ingredients underlying the basic instability giving rise to finger formation. This question has been addressed in several recent papers [118, 120, 121] with several issues still remaining to be clarified.

Acknowledgements This work was supported by NSF (PHY-1935762) and NSF (PHY-2019745).

References

1. R. Alert, X. Trepat, Ann. Rev. Condens. Matter Phys. **11**, 77 (2020)
2. B.A. Camley, W.J. Rappel, J. Phys. D Appl. Phys. **50**(11), 113002 (2017)
3. S. Banerjee, M.C. Marchetti, Cell migrations: causes and functions (2019), pp. 45–66
4. S. Jain, V.M. Cachoux, G.H. Narayana, S. de Beco, J. D'alessandro, V. Cellerin, T. Chen, M.L. Heuzé, P. Marcq, R.M. Mège et al., Nat. Phys. **16**(7), 802 (2020)
5. L. Li, E.C. Cox, H. Flyvbjerg, Phys. Biol. **8**(4), 046006 (2011)
6. N.G. Van Kampen, *Stochastic Processes in Physics and Chemistry*, vol. 1 (Elsevier, Amsterdam, 1992)
7. H.C. Berg, D.A. Brown, Nature **239**(5374), 500 (1972)
8. M.J. Schnitzer, Phys. Rev. E **48**(4), 2553 (1993)
9. H. Levine, W.J. Rappel, Phys. Today **66**(2) (2013)
10. R.c. Gao, X.d. Zhang, Y.h. Sun, Y. Kamimura, A. Mogilner, P.N. Devreotes, M. Zhao, Eukaryotic Cell **10**(9), 1251 (2011)
11. C.R. Doering, X. Mao, L.M. Sander, Phys. Biol. **15**(6), 066009 (2018)
12. E.F. Keller, L.A. Segel, J. Theor. Biol. **30**(2), 225 (1971)
13. R.T. Tranquillo, in *Biochemical Society Symposium*, vol. 65 (1999), pp. 27–42
14. X. Sun, M.K. Driscoll, C. Guven, S. Das, C.A. Parent, J.T. Fourkas, W. Losert, Proc. Natl. Acad. Sci. **112**(41), 12557 (2015)
15. G. Yu, J. Feng, H. Man, H. Levine, Phys. Rev. E **96**(1), 010402 (2017)
16. B.A. Camley, J. Zimmermann, H. Levine, W.J. Rappel, Phys. Rev. Lett. **116**(9), 098101 (2016)
17. B.A. Camley, J. Zimmermann, H. Levine, W.J. Rappel, PLoS Comput. Biol. **12**(7), e1005008 (2016)
18. R. Sunyer, V. Conte, J. Escribano, A. Elosegui-Artola, A. Labernadie, L. Valon, D. Navajas, J.M. García-Aznar, J.J. Muñoz, P. Roca-Cusachs et al., Science **353**(6304), 1157 (2016)
19. E.A. Novikova, M. Raab, D.E. Discher, C. Storm, Phys. Rev. Lett. **118**(7), 078103 (2017)
20. W.F. Loomis, *The Development of Dictyostelium discoideum* (Academic, New York, 1982)
21. B. Hu, W. Chen, W.J. Rappel, H. Levine, Phys. Rev. E **83**(2), 021917 (2011)
22. G. Casella, R.L. Berger, *Statistical Inference* (Cengage Learning, Boston, 2021)
23. D. Fuller, W. Chen, M. Adler, A. Groisman, H. Levine, W.J. Rappel, W.F. Loomis, Proc. Natl. Acad. Sci. **107**(21), 9656 (2010)
24. R.G. Endres, N.S. Wingreen, Proc. Natl. Acad. Sci. **105**(41), 15749 (2008)
25. C.A. Parent, P.N. Devreotes, Science **284**(5415), 765 (1999)
26. A. Levchenko, P.A. Iglesias, Biophys. J . **82**(1), 50 (2002)
27. K. Takeda, D. Shao, M. Adler, P.G. Charest, W.F. Loomis, H. Levine, A. Groisman, W.J. Rappel, R.A. Firtel, Sci. Signal.**5**(205), ra2 (2012)
28. W.J. Rappel, A. Nicol, A. Sarkissian, H. Levine, W.F. Loomis, Phys. Rev. Lett. **83**(6), 1247 (1999)
29. L. Song, S.M. Nadkarni, H.U. Bödeker, C. Beta, A. Bae, C. Franck, W.J. Rappel, W.F. Loomis, E. Bodenschatz, Eur. J. Cell Biol. **85**(9–10), 981 (2006)

30. H. Levine, W. Loomis, W. Rappel, in *Fields Institute Communications*, ed. by S. Sivaloganathan (American Mathematical Society, Providence, RI, 2010), pp. 1–20
31. J.E. Ferrell Jr., S.H. Ha, Trends Biochem. Sci. **39**(10), 496 (2014)
32. V. Sourjik, N.S. Wingreen, Curr. Opin. Cell Biol. **24**(2), 262 (2012)
33. H.C. Berg, E.M. Purcell, Biophys. J . **20**, 193 (1977)
34. W. Bialek, S. Setayeshgar, PNAS **102**(29), 10040 (2005)
35. K. Wang, W.J. Rappel, R. Kerr, H. Levine, Phys. Rev. E **75**(6), 061905 (2007)
36. Y. Tu, T.S. Shimizu, H.C. Berg, Proc. Natl. Acad. Sci. **105**(39), 14855 (2008)
37. W. Ma, A. Trusina, H. El-Samad, W.A. Lim, C. Tang, Cell **138**(4), 760 (2009)
38. E.L. Barnhart, K.C. Lee, K. Keren, A. Mogilner, J.A. Theriot, PLoS Biol. **9**(5), e1001059 (2011)
39. P.J. Van Haastert, PLoS ONE **6**(11), e27532 (2011)
40. F. Graner, J.A. Glazier, Phys. Rev. Lett. **69**(13), 2013 (1992)
41. D. Shao, W.J. Rappel, H. Levine, Phys. Rev. Lett. **105**(10), 108104 (2010)
42. F. Ziebert, S. Swaminathan, I.S. Aranson, J. R. Soc. Interface **9**(70), 1084 (2012)
43. J.C. Del Alamo, R. Meili, B. Alonso-Latorre, J. Rodríguez-Rodríguez, A. Aliseda, R.A. Firtel, J.C. Lasheras, Proc. Natl. Acad. Sci. **104**(33), 13343 (2007)
44. J. Feng, H. Levine, X. Mao, L.M. Sander, Soft Matter **15**(24), 4856 (2019). https://doi.org/10.1039/c8sm02564a
45. M. Buenemann, H. Levine, W.J. Rappel, L.M. Sander, Biophys. J . **99**(1), 50 (2010)
46. J. Kim, Y. Cao, C. Eddy, Y. Deng, H. Levine, W.J. Rappel, B. Sun, Proceedings of the National Academy of Sciences **118**(10) (2021)
47. T.D. Pollard, Nature **422**(6933), 741 (2003)
48. J.B. Collins, H. Levine, Phys. Rev. B **31**(9), 6119 (1985)
49. P.C. Hohenberg, B.I. Halperin, Rev. Mod. Phys. **49**(3), 435 (1977)
50. P.M. Chaikin, T.C. Lubensky, T.A. Witten, *Principles of Condensed Matter Physics*, vol. 10 (Cambridge University Press, Cambridge, 1995)
51. J.S. Langer, Rev. Mod. Phys. **52**(1), 1 (1980)
52. D. Shao, H. Levine, W.J. Rappel, Proc. Natl. Acad. Sci. **109**(18), 6851 (2012)
53. E. Tjhung, A. Tiribocchi, D. Marenduzzo, M. Cates, Nat. Commun. **6**(1), 1 (2015)
54. S. Sedwards, T. Mazza, Bioinformatics **23**(20), 2800 (2007)
55. K. Popov, J. Komianos, G.A. Papoian, PLoS Comput. Biol. **12**(4), e1004877 (2016)
56. Y. Mori, A. Jilkine, L. Edelstein-Keshet, Biophys. J . **94**(9), 3684 (2008)
57. B.A. Camley, Y. Zhao, B. Li, H. Levine, W.J. Rappel, Phys. Rev. E **95**(1), 012401 (2017)
58. Y. Cao, R. Karmakar, E. Ghabache, E. Gutierrez, Y. Zhao, A. Groisman, H. Levine, B.A. Camley, W.J. Rappel, Soft Matter **15**(9), 2043 (2019)
59. Y. Cao, E. Ghabache, Y. Miao, C. Niman, H. Hakozaki, S.L. Reck-Peterson, P.N. Devreotes, W.J. Rappel, J. R. Soc. Interface **16**(161), 20190619 (2019)
60. V. Hakim, P. Silberzan, Rep. Prog. Phys. **80**(7), 076601 (2017)
61. R. Mayor, E. Theveneau, Development **140**(11), 2247 (2013)
62. J.A. Sherratt, J.C. Dallon, C.R. Biol. **325**(5), 557 (2002)
63. P. Friedl, D. Gilmour, Nat. Rev. Mol. Cell Biol. **10**(7), 445 (2009)
64. P. Friedl, J. Locker, E. Sahai, J.E. Segall, Nat. Cell Biol. **14**(8), 777 (2012)
65. T. Vicsek, A. Czirók, E. Ben-Jacob, I. Cohen, O. Shochet, Phys. Rev. Lett. **75**(6), 1226 (1995)
66. C.W. Reynolds, in *Proceedings of the 14th Annual Conference on Computer Graphics and Interactive Techniques* (1987), pp. 25–34
67. M.C. Marchetti, J.F. Joanny, S. Ramaswamy, T.B. Liverpool, J. Prost, M. Rao, R.A. Simha, Rev. Mod. Phys. **85**(3), 1143 (2013)
68. N.D. Mermin, H. Wagner, Phys. Rev. Lett. **17**(22), 1133 (1966)
69. J. Cardy, *Scaling and Renormalization in Statistical Physics*, vol. 5 (Cambridge University Press, Cambridge, 1996)
70. H. Chaté, F. Ginelli, G. Grégoire, F. Peruani, F. Raynaud, Eur. Phys. J. B **64**(3), 451 (2008)
71. F. Ginelli, Eur. Phys. J. Special Topics **225**(11), 2099 (2016)
72. E. Méhes, T. Vicsek, Integr. Biol. **6**(9), 831 (2014)

73. H. Levine, W.J. Rappel, I. Cohen, Phys. Rev. E **63**(1), 017101 (2000)
74. M. Basan, J. Elgeti, E. Hannezo, W.J. Rappel, H. Levine, Proc. Natl. Acad. Sci. **110**(7), 2452 (2013)
75. A. Czirók, E. Ben-Jacob, I. Cohen, T. Vicsek, Phys. Rev. E **54**(2), 1791 (1996)
76. Y. Katz, K. Tunstrøm, C.C. Ioannou, C. Huepe, I.D. Couzin, Proc. Natl. Acad. Sci. **108**(46), 18720 (2011)
77. K. Tanner, H. Mori, R. Mroue, A. Bruni-Cardoso, M.J. Bissell, Proc. Natl. Acad. Sci. **6**, 2012 (1973)
78. G. Malet-Engra, W. Yu, A. Oldani, J. Rey-Barroso, N.S. Gov, G. Scita, L. Dupré, Curr. Biol. **25**, 242 (2015)
79. A. Bianco, M. Poukkula, A. Cliffe, J. Mathieu, C.M. Luque, T.A. Fulga, P. Rørth, Nature **448**(7151), 362 (2007)
80. M. Abercrombie, Nature **281**(5729), 259 (1979)
81. R. Mayor, C. Carmona-Fontaine, Trends Cell Biol. **20**(6), 319 (2010)
82. B. Lin, T. Yin, Y.I. Wu, T. Inoue, A. Levchenko, Nat. Commun. **6**(1), 1 (2015)
83. E. Ben-Jacob, I. Cohen, H. Levine, Adv. Phys. **49**(4), 395 (2000)
84. T.S. Deisboeck, Z. Wang, P. Macklin, V. Cristini, Annu. Rev. Biomed. Eng. **13**, 127 (2011)
85. Z. Wang, J.D. Butner, R. Kerketta, V. Cristini, T.S. Deisboeck, in *Seminars in Cancer Biology*, vol. 30 (Elsevier, 2015), pp. 70–78
86. S.A. Sandersius, T.J. Newman, Phys. Biol. **5**(1), 015002 (2008)
87. B. Sabass, M.L. Gardel, C.M. Waterman, U.S. Schwarz, Biophys. J . **94**(1), 207 (2008)
88. P. Vallotton, G. Danuser, S. Bohnet, J.J. Meister, A.B. Verkhovsky, Mol. Biol. Cell **16**(3), 1223 (2005)
89. J. Zimmermann, B.A. Camley, W.J. Rappel, H. Levine, Proc. Natl. Acad. Sci. **113**(10), 2660 (2016)
90. J.H. Kim, X. Serra-Picamal, D.T. Tambe, E.H. Zhou, C.Y. Park, M. Sadati, J.A. Park, R. Krishnan, B. Gweon, E. Millet et al., Nat. Mater. **12**(9), 856 (2013)
91. A. Brugués, E. Anon, V. Conte, J.H. Veldhuis, M. Gupta, J. Colombelli, J.J. Muñoz, G.W. Brodland, X. Ladoux, X. Trepat, Nat. Phys. **10**(9), 683 (2014)
92. M. Poujade, E. Grasland-Mongrain, A. Hertzog, J. Jouanneau, P. Chavrier, B. Ladoux, A. Buguin, P. Silberzan, Proc. Natl. Acad. Sci. **104**(41), 15988 (2007)
93. Y. Yang, H. Levine, Soft Matter **14**(23), 4866 (2018)
94. D. Staple, R. Farhadifar, J.C. Röper, B. Aigouy, S. Eaton, F. Jülicher, European Physical Journal E **33**, 117 (2010)
95. X. Du, M. Osterfield, S.Y. Shvartsman, Phys. Biol. **11**(6), 066007 (2014)
96. T. Nagai, H. Honda, Philos. Mag. B **81**(7), 699 (2001)
97. D. Bi, J. Lopez, J. Schwarz, M. Manning, Nat. Phys. **11**, 1074 (2015)
98. D. Bi, X. Yang, M.C. Marchetti, M.L. Manning, Phys. Rev. X **6**(2), 021011 (2016)
99. A.J. Liu, S.R. Nagel, Annu. Rev. Condens. Matter Phys. **1**(1), 347 (2010)
100. J.A. Park, J.H. Kim, D. Bi, J.A. Mitchel, N.T. Qazvini, K. Tantisira, C.Y. Park, M. McGill, S.H. Kim, B. Gweon et al., Nat. Mater. **14**(10), 1040 (2015)
101. R. Farhadifar, J.C. Röper, B. Aigouy, S. Eaton, F. Jülicher, Curr. Biol. **17**(24), 2095 (2007)
102. S. Koride, A.J. Loza, S.X. Sun, APL Bioeng. **2**(3), 031906 (2018)
103. M.F. Staddon, D. Bi, A.P. Tabatabai, V. Ajeti, M.P. Murrell, S. Banerjee, PLoS Comput. Biol. **14**(10), e1006502 (2018)
104. F. Graner, J. Theor. Biol. **164**, 455 (1993)
105. G. Schaller, M. Meyer-Hermann, Phys. Rev. E **71**, 051910 (2005)
106. E. Teomy, D.A. Kessler, H. Levine, Phys. Rev. E **98**(4), 042418 (2018)
107. S.E. Leggett, Z.J. Neronha, D. Bhaskar, J.Y. Sim, T.M. Perdikari, I.Y. Wong, Proc. Natl. Acad. Sci. **116**(35), 17298 (2019)
108. M.H. Swat, G.L. Thomas, J.M. Belmonte, A. Shirinifard, D. Hmeljak, J.A. Glazier, Methods Cell Biol. **110**, 325 (2012)
109. B.A. Camley, Y. Zhang, Y. Zhao, B. Li, E. Ben-Jacob, H. Levine, W.J. Rappel, Proc. Natl. Acad. Sci. **111**(41), 14770 (2014)

110. F.J. Segerer, F. Thüroff, A.P. Alberola, E. Frey, J.O. Rädler, Phys. Rev. Lett. **114**(22), 228102 (2015)
111. S. Wang, J. Sun, Y. Xiao, Y. Lu, D.D. Zhang, P.K. Wong, Adv. Biosyst. **1**(1–2), 1600019 (2017)
112. N. Murisic, V. Hakim, I.G. Kevrekidis, S.Y. Shvartsman, B. Audoly, Biophys. J . **109**(1), 154 (2015)
113. J. Toner, Y. Tu, Phys. Rev. E **58**(4), 4828 (1998)
114. J. Toner, Y. Tu, S. Ramaswamy, Ann. Phys. **318**(1), 170 (2005)
115. S. Banerjee, K.J.C. Utuje, M.C. Marchetti, Phys. Rev. Lett. **114**(22), 228101 (2015)
116. M. Basan, T. Risler, J.F. Joanny, X. Sastre-Garau, J. Prost, HFSP J. **3**(4), 265 (2009)
117. M. Basan, J.F. Joanny, J. Prost, T. Risler, Phys. Rev. Lett. **106**(15), 158101 (2011)
118. Y. Yang, H. Levine, Phys. Biol. **17**(4), 046003 (2020)
119. L. Petitjean, M. Reffay, E. Grasland-Mongrain, M. Poujade, B. Ladoux, A. Buguin, P. Silberzan, Biophys. J . **98**(9), 1790 (2010)
120. J.J. Williamson, G. Salbreux, Phys. Rev. Lett. **121**(23), 238102 (2018)
121. R. Alert, C. Blanch-Mercader, J. Casademunt, Phys. Rev. Lett. **122**(8), 088104 (2019)

Chapter 8
Modeling Biological Information Processing Networks

Xiao Gan and Réka Albert

Abstract Higher-level functions of complex biological systems are emergent properties that arise from the totality of lower-level elements and interactions. Network models of these systems can provide valuable insight into how the underlying lower-level interactions lead to higher-level emergent properties, and can help predict not-yet-characterized behaviors. This chapter describes the methodologies of network analysis and network-based discrete dynamic modeling and exemplifies them in the context of within-cell information processing networks and their determination of cellular behaviors. In addition to the specific predictions offered by models of individual systems, general insights can be gained by an expanded network representation that integrates the network structure and regulatory logic. This expanded network reveals the connectivity patterns that underlie the system's functional repertoire, and enables the characterization of their stability and control.

8.1 Introduction

Interacting systems abound at every level of biological organization (molecular, cellular, organ, organism, or population). For example, molecular interacting systems consist of genes, their transcripts (mRNAs), proteins, and small molecules; their interactions include gene transcription, protein translation, protein-protein interactions, and chemical reactions. A fundamental goal of biology is to understand why biological systems behave the way they do. One promising avenue toward this goal is to realize that interacting biological systems at each level can determine the behavior at the next level. For example, cellular decisions, behaviors, and phenotypes arise from the interactions of numerous molecular components. Similarly, interactions among cells determine how multi-cellular organisms develop and how tissues and organs

X. Gan (✉) · R. Albert
Pennsylvania State University, State College, PA, USA
e-mail: xxg114@psu.edu

R. Albert
e-mail: rza1@psu.edu

© Springer Nature Switzerland AG 2022
K. B. Blagoev and H. Levine (eds.), *Physics of Molecular and Cellular Processes*,
Graduate Texts in Physics, https://doi.org/10.1007/978-3-030-98606-3_8

function; interactions among individuals form the basis of social communities, and interactions among species underlie ecological communities.

A higher-level function, behavior, or phenotype is an emergent property that arises from the totality of the lower-level elements and interactions. That is, one usually cannot attribute a cell behavior to a single gene or protein. This does not necessarily mean, however, that all the elements and interactions are equally important in determining a higher-level behavior. Biological networks offer a visual and effective way to represent the lower-level elements and interactions; the analysis of these networks is a key step toward the elucidation of higher-level emergent properties. Specifically, network analysis and network-based dynamic modeling can be used to determine the repertoire of cellular behaviors associated with a within-cell network, and to identify the sub-networks that play a key role in the cell adopting a certain behavior.

Another aspect of understanding biology is that, despite the vast amounts of recent information about regulatory relationships among genes, proteins, and small molecules, many knowledge gaps still exist. Networks and network-based modeling can integrate fragmentary and qualitative interaction information, and can make powerful predictions about undiscovered biology.

In this chapter, we focus on the application of network-related methods and techniques in understanding biology. A variety of networks can be defined at the cellular, organismal, and ecosystem levels. The examples in this chapter will focus on the molecular to the cellular level. We aim to illustrate how to connect the properties of within-cell information processing networks to cellular phenotypes. We start by introducing network concepts and measures such as paths, cycles, centrality measures, strongly connected components, network motifs, and their biological interpretation. Then we introduce network-based dynamic modeling, which offers in-depth insights into dynamical processes and the effect of perturbations. We describe the construction of dynamic models, and demonstrate their predictive power through two examples. We also introduce methodologies that reveal structure-dynamics connections through the construction of so-called expanded networks.

8.2   Representing Biological Networks and Analyzing their Topology

A network (also called graph) consists of nodes (also called vertices) and edges that connect pairs of nodes. In a biological network, nodes represent biological elements, for example, proteins and molecules in a cell signaling process; edges represent interactions or regulatory effects between these elements. The edges of a network may be directed or undirected. An undirected edge connects a node pair without order, that is, edge (x, y) is identical to edge (y, x). For a directed edge, the order of the node pair matters: a directed edge (x, y) starts from x and ends in y. One can refer to x as the head or source of the edge and refer to y as the tail or target of the edge; y is also said to be a direct successor of x and x is said to be a direct predecessor of y. Edges can also be characterized by positive or negative signs. In biological networks, the sign of an edge represents the effect of the regulation. A positive edge stands for positive regulation (i.e. activation); a negative edge stands for negative regulation (i.e. inhibition). Biological networks are often directed and signed; this way the network is a reflection of the flow of mass and information in the system. During the construction of the network, certain nodes may be designated as markers or proxies for higher-level behavior. For example, certain genes or proteins can be used as markers of cell types (as it is also done in experimental investigation), and abstract nodes can be added as proxies of the phenotypic outcome of a signal transduction network.

We use as the first illustration a signal transduction network inside plant guard cells. Guard cells border the stomata, which are pores on leaf surfaces that allow the plant to exchange carbon dioxide (CO_2) and oxygen with the atmosphere. The shape change of the guard cells determines stomatal opening (increased aperture) or closure. This shape change is elicited by environmental signals, including light of different wavelength, CO_2 concentration, as well as internal signals (hormones) such as abscisic acid (ABA). Thus, the within-guard cell signal transduction network can be defined as the elements and interactions that respond to the external and internal signals and yield stomatal opening (or closure). Sun et al. constructed a signaling network of light-induced stomatal opening, which contained more than 70 nodes and 150 directed and signed edges [2]. Figure 8.1 shows a reduced version of this network, with 32 nodes, including the outcome node Stomatal Opening, and 81 edges [1].

The organizational features of a network reflect the properties that are critical for emergent behavior. One way to connect the micro-scale (node) properties to the macro-scale (network) properties is to look at the connectivity patterns of a network. First, one can analyze the patterns of how the edges are distributed among nodes. A local measure of this is the node degree, which in directed networks can be separated into in- and out-degree. The in-degree of a node is the number of its incoming edges; the out-degree of a node is the number of its outgoing edges. Nodes with a high in-degree have many regulators and nodes with a high out-degree regulate many other nodes. A node can also have a high degree (sum of in-and out-degree) by having intermediary values of in- and out-degree. All of these types of high-degree nodes (also called hub nodes) have biological meanings.

A node with only outgoing edges and no incoming edges is called a source node. These nodes represent external signals. A node with only incoming edges and no outgoing edges is called a sink node; these nodes represent the outcomes of the network. In the reduced stomatal opening network, there are four source nodes, each representing a signal, namely CO_2, Blue Light, Red Light, and ABA. There is a single sink node, Stomatal Opening. The highest-degree (hub) nodes include AnionCh (referring to multiple anion channels) with in-degree 6 and out-degree 2, and $[Ca^{2+}]_c$ (cytosolic Ca^{2+} concentration) with in-degree 4 and out-degree 7.

In order to characterize the flow of information from source nodes (signals) to sink nodes (outcomes), we can use the concept of path. A path is a sequence of distinct nodes in which each node is adjacent (connected by an edge) to the next

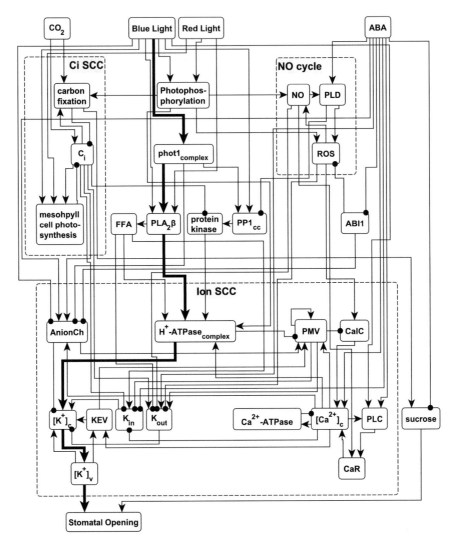

Fig. 8.1 Signal transduction network corresponding to the process of stomatal opening in plants, adapted from [1]. This network has 32 nodes and 81 directed edges. Arrows represent positive edges, and terminal-filled circles represent negative edges. The network contains three strongly connected components (SCCs), marked with dotted lines. The thick edges indicate a path from the source node "Blue light" to the sink node "Stomatal Opening"

one. In a directed network, a path needs to follow the direction of the edge(s). If there is a path from node A to node B, B is said to be reachable from A, meaning that information may be transmitted from A to B. For example, the thick edges in the reduced stomatal opening network form a path from the input signal Blue Light to the outcome node Stomatal Opening. There is no path from Stomatal Opening to Blue light, meaning that the signal transduction has a single direction. All the edge signs in the Blue Light to Stomatal Opening path are positive, making the path positive. Another way in which a path is positive is if it contains an even number of negative edges. Conversely, a path is negative if it contains an odd number of negative edges. The indirect connection between two nodes can be characterized by the distance between them, defined as the number of edges along the shortest path connecting them. The thick path shown above has a length of 6. However, it is not the shortest path from Blue Light to Stomatal Opening. The shortest path has four edges, thus the distance between these two nodes is 4. We leave it to the reader to find which path(s) is(are) the shortest.

A pair of nodes can be connected by multiple paths (as we have seen for Blue Light and Stomatal Opening). If all these paths have the same sign, the regulatory relationship between the two nodes can be unambiguously characterized as positive or negative. A network wherein the relationship between all pairs of nodes is unambiguous is called sign-consistent (or structurally balanced). It is also possible that paths of both signs exist between a pair of nodes, making their relationship ambiguous. This ambiguity can be resolved by additional, dynamic information (which will be described later). Finally, it is possible that two nodes are disconnected (there are no paths between them); these nodes do not influence each other. Considering connectivity at the network level, an undirected network is said to be connected if there is a path between any pair of nodes. A directed network is strongly connected if there is a path between any ordered pair of nodes (i.e. there is a path from node A to node B and another path from node B to node A). On the other hand, a directed network is weakly connected if the undirected network that results when disregarding edge directions is connected.

A special type of path is the cycle: it starts and ends at the same node and does not revisit any nodes. Another way to refer to a directed cycle is a feedback loop. For example, in the stomatal opening network, the nodes NO, PLD, and ROS form the NO cycle, circumscribed by the dotted rectangle in Fig. 8.1. In a directed and signed network, one can define the sign of a cycle (feedback loop) depending on whether the number of negative (inhibitory) edges is odd or even. If a feedback loop has an even number of negative edges, it is a positive feedback loop. Thus, mutual inhibition between two nodes is an example of a positive feedback loop. If a directed cycle has an odd number of negative edges, it is a negative feedback loop. The NO cycle is a positive feedback loop. In the stomatal opening network, there is a negative feedback loop between Ca_c^{2+} and the node Ca^{2+} ATPase, which represents the pumps and transport mechanisms aiming to prevent a sustained high cytosolic Ca^{2+} concentration, which would be detrimental to the cell. The sign of feedback loops has significant meaning in predicting emergent properties of a network: positive feedback loops are necessary for multi-stability, and negative feedback loops are necessary for sustained oscillations [3]. We will talk about dynamics in detail in the next section.

With the above network metrics introduced, one can evaluate node centrality, which identifies the most important node(s) within a graph. Centrality can be measured in various ways, depending on the context of a network. The most straightforward measure is degree centrality, i.e. nodes with the highest degrees are defined as most important. Another common centrality measure is betweenness centrality, defined as the (normalized) number of shortest paths that pass through the node. Betweenness centrality can complement degree centrality by capturing the importance of "bridge nodes", which have low node degree, but mediate many shortest paths between other nodes. In biological networks, degree centrality and betweenness centrality often point to similar nodes. For example, in the stomatal opening network, node $[Ca^{2+}]_c$ (cytosolic Ca^{2+} concentration) has both the highest degree centrality and the highest betweenness centrality.

Connectivity pattern is an important property of a network. The simplest yet most important connectivity pattern of a network is its strongly connected component (SCC). An SCC is a sub-network in which every node is reachable from every other node. As each SCC is made up of cycles, it can serve as an information processing, decision-making unit [4]. For a strongly connected component, one can identify its in-component as the nodes that can reach the SCC (but cannot be reached from the SCC), and its out-component as the nodes that can be reached from the SCC (but cannot reach the SCC). These components often have functional interpretations. For example, many biological networks contain a dominant SCC. The in-component of this SCC contains the signal(s) and its out-component contains the outcome(s); most of the paths from signal(s) to outcome(s) pass through the SCC. In the stomatal opening network, there are three strongly connected components, namely the C_i SCC, the NO cycle, and the ion SCC, as shown in Fig. 8.1. The ion SCC is the dominant SCC. Its in-component includes the four signals, the other two smaller SCCs, and 7 other nodes (i.e. all nodes above the ion SCC in the figure). Its out-component is a single node, Stomatal Opening. The node sucrose is neither in the in-component nor in the out-component of the ion SCC.

Network motifs are another important connectivity pattern. The term "network motif" does not have a universal definition, but instead refers to a small sub-network that is of specific interest. For example, a common definition of network motif is a sub-network that occurs much more frequently in a biological network than in randomized versions (null models) of the network. Small feedforward loops (such as the sub-network formed by Blue Light, phot1$_{complex}$, and PLA2β in the stomata opening network) and feedback loops (such as the sub-network formed by NO, PLD, and ROS) were found to be over-represented in biological networks. These network motifs were proposed to serve as functional units of the larger network [5]. Feedforward loops enable specific dynamic behaviors such as noise filtering, while feedback loops enable bistability. Later in this chapter, we will introduce the related concept of stable motifs, which are a type of positive feedback loop. Stable motifs are identified based on a dynamic model and determine the dynamical repertoire of the system.

A variety of software for network visualization and analysis exists. For example, yEd excels in visualizing mid-size networks using a number of effective layouts. Cytoscape is an open-source software platform for visualizing molecular interaction networks and integrating these networks with multiple types of data [6]. NetworkX is a Python package for the creation and analysis of complex networks [7].

8.3 Dynamic Modeling

Networks defined in the previous section indicate which biological entities interact with and regulate each other, but do not provide details about the results of multiple regulatory relationships that are incident on the same node. This is especially problematic if the network is not sign-consistent, meaning that the regulatory relationships of a subset of the nodes are ambiguous. For example, in the Stomatal Opening network, the node CO_2 positively and directly regulates C_i, but it also has an indirect negative effect on C_i via carbon fixation. If one wants to evaluate the aggregated result of multiple interactions like this, one must consider the temporal and quantitative aspects of information propagation on the network. This is done by network-based dynamic modeling.

After a network is established, one associates each node with a variable to represent its state. For example, if the node represents a protein in a cell signaling network, the state variable can represent this protein's concentration or activation level; if a node represents a species in a food web, the state variable can represent the population of this species. Then one constructs a regulatory function for this variable, based on the regulators of the node indicated in the network. In this way, information (realized as a state change of a node) propagates through the network. Each node's state will evolve over time, eventually converging into a long-term behavior such as a steady state or a sustained oscillation. The phenotype of the system can then be characterized by the long-term states of all nodes, or of a subset of the nodes. For example, if there is a sink node that represents a phenotypic outcome, the long-term state of this node may be a sufficient proxy to describe the whole system.

To construct a dynamic model, the modeler needs to start by identifying the process to be modeled, which will specify the signals and outcomes of the network. Next is to identify additional nodes involved in the process, and the interactions among them. Experimental interaction data, such as physical interactions, chemical reactions, post-translational modifications, and causal effects of knockouts, are used in this process. Then the regulatory function for each node needs to be determined, and is usually parameterized using experimental interaction data. In the vast majority of cases, there isn't enough information to fully characterize and parameterize each regulatory function.

Once a model is established, one can validate it by simulating the model and comparing the results with experimental data. A simulation starts at an initial state that represents the resting (pre-stimulus) status of the system, and it identifies the consecutive states by applying the regulatory functions. The simulation result should agree with the experimentally known response of the system to the signal(s). Intervention or perturbation scenarios can also be simulated and analyzed. Comparing the model's results with existing experimental results in these scenarios is an additional test of the model. If there are discrepancies, one or more regulatory functions need to be adjusted until a reasonable percentage of simulations is consistent with experiments. This adjustment process decreases the uncertainty of the regulatory functions.

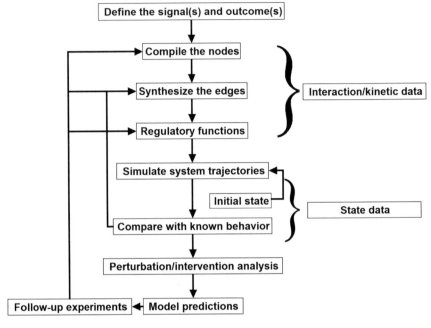

Fig. 8.2 Flowchart of the main steps of constructing and analyzing a dynamic model of a signal transduction network. The key to the construction and validation of the model is experimental data. Different types of data are used for model construction and model validation: interaction data and initial state data are used as inputs; and time-course or long-term state data are used for model validation. This separation of information helps avoid overfitting

After the model is validated, it can be used to make predictions about situations that were not studied before. For example, the model can identify key nodes, whose perturbation disrupts a certain behavior of the system. If this behavior is undesired (e.g. it represents uncontrolled growth of cancer cells), these key nodes serve as intervention targets. The model's predictions should be tested by follow-up experiments, which may confirm the predictions or contradict them. Both cases represent a gain of knowledge. An invalidated prediction spurs further revisions and increased certainty of the regulatory functions. Figure 8.2 presents a flowchart of the modeling process.

There are multiple frameworks for dynamic modeling, categorized by the type of their state variables, the type of their time variable, or by the incorporation of stochasticity in the model. For example, continuous models follow continuous-valued variables in continuous time, and the regulatory functions describe the rate of change of the state variables by differential equations. In discrete modeling, the state variables are discrete, with regulatory functions that indicate the value of the state variables after a time delay (which usually is given as a multiple of a discrete time unit). The major advantage of discrete modeling is that it can reflect the functional repertoire of a biological system without the need for a lot of kinetic information. If one wants to construct a continuous regulatory function as part of a differential equation model,

Boolean truth table		
A* = B or C		
B	C	A*
0	0	0
1	0	1
0	1	1
1	1	1

Multi-level truth table		
A* = B + C		
B	C	A*
0	0	0
1	0	1
0	1	1
1	1	2

Fig. 8.3 Example of Boolean and multi-level regulatory functions in truth table representation. A truth table is generated by enumerating all input combinations and indicating the corresponding outputs. The output of the regulatory function will become the next state of the target node. Here and throughout the chapter, we represent the next state of node A as A*

few parameter values are known beforehand, and one needs to estimate a lot of parameters by fitting experimental data. However, such experimental data is difficult to obtain, and much less such data is available than what would be enough for the construction and validation of continuous models. On the other hand, discrete models, especially Boolean models, require a minimal number of parameters, and are shown to be useful in biological modeling, especially in modeling large systems [8–12]. This chapter will focus on discrete dynamic modeling using discrete time.

The simplest discrete dynamic model is the Boolean model, where each node can only employ two states. The 0 (OFF) state refers to a concentration or activity insufficient (below threshold) to initiate downstream processes; conversely, the 1 (ON) state represents a sufficient (above-threshold) concentration or activity. The Boolean regulatory functions can be expressed with the Boolean operators "AND", "OR", and "NOT"; they can also be expressed as truth tables. When a node has more than two levels, the 0 state refers to inactivity, and different levels of activity are usually represented by positive integers, e.g. 1, 2, until the maximal activity. The choice of number of levels to use is determined by experimental evidence (e.g. if there are different outcomes when a node has an intermediate activity compared to when it is fully active). For example, the stomatal opening model has more than two levels for about one-third of the nodes, informed by observations of additive or synergistic relationships between blue and red light in regulating nodes of the network. The rest of the nodes, for which no such evidence exists, are binary. The regulatory functions of multi-level discrete dynamic models can be expressed in multiple ways [9, 13], including a truth table, as shown in Fig. 8.3.

Discrete models use different implementations of time evolution, called update schemes. A synchronous update scheme is where all nodes are evaluated at once, and each node will take its regulatory function-given value as its state in the next time step (e.g. the next state of node A, denoted as A*, is given in the last column of the truth table in Fig. 8.3). This update scheme is realistic if the synthesis and decay processes of each node are the same; for example, this may apply to certain gene regulatory networks, as the timing of gene transcription and mRNA degradation is similar, on the order of minutes. Asynchronous update schemes allow different nodes to update at different rates, which is necessary for networks that include both pre- and post-translational events. There are many ways to implement an asynchronous update. Some are deterministic, for example, updating nodes according to a fixed order; others are stochastic, for example, in general asynchronous update, at each time step a randomly chosen node is updated.

Given an initial state and an update scheme, the system's state will eventually evolve into an attractor. An attractor is a minimal set of states of the system, from which only states in the same set can be reached. The simplest attractor, called a fixed point, consists of a single state. This state is also referred to as a steady state (in analogy with continuous models). Attractors consisting of more than one state, which the system keeps revisiting, are called complex attractors or oscillating attractors.

The evolution of a system can be effectively summarized into a state transition graph (STG), whose nodes are the states of the system, and whose edges represent allowed state transitions. The state transition graph indicates fixed point attractors as sink states and complex attractors as terminal SCCs (SCCs that do not have any successor nodes). The intuition of this is simple: if the system gets into an attractor, it cannot escape from it as there are no outgoing state transitions. Since discrete dynamic models of biological systems have a finite number of nodes and a finite number of states, the system will eventually evolve into an attractor, and then stay in this attractor unless disrupted by a change in external signals or an internal perturbation. The biological significance of this is that the attractors represent biological phenotypes. For example, in the stomatal opening model, one attractor represents open stomata, while another attractor represents closed stomata.

State transitions depend on the update scheme. For example, in synchronous update, one state can only transit into one state, i.e. each state has one and only one outgoing edge in the STG, while in some stochastic asynchronous update schemes one state can transit into different states. This means that the attractors that involve state transitions, i.e. complex attractors, will depend on the update scheme, too. Fixed points are the same under different update schemes. Figure 8.4 demonstrates an example where a Boolean model's complex attractor depends on the update scheme.

In order to determine the complete repertoire of dynamic trajectories of a network-based model, one needs to identify all possible state transitions. This is computationally challenging, as the number of states increases exponentially with the number of nodes (e.g. 2^N for a Boolean network with N nodes). An effective way to reduce the state space is network reduction; of course, this reduction needs to preserve the dynamic repertoire of the system. Two types of nodes can be reduced (eliminated or merged): source nodes that have a sustained state, and simple mediator nodes that have one incoming and/or one outgoing edge. In the reduction, the source node's state is directly plugged into the regulatory function of all of its direct successor nodes; then the source node is eliminated. For a simple mediator node with one direct predecessor (regulator) and/or one direct successor (target), its regulator is connected directly to its target, and the mediator node is merged into the regulator. This reduction method is proven to conserve attractors [14, 15].

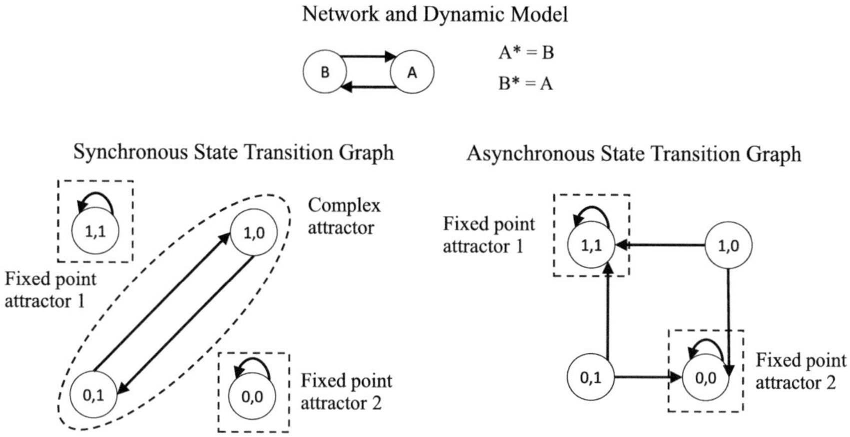

Network and Dynamic Model

$$A^* = B$$
$$B^* = A$$

Synchronous State Transition Graph   **Asynchronous State Transition Graph**

Fig. 8.4 Example of a toy Boolean network model and its dynamics under synchronous update (when both nodes are updated simultaneously) and under general asynchronous update (when one node is updated at each time). The dynamics of the model is represented by a state transition graph (STG), in which system states are represented by nodes and state transitions are represented by edges. Terminal strongly connected components (including nodes with only a self-loop) in an STG are attractors of the system. This model exemplifies that complex attractors may depend on update schemes. Specifically, under synchronous update, there is a complex attractor formed by two states that differ in the values of both nodes. As state transitions that change the value of two nodes are not possible under general asynchronous update, this complex attractor disappears under general asynchronous update

A variety of software exists to facilitate discrete dynamical modeling. Model and software development efforts are coordinated by the Consortium of Logical Models and Tools (CoLoMoTo), an international open community that aims to develop standards for model representation and interchange, establish criteria for the comparison of methods, models, and tools, and promote these methods, tools, and models [16]. CoLoMoTo members have developed the Qualitative Models Package ("qual") of the Systems Biology Markup Language (SBML) [17]. The Cell Collective is a web-based platform that enables collective model construction and real-time model simulation [18]; GINsim allows asynchronous and/or multi-level dynamics and STG construction [19]. The Python library BooleanNet allows simulation of Boolean models with different update schemes [20]; SimBoolNet is a Cytoscape app that benefits from the functionalities and friendly graphic user interface of Cytoscape [21]; the R package BoolNet can construct and simulate Boolean models and analyze attractors using exhaustive or heuristic search methods [22].

In the following, we present two published models to demonstrate the power of the dynamical modeling of biological networks.

8.3.1 Modeling T Cell Survival

This model reflects the survival and proliferation of cytotoxic T cells in the context of the disease T-LGL leukemia (Fig. 8.5). Cytotoxic T cells are generated to fight infection by eliminating infected cells. After the infection is over, they usually undergo the process of activation-induced cell death. However, in T-LGL leukemia they survive, adopt a cell state different both from resting and from activated T cells, and start attacking healthy cells. Zhang et al. synthesized the pathways involved in activation-induced cell death, cell proliferation, as well as the pathways that are known to be different in T-LGL cells compared to normal cytotoxic T cells [23]. They formulated a Boolean model of the process and simulated its trajectories, starting from a just-stimulated T cell, using stochastic timing. The model reproduces the survival of a fraction of the initial stimulated cells and the known markers of this process, for example, the activation of JAK in every surviving cell. The model has two fixed points: the normal fixed point that corresponds to programmed cell death, and the disease fixed point that reproduces the T-LGL survival state. The model predicts that a small subset of the known deregulations (abnormal node states) is sufficient to cause all the others, thus preventative efforts should focus on this subset. The model predicts 12 additional nodes whose states stabilize in the T-LGL state. The model also predicts several key nodes whose state changes can ensure the apoptosis of the whole population; these key nodes are potential therapeutic targets for T-LGL leukemia. Several of these predictions have been verified experimentally.

In a follow-up project, Saadatpour et al. reduced the system to 6 nodes (in a way that preserves the attractor repertoire) and determined its state transition graph [24]. They found that the basin of attraction of the normal fixed point is larger than the basin of the T-LGL fixed point, but there is a significant overlap between the basins, meaning that there exist states from which certain trajectories lead to the normal fixed point and other trajectories lead to the T-LGL fixed point, depending on the order of events. They also performed a systematic single-node perturbation analysis starting from the T-LGL state, wherein a node is driven into and maintained in the state opposite of its state in the T-LGL survival state. They found that the perturbation of any one of 19 nodes leads to the disappearance of the T-LGL attractor, meaning that the only possible long-term outcome is apoptosis. Thus, these 19 nodes are potential therapeutic targets, whose control (knockout or constitutive activity) leads to apoptosis of the T-LGL cells. The majority (68%) of these predictions are corroborated by experimental evidence; the rest have not yet been assessed. This work illustrates how network-based modeling can be used for predictions that can potentially lead to identifying new therapeutic targets.

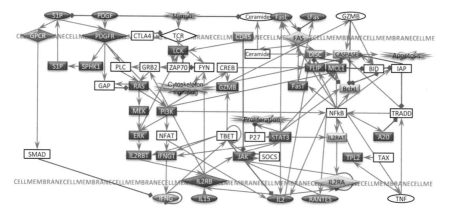

Fig. 8.5 T-LGL survival signaling network by Zhang et al., reproduced from [23], copyright (2008) National Academy of Sciences, U.S.A. The network contains 58 nodes and 123 edges. Up-regulated or constitutively active nodes are in red, down-regulated or inhibited nodes are in green, nodes that have been suggested to be deregulated (either by up-regulation or down-regulation) are in blue, and the states of white nodes are unknown or unchanged compared with normal. Blue edges with arrowheads indicate activation and red edges that terminate in diamonds indicate inhibition. The shape of the nodes indicates the cellular location of the corresponding proteins, transcripts, or molecules: rectangles indicate intracellular components, ellipses indicate extracellular components, and diamonds indicate receptors. Conceptual nodes (Stimuli, Cytoskeleton signaling, Proliferation, and Apoptosis) are orange

8.3.2 Modeling Epithelial to Mesenchymal Transition (EMT)

The epithelial to mesenchymal transition (EMT) is the process where epithelial cells lose their cell polarity and cell-cell adhesion, and gain migratory and invasive properties, to ultimately become mesenchymal cells. The loss of the expression of the protein E-cadherin is considered the hallmark of the EMT transition. This cell fate change is beneficial during embryonic development and wound healing, but it also is the first step of cancer metastasis. Steinway et al. constructed a signal transduction network and Boolean model of this process [25], see Fig. (8.6). The model uses stochastic update with separate update probabilities (and thus separate time-scales) for nodes regulated at the protein and mRNA levels.

Simulations of the model start from the epithelial state, after which a sustained input signal, TGFβ, is provided. During the simulation, most nodes in the model change states, and the system converges into a fixed point attractor that recapitulates the mesenchymal state, including the inactivity of E-cadherin. The model reproduces known molecular markers of the transition and captures the importance of known key mediators, for example, the transcription factors that down-regulate the E-cadherin mRNA. The model also predicts that several pathways which were previously thought to be independent of TGFβ are also activated through the process. In the sustained presence of TGFβ, the EMT network can be simplified to 16 nodes, which enables the determination of a state transition graph (STG). Model simulations and the STG

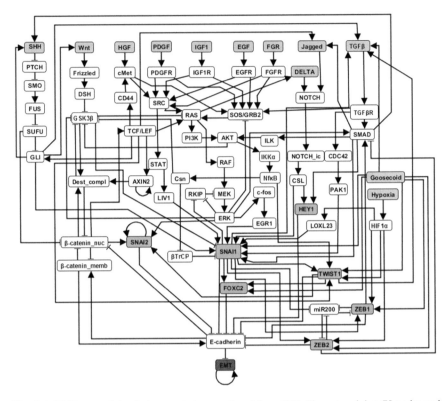

Fig. 8.6 EMT network by Steinway et al., reproduced from [25]. The network has 70 nodes and 135 edges. Nodes that represent extracellular signals are shown in blue, green nodes are transcription factors, and the single output node EMT is shown in red. Multiple molecules that serve as extracellular signals are also produced by the cell, thus these nodes have incoming edges

both indicate that despite the timing (update) stochasticity, all the trajectories end in the mesenchymal state, indicating that the EMT transition is a robust process. Based on the model, the authors predicted interventions that can block the transition, and validated several of these predictions experimentally [26]. This work is important because EMT is the first step of cancer metastasis, so therapies that block it have high clinical potential.

8.4 Integration of the Interaction Network and Regulatory Rules

As we have seen in the previous section, the determination of the attractor repertoire of a dynamical system, and of the ways in which this attractor repertoire changes in response to perturbations and interventions, is a key step in connecting molecular interaction networks with cellular behaviors. One of the methods to determine the attractor repertoire is to use the state transition graph, which contains all the trajectories of the system. However, the STG can have an enormous number of nodes and edges if the biological system is large. An alternative way to determine the attractor repertoire of a system is to exploit the connectivity patterns of the network. Indeed, it has been shown in multiple dynamic frameworks, including discrete dynamic systems, that positive feedback loops are necessary for multi-stability, while negative feedback loops are necessary for sustained oscillations [3, 27–29]. A recently proposed family of methods to connect structural and dynamic analysis is based on integrating the signal transduction network with its regulatory functions, into an expanded network. By using this approach, one can determine elementary and independent signal transduction pathways, find centers of stability in the network, reveal the attractor repertoire, and drive the system into beneficial attractors or away from undesired ones [30–33].

The regulatory logic is integrated into the signaling network in two steps. First, one creates a virtual node for each state of a node. This virtual node will be Boolean, with the 1 (True) value indicating that the original node is in this state and the 0 (False) value indicating that the original node is not in this state. One can construct this virtual node's regulatory function by summarizing the corresponding input combinations. In the Boolean case, the virtual nodes' regulatory functions can be straightforwardly obtained from the original node's regulatory function. For example, a Boolean function $A* = $ not B will now be represented as two functions, $A1* = B0$ and $A0* = B1$. For multi-level models, the regulatory functions can be constructed from the truth table by summing up the corresponding input combinations. For example, the virtual nodes and regulatory functions of the multi-level truth table of Fig. 8.3 are $A0* = B0$ and $C0$, $A1* = B1$ and $C0$ or $B0$ and $C1$, $A2* = B1$ and $C1$. The resulting regulatory functions are in a Boolean disjunctive form [34]. Second, one eliminates AND/OR ambiguity by representing each "AND" clause with a composite node. The nodes in the clause will have edges pointing to the composite node, and the composite node will have an edge pointing to the regulated node. This expanded network contains positive edges only, and explicitly identifies interactions of a combinatorial nature. Examples of expanded network construction for both Boolean and multi-level functions are shown in Fig. 8.7.

The expanded network makes it easy to identify a sufficient condition to activate a virtual node (i.e. to make the original node attain the state represented by the virtual node): a virtual node will have state 1 if any of its regulator virtual nodes has state 1, or if any of its regulator composite nodes has all its input virtual nodes in state 1. If either of these conditions is satisfied, the target virtual node will have state 1, regardless of the state of other regulators of the target node. Following this intuition to more distant virtual nodes, one can see that a path or subgraph in the expanded network satisfying the above criterion allows signal propagation from the first node of the path/subgraph to the last node of the path/subgraph, independent of other nodes; and a cycle in the expanded network satisfying the above criterion will be self-sufficient to stabilize.

Boolean:

A0* = B1 and C0

A1* = B0 or (C1 and B1)

Multi-level:

A0* = B0 or (C1 and B1)

A1* = B1 and C0

A2* = B2

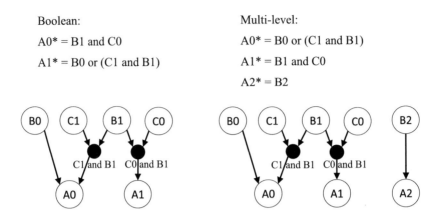

Fig. 8.7 Examples of expanded network construction in the Boolean and multi-level case. Each virtual node is labeled with the state it represents. Each composite node is black, with a label indicating which node combination it represents. The complete expanded network is obtained by expanding all regulatory functions of the original model

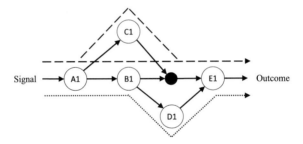

Fig. 8.8 Example of elementary signaling modes (ESMs) in a partially expanded network. The labeled virtual nodes correspond to the ON state of the respective nodes in the original signal transduction network; the black node is a composite node. There are two ESMs in the network: the path A1 B1 D1 E1, shown as the dotted line, and the subgraph that contains A1, C1, B1, the composite node, and E1, shown with a dashed line. Each is sufficient for the signal to activate the outcome. This figure was adapted from [30]

Connectivity patterns of the expanded network lead to the definition of elementary signal modes and stable motifs, which reveal important dynamical properties of the system. An elementary signaling mode (ESM) is defined as a minimal set of components that can perform signal transduction from signals (source nodes) to outcome nodes (proxies for cellular responses) [30, 33]. A key property of an elementary signaling mode is that if it includes a composite node, it must include all the regulators of the composite node as well (see Fig. 8.8). There are many applications of the ESMs: one can evaluate the importance of signaling components by the effect of their perturbation on the ESMs of the network; the number of node-independent elementary signaling modes also shows the redundancy of a network. In many signaling networks, the number of node-independent elementary signaling modes is

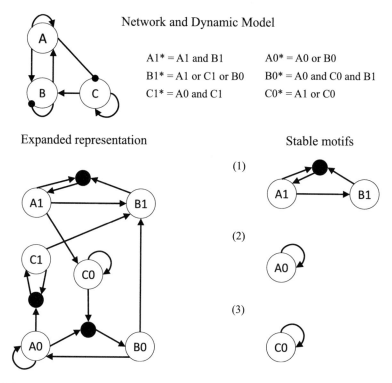

Network and Dynamic Model

$A1^* = A1$ and $B1$ $A0^* = A0$ or $B0$

$B1^* = A1$ or $C1$ or $B0$ $B0^* = A0$ and $C0$ and $B1$

$C1^* = A0$ and $C1$ $C0^* = A1$ or $C0$

Expanded representation   Stable motifs

(1)

(2)

(3)

Fig. 8.9 Example of stable motif identification in a three-node Boolean dynamic model. The regulatory functions of the virtual nodes are given. The black nodes in the expanded network are composite nodes. Three stable motifs can be identified from the expanded representation of the network. The first stable motif represents the simultaneous activation (state 1) of nodes A and B. The second and third stable motifs represent the sustained inactivation (state 0) of A and C, respectively. Notice that a stable motif corresponds to a positive feedback loop (or SCC) in the original network, but not all positive feedback loops are stable motifs

one, meaning that there is no more than one independent modality of signaling, and the loss of a single node can disrupt signaling.

A stable motif is defined as a strongly connected component in the expanded network that satisfies two criteria: 1. It does not contain multiple virtual nodes that correspond to the same original node; 2. If it contains composite nodes, it also contains these nodes' inputs [31, 34]. Such definition guarantees that a stable motif is a self-sufficient cycle, so that it can stabilize on its own, regardless of the rest of the network. Figure 8.9 is an example of stable motif identification in a three-node model. It is important to note that a stable motif is both a network motif and an associated state, encoded in the names of the virtual nodes that form the stable motif. For example, the first stable motif in Fig. 8.9 indicates that the positive feedback loop between A and B is sufficient to sustain both nodes in the ON state. Stable motifs are centers of stability in the system and have a one-to-one correspondence to the partially fixed points of the system. Specifically, each stable motif determines a partial fixed point in which the nodes of the stable motif, and potentially additional nodes, stabilize. Conversely, each partial fixed point (i.e. fixed state of a subset of the nodes) corresponds to one or more stable motif(s).

This one-to-one correspondence indicates that identifying stable motifs is enough to determine the stabilized part of any attractor of the system. A node must either stabilize or oscillate in an attractor. Since stabilized nodes are associated with stable motifs, the nodes not associated with stable motifs must be oscillating or influenced by an oscillating regulator. In this way, one can identify the attractors of the system by finding stable motifs. The main advantage of this method is that it allows the identification of all attractors without enumerating the entire state space. As the size of the expanded network is smaller than the size of the state space, stable-motif-based attractor identification is more efficient computationally than state-space-based attractor identification.

The implementation of the attractor identification is an iterative network reduction based on stabilized components. The idea is simple: if a node is known to stabilize, one can plug its state into the regulatory functions of its direct successors and eliminate the node. Similarly, after identifying a stable motif, one can plug in the corresponding states, identify additional stabilized nodes, and reduce them until no more nodes stabilize. After each step of reduction, new stable motifs may be found and can be plugged in. If at the end of this iterative process there are any nodes left that cannot be reduced, they must be related to oscillations. The stable motif sequence (regardless of order) found in the reduction process determines the attractor [32]. Figure 8.10 demonstrates the complete attractor identification process of the network example presented in Fig. 8.9. Note that this process is the same for both Boolean and multi-level models. The resulting diagram is referred to as a stable motif succession diagram. This diagram also reflects the system's natural dynamical repertoire: starting from an arbitrary initial condition, sooner or later one of the possible stable motifs will stabilize, which will make other nodes stabilize, and so on. When a system allows multiple stable motifs, the timing of events determines which stable motif stabilizes first, which may make other stable motifs unattainable. For example, in Fig. 8.10, the initial condition and timing determine whether both A and B stabilize at 1 (first row) or A stabilizes at 0 (second row). These two stable motifs are mutually exclusive. The system wherein A stabilized at 0 may achieve stabilization of C at 1 or at 0, reaching attractor 2, or attractor 3, respectively.

We illustrate stable motif and ESM analysis on our two previously introduced examples, the T-LGL network and the EMT network (Figs. 8.11, 8.12, and 8.13). The complete expanded networks are too large and complex to be visually parsed, so we illustrate the stable motifs in each network. Figure 8.11 is a part of the stable motif succession diagram of the T-LGL network, illustrating one motif sequence whose stabilization leads to the normal, apoptosis attractor, and a motif whose stabilization leads to the T-LGL leukemia attractor. The complete succession diagram contains more motif sequences. Note that the first stable motif in the apoptosis-inducing sequence and the T-LGL-causing stable motif contain opposite states of the nodes S1P, PDGFR, and SPHK1. This suggests that the positive feedback among these

Motif Succession Diagram

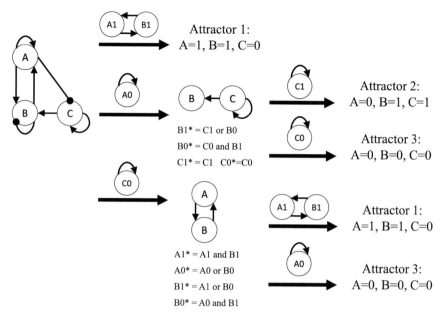

Fig. 8.10 Example of attractor identification with iterative stable motif guided network reduction using the same model as in Fig. 8.9. There are three stable motifs in the model. In the iterative reduction process, each of them is plugged into the regulatory functions (represented by indicating the stable motif above an arrow), resulting in a reduced model (indicated by the interaction network and regulatory functions), where further stable motif analysis is performed. For simplicity of representation of the A1, B1 stable motif, we do not show the composite node. When all nodes' states are identified in the process, the reduction is complete and an attractor is obtained

nodes, coupled with the mutual inhibition between S1P and Ceramide, is an attractor-determining connectivity pattern in the T-LGL leukemia network.

There are 8 stable motifs associated with the mesenchymal state in the EMT network, ranging in size from four to eleven nodes. Stabilization of any one of these stable motifs can independently drive the system into the mesenchymal state. Figure 8.12 shows the logic backbone of the EMT network, where stable motifs are represented with blue nodes [35]. All edges of the backbone represent sufficient activation, mediated by a path or subgraph of the EMT network. The figure indicates that any input signal is sufficient to drive all stable motifs, any of which is sufficient to drive EMT. An example ESM is given in Fig. 8.13.

The existence of eight stable motifs and their connectivity illustrated in Fig. 8.14 indicates that EMT is a very robust process. Steinway et al. analyzed the effect of single- and multiple-node knockout (sustained OFF state) on TGFβ-driven EMT, focusing on the status of the outcome node EMT. They found that knockout of the TGFβ receptor or of one of the seven transcription factors that down-regulate E-

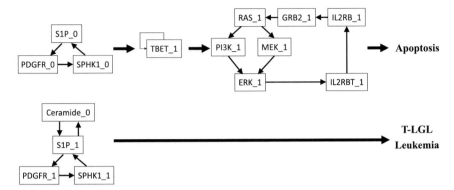

Fig. 8.11 Part of the stable motif succession diagram of the T-LGL network, adapted from [32]. The state of the nodes in each motif is indicated by a number, separated from the node name by an underscore (e.g. S1P_0 represents S1P at state 0). A stable motif sequence determines the attractor, i.e. Apoptosis or T-LGL leukemia (cancer). For example, the activation of the Ceramide = 0, S1P = 1, PDGFR = 1, SPHK1 = 1 motif leads to the reduction of the whole network and convergence into the T-LGL leukemia attractor

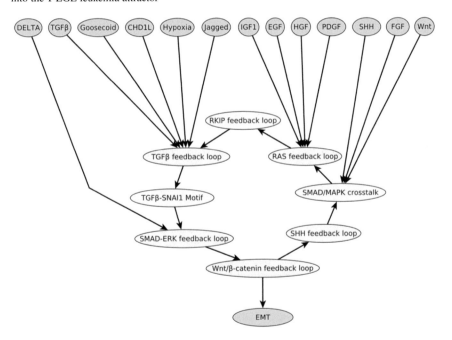

Fig. 8.12 The logic backbone of the EMT model, reproduced from [35]. This is a condensed version of the EMT network, where each stable motif of the model is represented by a single node (in blue), and its causal relationships with the signals and the outcome node EMT (in yellow) are visualized. All edges are sufficient activations, i.e. the activity (sustained ON state) of the input node/motif will activate the target node or motif. Any signal, or any stable motif, is sufficient to drive EMT

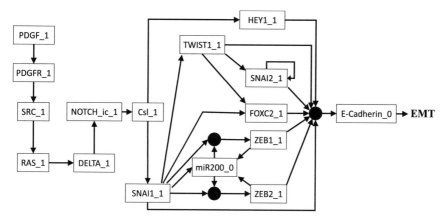

Fig. 8.13 An example ESM from the EMT network, from the signal PDGF to output EMT. The states of the nodes are marked at the end of each node label (e.g. PDGF_1 means PDGF at state 1). The existence of this ESM indicates that the sustained presence of the PDGF signal alone is sufficient to drive EMT. Note that this ESM contains three composite nodes

cadherin are the only EMT-blocking single node interventions. The effective double-node interventions include knockout of SMAD combined with a knockout of another node out of nine, marked with blue color on Fig. 8.14.

Stable motifs also offer a way to control the network. Generally, control can have two meanings: 1. to be able to drive the system into an arbitrary state (but the system may not necessarily stay there); 2. to be able to drive the system into an arbitrary attractor. The most appropriate control interventions depend on the control objectives as well as the structure and dynamics of the system, as reviewed in [36]. Because the cellular phenotypes are the attractors of molecular interaction systems, we set the second control objective. Since stable motifs correspond to (partial) fixed points of the system, a sequence of stable motifs will determine an attractor. Therefore, controlling one or more stable motifs (i.e. eliciting their stabilization by maintaining one or more nodes in a fixed state) is enough to drive the system into one of its attractors. The number of nodes that need to be controlled (maintained in a fixed state) can be minimized in two ways: First, not all stable motifs in a sequence need to be controlled. Specifically, stable motifs whose stabilization inevitably follows from the stabilization of a previous motif do not need independent control. Furthermore, to control a stable motif, one does not need to control all of its nodes, but only a subset of nodes called driver nodes. These two criteria can be used to predict a small set of driver nodes that can drive the entire system into the desired attractor.

Let's consider the EMT network again, but now focusing on the epithelial state. The stable motif associated with the epithelial state, shown in Fig. 8.14, is quite large (it is the entire SCC of the EMT network). Yet to control this motif, one only needs to control as few as five nodes: one node in each yellow rectangle. Maintaining these five nodes in their epithelial states is enough to ensure convergence to the epithelial state from any initial state of the system. Taken together, stable motif analysis of

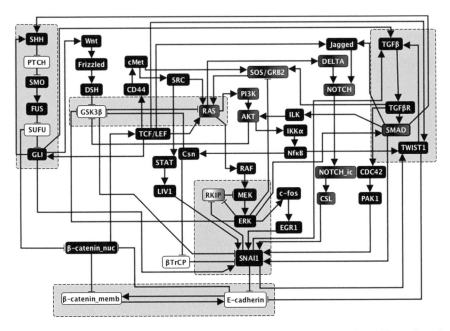

Fig. 8.14 Stable motif associated with the epithelial state in the EMT network and illustration of control sets that guarantee convergence to the epithelial state, reproduced from [26]. The entire graph is the epithelial stable motif. Nodes in black are OFF, and nodes in white are ON. Controlling of one node in each yellow rectangle, e.g. SMAD, SNAI1, RAS, SHH knockout combined with β-catenin_memb constitutive activation, ensures convergence to the epithelial state. The nodes highlighted in blue represent SMAD and the nine nodes whose knockout in combination with SMAD is able to prevent TGFβ-driven EMT. The fact that these blue nodes are either part of a yellow rectangle (SMAD, RAS), on a path that ends in a node of a yellow rectangle (DELTA, NOTCH, NOTCHic, CSL), or on a path that starts with a node of a yellow rectangle (PI3K, AKT) indicates the inclusive relationship between node sets whose control prevents or, respectively, reverses EMT

the EMT model allowed the prediction of two types of interventions: interventions that block TGFβ-driven EMT, thus suppressing features of invasive tumors, and interventions that revert mesenchymal cells to their epithelial state.

8.5   Conclusions

In this chapter, we aimed to show how network analysis and network-based dynamic modeling can be used to determine the repertoire of cellular behaviors associated with a within-cell network, and to identify the sub-networks that play a key role in the cell adopting a certain behavior. Overall, we found that the expanded network representation, an integration of the network topology with regulatory functions, reveals the indirect and self-sustaining influences in the system, which ultimately determine its repertoire of behaviors. The emerging answers indicate that stable motifs are a key information processing, decision-making connectivity pattern. Stable motifs receive information from external signals and internal perturbations, and their stabilization serves as a point of no return in the system's dynamics. One can characterize attractors by the stable motifs they are determined by, and one can control the system's outcome by controlling stable motifs.

Network analysis and network-based discrete dynamic modeling are an insightful first step that can be followed up by more detailed, quantitative approaches. By focusing on the connectivity patterns identified by the foundational analysis, these quantitative models can identify the exact tuning of the variables and kinetic parameters that can achieve the desired control. Ultimately, network modeling will become a tool in the toolbox of physicists studying life.

Acknowledgements The Albert group's research reviewed in this chapter was funded by NSF grants IIS 1161001, PHY 1205840, PHY 1545839.

References

1. X. Gan, R. Albert, BMC Syst. Biol. **10**(1), 78 (2016)
2. Z. Sun, X. Jin, R. Albert, S.M. Assmann, PLoS Comput. Biol. **10**(11), e1003930 (2014)
3. R. Thomas, E.M.B. Organization., *Kinetic logic : a Boolean approach to the analysis of complex regulatory systems : proceedings of the EMBO course "Formal analysis of genetic regulation," held in Brussels, September 6-16, 1977*. Lecture notes in biomathematics (Springer, Berlin, New York, 1979)
4. R. Albert, B.R. Acharya, B.W. Jeon, J.G.T. Zañudo, M. Zhu, K. Osman, S.M. Assmann, PLOS Biol. **15**(9), e2003451 (2017)
5. U. Alon, *An Introduction to Systems Biology: Design Principles of Biological Circuits*. Chapman & Hall/CRC Mathematical and Computational Biology (Taylor & Francis, 2006)
6. P. Shannon, A. Markiel, O. Ozier, N.S. Baliga, J.T. Wang, D. Ramage, N. Amin, B. Schwikowski, T. Ideker, Genome Res. **13**(11), 2498 (2003)
7. A.A. Hagberg, D.A. Schult, P.J. Swart, in ed. by G. Varoquaux, T. Vaught, J. Millman *Proceedings of the 7th Python in Science Conference (SciPy2008)*, pp. 11 – 15
8. M.K. Morris, J. Saez-Rodriguez, P.K. Sorger, D.A. Lauffenburger, Biochemistry **49**(15), 3216 (2010)
9. W. Abou-Jaoudé, P. Traynard, P.T. Monteiro, J. Saez-Rodriguez, T. Helikar, D. Thieffry, C. Chaouiya, Front. Genetics **7**(94) (2016)
10. M.L. Wynn, N. Consul, S.D. Merajver, S. Schnell, Integr. Biol. (Camb.) **4**(11), 1323 (2012)
11. R. Laubenbacher, F. Hinkelmann, D. Murrugarra, A. Veliz-Cuba, *Algebraic Models and Their Use in Systems Biology* (Springer, Berlin, 2014), pp. 443–474
12. R.S. Wang, A. Saadatpour, R. Albert, Phys. Biol. **9**(5), 055001 (2012)
13. A. Veliz-Cuba, A.S. Jarrah, R. Laubenbacher, Bioinformatics **26**(13), 1637 (2010)
14. A. Saadatpour, R. Albert, T.C. Reluga, SIAM J. Appl. Dyn. Syst. **12** (2013)
15. A. Naldi, E. Remy, D. Thieffry, C. Chaouiya, Theor. Comput. Sci. **412**(21), 2207 (2011)
16. A. Naldi, P.T. Monteiro, C. Müssel, H.A. Kestler, D. Thieffry, I. Xenarios, J. Saez-Rodriguez, T. Helikar, C. Chaouiya, Bioinformatics **31**(7), 1154 (2015)
17. C. Chaouiya, D. Berenguier, S.M. Keating, A. Naldi, M.P. van Iersel, N. Rodriguez, A. Drager, F. Buchel, T. Cokelaer, B. Kowal, B. Wicks, E. Goncalves, J. Dorier, M. Page, P.T. Monteiro, A. von Kamp, I. Xenarios, H. de Jong, M. Hucka, S. Klamt, D. Thieffry, N. Le Novere, J. Saez-Rodriguez, T. Helikar, BMC Syst. Biol. **7**, 135, 055001 (2013)

18. T. Helikar, B. Kowal, S. McClenathan, M. Bruckner, T. Rowley, A. Madrahimov, B. Wicks, M. Shrestha, K. Limbu, J.A. Rogers, BMC Syst. Biol. **6**(1), 96 (2012)
19. C. Chaouiya, A. Naldi, D. Thieffry, Methods Mol. Biol. **804**, 463 (2012)
20. I. Albert, J. Thakar, S. Li, R. Zhang, R. Albert, Source Code for Biol. Med. **3**(1), 16 (2008)
21. J. Zheng, D. Zhang, P.F. Przytycki, R. Zielinski, J. Capala, T.M. Przytycka, Bioinformatics **26**(1), 141 (2010)
22. C. Müssel, M. Hopfensitz, H.A. Kestler, Bioinformatics **26**(10), 1378 (2010)
23. R. Zhang, M.V. Shah, J. Yang, S.B. Nyland, X. Liu, J.K. Yun, R. Albert, J. Loughran, T.P., Proc Natl. Acad. Sci. U S A **105**(42), 16308 (2008)
24. A. Saadatpour, R.S. Wang, A. Liao, X. Liu, T.P. Loughran, I. Albert, R. Albert, PLOS Comput. Biol. **7**(11) (2011)
25. S.N. Steinway, J.G. Zanudo, W. Ding, C.B. Rountree, D.J. Feith, J. Loughran, T. P., R. Albert, Cancer Res. **74**(21), 5963 (2014)
26. S.N. Steinway, J.G.T. Zañudo, P.J. Michel, D.J. Feith, T.P. Loughran, R. Albert, NPJ Syst. Biol. Appl. **1**, 15014 (2015)
27. E. Remy, P. Ruet, D. Thieffry, Adv. Appl. Math. **41**(3), 335 (2008)
28. A. Richard, J.P. Comet, Discret. Appl. Math. **155**(18), 2403 (2007)
29. A. Richard, Adv. Appl. Math. **44**(4), 378 (2010)
30. R.S. Wang, R. Albert, BMC Syst. Biol. **5**(1), 44 (2011)
31. J.G. Zanudo, R. Albert, Chaos **23**(2), 025111 (2013)
32. J.G. Zanudo, R. Albert, PLoS Comput. Biol. **11**(4), e1004193 (2015)
33. Z. Sun, R. Albert, Netw. Sci. **4**(3), 273 (2016)
34. X. Gan, R. Albert, Physical Review E **97**(4), 042308 (2018)
35. P. Maheshwari, R. Albert, BMC Syst. Biol. **11**(1), 122 (2017)
36. Y.Y. Liu, A.L. Barabási, Rev. Mod. Phys. **88**, 035006 (2016)

Chapter 9
Introduction to Evolutionary Dynamics

David A. Kessler and Herbert Levine

Abstract The theory of Darwinian evolution can be treated using the formalism of chemical systems where the reactions consist of birth, death and genomic changes. In this chapter we introduce this approach and focus on three important problems that yield insight into evolutionary dynamics. First, we focus on the fixation of single mutations in a population of fixed size. The leads to the notion of there being a tradeoff between the size of the population and the smallest fitness advantage that can be resolved; at smaller N, the population suffers genetic drift and mutations are effectively neutral. A second example focuses on the balance between mutation and selection for rapidly evolving populations where many genotypes compete. Here we encounter the phenomenon of clonal interference, which dramatically affects the rate of evolutionary advance. Finally, we study the Delbruck-Luria model of the emergence of antibiotic resistance in a growing bacterial population. Its predictions concerning the anomalously broad distribution of resistant population sizes were critical in establishing the independence of the mutational event from the application of the drug, demonstrating the principle of natural selection in its simplest form.

9.1 Birth-Death Processes

A famous biologist once quipped that nothing in biology makes sense except in the context of evolution [1]. It is therefore surprising that most courses and textbooks of biophysics have nothing to say about this extremely important topic. This is even more surprising given that many of the techniques in common use for studying biochemical reaction dynamics can be used to derive important features of Darwinian evolution. The purpose of this chapter is to partially remedy this educational failing.

D. A. Kessler (✉)
Department of Physics, Bar-Ilan University, Ramat-Gan, Israel
e-mail: kessler@dave.ph.biu.ac.il

H. Levine
Northeastern University, Boston, MA, USA
e-mail: h.levine@northeastern.edu

© Springer Nature Switzerland AG 2022
K. B. Blagoev and H. Levine (eds.), *Physics of Molecular and Cellular Processes*,
Graduate Texts in Physics, https://doi.org/10.1007/978-3-030-98606-3_9

We will be treating population dynamics, that is, the study of the temporal dynamics of populations of organisms, as a special type of chemical reaction system. Specifically we will consider the "reactions" of birth, death, mutation and (possibly) recombination. The simplest example would be a pure continuous-time birth/death process [2]

$$A \to 2A \quad \text{rate } b \qquad A \to 0 \quad \text{rate } d$$

At the level of rate equations we get pure exponential growth or decay

$$\frac{dA}{dt} = (b - d)A. \tag{9.1}$$

What about the stochastic version, i.e. the treatment that takes into account the discreteness of individuals and hence the possibility of going extinct? Here we use the standard approach of deriving a master equation governing the probability that there exist n individuals at time t. Specifically

$$\frac{dP_n(t)}{dt} = b\left((n-1)P_{n-1}(t) - nP_n(t)\right) + d\left((n+1)P_{n+1}(t) - nP_n(t)\right). \tag{9.2}$$

These terms arise from the birth process where there can be transitions in and out of the n-particle state as well as the death process with the same two possibilities. Note the n-dependent factors representing the fact that b and d are the rates of individual birth and death respectively

In this simple case, we can solve this difference equation exactly by using the generating function method; this method is analogous to using a Laplace transform to solve a differential equation. We define

$$W(\lambda) \equiv \sum_{n=0}^{\infty} \lambda^n P_n(t)$$

Note that P_0, i.e. the extinction probability, is just $W(0)$. We multiply both sides of the master equation by λ^n and sum over n. This gives

$$\frac{\partial W}{\partial t} = b \sum_n \lambda^n \left[((n-1)P_{n-1}(t) - nP_n(t)\right] + d \sum_n \lambda^n \left[(n+1)P_{n+1}(t) - nP_n(t)\right]. \tag{9.3}$$

Shifting the various summations, the right hand side of the equation becomes

$$\left[b(\lambda - 1) + d(\frac{1}{\lambda} - 1)\right] \sum_n n\lambda^n P_n = (b\lambda^2 - (b+d)\lambda + d)\frac{\partial W}{\partial \lambda} = (b\lambda - d)(\lambda - 1).\frac{\partial W}{\partial \lambda} \tag{9.4}$$

We have now converted the master equation into a first-order PDE for the generating functional.

This equation can now be solved by the method of characteristics [3]. Let us define a new variable q via the requirement that

$$\frac{\partial W}{\partial q} = -(b\lambda - d)(\lambda - 1)\frac{\partial W}{\partial \lambda} \tag{9.5}$$

This of course necessitates $\frac{d\lambda}{dq} = -(b\lambda - d)(\lambda - 1)$ which leads to

$$q = \frac{1}{b-d}\ln\left[\frac{b\lambda - d}{d(\lambda - 1)}\right]. \tag{9.6}$$

In terms of q, the PDE is now quite trivial as $\frac{\partial W}{\partial t} = -\frac{\partial W}{\partial q}$ with the obvious solution $W = \tilde{W}_0(q - t)$ where $\tilde{W}_0$ is the generating function at time 0 as a function of the auxiliary variable q. To see what this means, we can focus on the case where we initially have exactly one individual in the population. This means that

$$W(t = 0) = \lambda = \frac{d\left(e^{(b-d)q} - 1\right)}{\left(de^{(b-d)q} - b\right)} \equiv \tilde{W}_0(q). \tag{9.7}$$

Then, substituting in $q - t$ for q and re-expressing the answer in terms of λ gives the final result

$$W(\lambda, t) = \frac{\frac{b\lambda - d}{\lambda - 1}\left(e^{-(b-d)t}\right) - d}{\frac{b\lambda - d}{\lambda - 1}\left(e^{-(b-d)t}\right) - b}. \tag{9.8}$$

We can use this result to obtain P_n, using

$$P_n(t) = \frac{1}{n!}\left(\frac{d}{d\lambda}\right)^n W(\lambda, t)\Big|_{\lambda=0}.$$

In particular, the extinction probability is

$$W(0, t) = \frac{d\left(e^{-(b-d)t} - 1\right)}{de^{-(b-d)t} - b}. \tag{9.9}$$

Note that it is proportional to d and vanishes at $t = 0$ as it must. Asymptotically for long time, P_0 approaches d/b if $b > d$; conversely, if the death rate is bigger than the birth rate, the population always goes extinct. The case of equal birth and death is called critical, and a limiting procedure gives $P_0(t) = bt/(1 + bt)$; hence extinction is guaranteed but it can take a very long time. In fact, the mean extinction time is infinite. The rest of the probability distribution for the critical case

$$P_n = \frac{(bt)^{n-1}}{(1 + bt)^{n+1}}, \quad n \geq 1 \tag{9.10}$$

is geometric, with the ratio of terms being $bt/(1 + bt)$, which approaches unity for large time. Thus the mean population size conditioned on survival grows as bt for large time, which multiplied by the $1/bt$ probability of survival gives an overall constant mean population size. The mean population size for the general case $\bar{n}(t) = \frac{dW}{d\lambda}\big|_{\lambda=1} = e^{(b-d)t}$ and the long-time limit of the population size variance $\frac{b}{d-b}e^{2(b-d)t}$.

If we want to obtain the full distribution for the general case, it is useful to rephrase the relationship for obtaining P_n from the generating function as the contour integral formula

$$P_n = \frac{1}{2\pi i} \oint dz \, \frac{W(z,t)}{z^{n+1}}, \tag{9.11}$$

where the integral is taken around a small contour encircling the origin. We can make use of the fact that W just has a simple pole outside this contour to get an explicit expression for the probability. We leave as an exercise for the reader the derivation of the result for $n \neq 0$; The answer is

$$P_n = \frac{b^{n-1}(b-d)^2 e^{-(b-d)t} \left(e^{-(b-d)t} - 1\right)^{n-1}}{\left(de^{-(b-d)t} - b\right)^{n+1}}, \qquad n \geq 1 \tag{9.12}$$

giving a geometric distribution except for P_0. This geometric distribution for long times has mean $be^{(b-d)t}/(b-d)$. This times the probability of surviving to generate a large population size, $1 - P_0 = (b-d)/b$ gives the overall mean $e^{(b-d)t}$.

9.2   The Kimura Problem

The above warm-up exercise is not really about evolution, as we have birth and death but no mutation. Our first real example is the calculation of the probability that a mutant organism can take over an entire non-mutant (known as wild-type) population. We will assume that the overall population size is fixed, which means that every birth event is automatically followed by a corresponding death. The continuous-time Markov process version of this problem is known as the Moran model [4]. We will assume that the process of fixation is fast enough that we can ignore additional mutations and focus just on the fate of one mutant at the start of the process. The relevant dynamical variable is then $N_\mu(t)$ mutants out of N total individuals. The basic reaction steps are as follows: (1) A mutant individual gives birth at rate b_μ. replacing a wild-type individual. This occurs at overall rate

$$b_\mu N_\mu \frac{N - N_\mu}{N},$$

where the last factor arises due to the requirement that the number of mutants will change only if a wild-type is selected (at random) to die following a mutant cell birth;

(2) a wild-type individual gives birth at rate b_w and a mutant is selected to die; this will occur at rate

$$b_w(N - N_\mu)\frac{N_\mu}{N}.$$

We could proceed by writing down the master equation for $P(N_\mu)$. The tine-dependent distribution cannot easily be found. But it is much simpler to solve for the final distribution. Since both the no mutant and no wild-type states are absorbing (which means that we cannot leave them once we enter), the final distribution will just consist of some probability (referred to as the splitting probability) of being in each of these two possibilities. To study splitting probabilities, it is useful to derive a slightly modified form of the master equation, sometimes called the backwards Kolmogorov equation for reasons that will become apparent. Note that this derivation is a specific application of the theory of first passage problems [5]. Imagine we are interested in $\Pi_{N_\mu}(t)$, the probability that the population will become fixed in the fully mutant state assuming that it is currently in state N_μ. This probability can be written down as a sum of terms corresponding to what happens in a small interval dt following t. There are three terms on the right hand side of the equation, corresponding to the three possibilities: nothing happens; mutants increase by one; mutants decrease by one. In detail,

$$
\begin{aligned}
\Pi_{N_\mu} = {} & b_\mu dt \frac{N_\mu(N - N_\mu)}{N} \Pi_{N_\mu+1} \\
& + b_w dt \frac{N_\mu(N - N_\mu)}{N} \Pi_{N_\mu-1} \\
& + \left(1 - (b_\mu + b_w)dt \frac{N_\mu(N - N_\mu)}{N}\right) \Pi_{N_\mu}.
\end{aligned}
\tag{9.13}
$$

Cancelling Π_{N_μ} from both side, all the remaining terms have a common factor of $dt \frac{N_\mu(N-N_\mu)}{N}$ which can be divided out, leaving us with the equation

$$0 = b_\mu \Pi_{N_\mu+1} + b_w \Pi_{N_\mu+1} - (b_\mu + b_w)\Pi_{N_\mu}.
\tag{9.14}$$

Note that this is exactly the same equation we would get for a random walk with constant jump probabilities of $p_r = b_\mu/(b_\mu + b_w)$ to the right and $p_l = b_\mu/(b_\mu + b_w) = 1 - p_r$ to the right. Clearly one solution of this second-order difference equation is Π_{N_μ} constant. The other solution can be found by the ansatz $\Pi_{N_\mu} \sim \alpha^{N_\mu}$. Substituting into the steady-state equation leads to

$$(-b_\mu + b_w) + b_\mu\alpha + b_w/\alpha = 0,$$

which upon multiplication by α can be rewritten as

$$(\alpha - 1)(b_\mu\alpha - b_w) = 0,
\tag{9.15}$$

thus showing that the other choice is $\alpha = b_w/b_\mu$. Finally we need to impose the two boundary conditions $\Pi_0 = 0$ and $\Pi_N = 1$, because $N_\mu = 0$ means that the mutation cannot be fixed in the population, and conversely $N_\mu = N$ means that it is already fixed. We therefore obtain the final answer to this problem

$$\Pi_{N_\mu} = \frac{1 - \left(\frac{b_w}{b_\mu}\right)^{N_\mu}}{1 - \left(\frac{b_w}{b_\mu}\right)^{N}}. \tag{9.16}$$

Recall that the problem we wanted to study was the splitting probability starting from one mutant; hence, we set $N_\mu = 1$ in the preceding formula. There are now several very important limits of the formula we have derived when N is large. First, imagine that the mutant is significantly deleterious (i.e. has lower birth rate, aka fitness), namely that

$$\frac{b_w - b_\mu}{b_w} >> \frac{1}{N}$$

The second term in the denominator is now very large, and the entire expression is exponentially small

$$\Pi_1 \simeq \left(\frac{b_w}{b_\mu} - 1\right) e^{-N \ln(b_w/b_\mu)}. \tag{9.17}$$

Thus such a mutation is essentially always eliminated and in fact mutation problems can basically subsume deleterious mutations into just an overall reduction in birth rate and focus instead of beneficial mutations. In the same limit for beneficial mutations,

$$\frac{b_\mu - b_w}{b_w} >> \frac{1}{N},$$

and the exponential in the denominator is very small and can be neglected compared to 1. This gives us the result

$$\Pi_1 = \frac{b_\mu - b_w}{b_\mu}. \tag{9.18}$$

This is consistent with what we saw in the pure birth-death problem above that the population starting out as a single individual has a significant chance of dying out despite its positive net growth rate. Here too we see that the in principle faster growing subpopulation has in fact a relatively small overall advantage (exactly equal to the relative fitness change which is usually expected to be small) since we start with just one mutant, and there is quite a large probability that it will become extinct due to bad luck. The limit of very small fitness change (small compared to $1/N$) is also quite interesting. The above expression reduces to the extremely simple form

$$\Pi_1 = \frac{1}{N}.$$

This result is easy to understand. After a sufficient amount of time, all individuals in an asexually reproducing population will be descended from one ancestor, and the chance of any one of the original N individuals to be that sole progenitor is $1/N$, in the absence of fitness differences. Parenthetically, this is the underlying explanation for the popular notion of "mitochondrial Eve" [6], since mitochondria are inherited strictly through the maternal line. If the mutation is neutral, the relative choice of that ancestor being from the mutant subpopulation is just the initial fraction of mutants. The point of this calculation is that for this to hold, the mutation does not need to be strictly neutral, only that it confers fitness change which is small compared to $1/N$. This means for a deleterious mutation to be eliminated by selection, it must give rise to a fitness deficit which is large compared to $1/N$. Thus, small populations have a tendency to fix deleterious mutations via this stochastic effect (termed genetic drift); if there is a relative absence of compensating beneficial mutations they will see their fitness systematically decline, a process that is referred to as Muller's ratchet [7].

All else being equal, the rate at which mutations arise in a population scales as the population size. Thus, if most mutations are effectively neutral and each mutation has an independent chance of fixation, we would expect a constant rate of mutation accumulation. Using this idea, Kimura suggested [8, 9] that the number of mutations seen in a population could be used as a genetic clock, telling us how long it had been since that population diverged from an ancestor. This idea has been used extensively since to create a phylogenetic history of species alive today.

9.3 Selection-Mutation Equilibrium

In the small mutation limit considered so far, populations make transitions from a state of fitness b to one of fitness $b(1 + s)$ due to a beneficial mutation arriving at probability per birth μ_b or $b(1 - s)$ due to a deleterious one at probability μ_d. The model above can be thought of as having a trivial "fitness landscape"

$$b(n) = b_0 + ns, \tag{9.19}$$

with respective transition rates

$$R_+(n) = b(n)N\mu_b \frac{1 - e^{-s}}{1 - e^{-Ns}}; \qquad R_-(n) = b(n)N\mu_d \frac{1 - e^{-s}}{1 - e^{Ns}}. \tag{9.20}$$

Let us first imagine that the two mutation probabilities are equal. Then it is easy to show we can define an effective "energy" $E(n) = -(N - 1)ns - \ln(b_0 + ns)$ for a state with n beneficial mutations compared to some initial reference state, such that we can satisfy the detailed balance condition

$$R_+(n - 1)e^{-E(n-1)} = R_-(n)e^{-E(n)}. \tag{9.21}$$

For N large, the second term in the energy is negligible.

This model by itself, however, does not lead to equilibrium because one can continue finding and fixing beneficial mutations *ad infinitum*. There are two non-exclusive ways to remedy this lack of an equilibrium mutation-selection balance. The first way involves having a more complex relationship between mutations and the birth rate. Imagine that the gene encodes for a person's height $h(n) = h_0 + n\Delta_h$, and birth rate, aka fitness, takes the form

$$b(h) = b_{max} - \alpha (h - h_{max})^2 . \tag{9.22}$$

In other words there is a preferred value of this "quantitative trait" that in this case is the height with quadratic decrease of fitness away from this maximum. The transition probabilities now have s replaced by the fitness difference $(b(n) - b(n-1))/b(n) \simeq (\ln b)'$ and the energy in the detailed balance formula becomes $E(n) = -(N-1)\ln b(n)$. Close to the maximum, this gives a Gaussian distribution

$$P(n) \simeq P_0 e^{-\frac{N\alpha}{b_{max}} \Delta_h^2 (n - n_{max})^2} . \tag{9.23}$$

Thus, the distribution has a width proportional to $1/\sqrt{N}$ so that the distribution narrows as N increases.

The second possibility arises from relaxing the equal mutation probabilities assumption. We will focus on the simplest model which exhibits this effect. We posit a genome of a large number L of binary genes and define the fitness as

$$b = b_0 + \Delta b \sum_{i=1}^{L} S_i , \tag{9.24}$$

where $S_i = \{0, 1\}$. Every member of the population has a genome and a specific phenotype (here just its birth rate) determined by the number of its genes that have the beneficial allele, $S = 1$. There is then a maximum fitness which occurs when all alleles have $S = 1$. As we near this state, there are fewer and fewer beneficial mutations available and conversely it become much more likely that a randomly chosen mutational event is deleterious. Quantitatively, we have

$$\mu_b(n) = \mu\frac{L-n}{L}; \qquad \mu_d(n) = \mu\frac{n}{L}. \tag{9.25}$$

We then find the ratio in the absence of any selection effects

$$\frac{R_+(n-1)}{R_-(n)} = \frac{L+1-n}{n}, \tag{9.26}$$

which will satisfy detailed balance with an energy equal (up to a constant) to $E(n) = \ln(k!) + \ln((L-k)!)$. The lowest energy level is clearly $n = L/2$. If one now adds together both effects, i.e. adds back the feature that beneficial mutations increase the fitness, we obtain the final result that

$$E(n) = -(N-1)ks - \ln(b_0 + ns) + \ln(k!) + \ln((L-k)!). \qquad (9.27)$$

When N is large, fitness effects win and the system is peaked around $n = L$; as Ns is decreased, the "entropy" terms become important and the peak of the equilibrium distribution moves downward.

9.4 Clonal Interference

The above selection calculation dealt with a population of only two competing genotypes. This implicitly assumes that the time scale for the arrival of new mutations is sufficiently long that the fate of each mutant population is decided before the next mutant arrives. This assumption clearly breaks down in many systems. In bacterial colonies for example, there may be upwards of 10^9 cells and an error rate per base pair of $10^{-8} - 10^{-7}$; thus each cell will typically give rise to of 10–100 mutations at each division, and hence the rate of production of new mutant populations is gigantic. The assumption of independent fixation of separately generated mutants would give rise to a rate of fitness increase of approximately

$$\underbrace{b_W \mu_{bp} LN}_{\substack{\text{beneficial mutation} \\ \text{supply rate}}} \times \underbrace{\frac{b_\mu - b_W}{b_W}}_{\substack{\text{fixation} \\ \text{probability}}} \times \underbrace{b_\mu - b_W}_{\substack{\text{fitness} \\ \text{increase}}}.$$

Here L is the size of the genome, N is the size of the population, and μ_{bp} is the rate of beneficial mutations whose fitness advantage exceeds $1/N$. Even if the percentage of beneficial mutations is a small percentage of all mutations, this formula tremendously overestimates the actual rate of fitness advance. The reason for this is simple; different beneficial mutations arise simultaneously in different clones and cannot all be fixed. This phenomenon is known as clonal interference.

Let us return to the simple genotype-phenotype model discussed above and now focus on the dynamic evolution process. The apparently obvious approach to the behaviour of this population relies on the use of chemical rate equations, i.e. assumes a deterministic value for the number of individuals that have a specific value of b at some fixed time. Assuming again that the total population size is fixed and that mutations can only change one gene at a time, we can straightforwardly derive the coupled set of equations for this number

$$\frac{\partial P_b}{\partial t} = (b - \bar{b}) P_b + \mu \Big((b + \Delta b) P_{(b+\Delta b)} f(b + \Delta b) +$$
$$(b - \Delta b) P_{(b-\Delta b)} \tilde{f}(b - \Delta b) - 2b P_b \Big). \qquad (9.28)$$

The factor $f(b)$ is the chance that a mutation lowers the birth rate by Δb which just as before equals the percentage of beneficial alleles $(b - b_0)/(L\Delta b)$ and $\tilde{f} = 1 - f$ is the complementary probability that a mutation raises the birth rate. Finally, $\bar{b}$ is the average growth rate for the entire population and its presence in the equation clearly enforces the overall conservation of population size.

To proceed with the analysis, we will simplify the situation in a variety of ways. First, we will assume that $b_0 > L\Delta b$ which means that the overall percentage change in growth rate is small. This allows us to neglect the shifts in birth rates that occur in the mutations terns of the preceding equation. Also, we will assume that L is large compared to the width of the fitness distribution and hence $(b - b_0)/(L\Delta b) \simeq (\bar{b} - b_0)/(L\Delta b)$ and is the same for all phenotypes. Also, for a finite velocity the rate of change of this term is $O(1/L)$ and hence this number can be taken as constant in time. These assumptions reduce our equation to the form

$$\frac{\partial P_b}{\partial t} = (b - \bar{b})P_b + \mu b_0 \left(P_{(b+\Delta b)} f + P_{(b-\Delta b)}(1 - f) - 2P_b \right), \qquad (9.29)$$

or, if we take the continuum limit in phenotype space

$$\frac{\partial P}{\partial t}(x, t) = (x - \bar{x})P + \mu b_0 \left(2\Delta b(f - 1/2)\frac{\partial P}{\partial x} + (\Delta b)^2 \frac{\partial^2 P}{\partial x^2} \right). \qquad (9.30)$$

The various terms on the right hand side have simple interpretations. They are in turn selection due to birth rate above the mean, bias in motion on the fitness landscape because of an excess of deleterious versus beneficial mutations if $f > 1/2$, and diffusion due to the stochastic nature of mutation. The bias term can be absorbed by defining a new frame of reference $y = x - 2b_0\mu\Delta b(f - 1/2)t$ since

$$\left.\frac{\partial P}{\partial t}\right|_x = \left.\frac{\partial P}{\partial t}\right|_y + 2b_0\mu\Delta b\frac{\partial P}{\partial x}. \qquad (9.31)$$

We thus obtain the simplest possible final form

$$\frac{\partial P(y, t)}{\partial t} = (y - \bar{y})P(y, t) + D\frac{\partial^2 P(y, t)}{\partial y^2}, \qquad (9.32)$$

with $D = b_0\mu(\Delta b)^2$. We note again that this equation comes from the classic chemical reaction formalism and does not take into account fluctuations due to discreteness of the individuals making up the population. We will see shortly that this is an insufficient approach and that one has to go beyond this equation to obtain sensible results.

What is the problem with this equation? Essentially, a very small number of very fit individuals at the leading edge of the fitness distribution at large y will reproduce faster and rapidly take over the entire population. A direct simulation of this effect will show a "velocity", the rate of growth of the average fitness, diverging to infinity at finite time. This bad behaviour is not because of all the approximations we have made in deriving this simplified form of the reaction kinetics equation; a full simulation of the original equations shows that system rapidly accelerates to a very high speed and gets to the end of the landscape (where all alleles are beneficial) in times that scale as $\ln L$ as opposed to $O(L)$. Actually, the problem with equations of this form was first noticed in a completely different context, the growth of a diffusion-limited solid [10] confined to a channel [11], in the limit of very slow growth compared to the diffusion of needed material to incorporate into the growing phase. In that limit, the diffusing field obeys Laplace's equation and has a linear profile, driving the solid front to rapidly accelerate. The following calculation, which uses the heuristic idea of a reaction cutoff, will directly show this anomalous behaviour as the front velocity will diverge in the limit of zero cutoff.

We now introduce the cutoff idea [11–13]. As already argued, the problems with the previous formulation arise due to the high growth rate for large y, effectively rapidly amplifying extremely small values of $P(y)$. But, for discrete individuals making up the population, the smallest possible value of P for some large value of the birth rate b is one person. This gives a smallest possible non-zero value of the density of $O(1)$. This of course can be handled exactly if we trade in the reaction kinetic equation for a full master equation formulation, but this leads to an unsolvable problem. Instead, we will by hand restrict the growth term to be operative only when $P > P_{cut}$, transforming our previous equation to the new form

$$\frac{\partial P(y, t)}{\partial t} = \theta(P - P_{cut})(y - \bar{y})P(y, t) + D\frac{\partial^2 P(y, t)}{\partial y^2}. \tag{9.33}$$

We now look for a traveling pulse solution of this equation $P(y, t) = P(y - vt)$ and fix translation invariance by assuming that $\bar{y} = vt$. This leads to

$$0 = v\frac{\partial P(y, t)}{\partial y} + \theta(P - P_{cut})yP(y, t) + D\frac{\partial^2 P(y, t)}{\partial y^2}. \tag{9.34}$$

Next, let us assume that $P = P_{cut}$ at $y = y_c$. For $y > y_c$ taking into account the fact that the P must vanish as $y \to \infty$, we have

$$P(y) = P_{cut}e^{-\left(\frac{v}{D}(y - y_c)\right)}. \tag{9.35}$$

This needs to match the solution for $y < y_c$. In this region, we write

$$P(y) = e^{-vy/2D}\psi(y),$$
$$P'(y) = e^{-vy/2D}\left(\psi' - \frac{v}{2D}\psi\right),$$
$$P''(y) = e^{-vy/2D}\left(\psi'' - \frac{v}{D}\psi' + \frac{v^2}{4D^2}\psi\right). \tag{9.36}$$

Substituting, we arrive at the equation

$$0 = D\psi'' + \left(y - \frac{v^2}{4D^2}\right)\psi. \tag{9.37}$$

This equation has a solution in terms of the well-known Airy function, $\psi = C\text{Ai}(\frac{v^2}{4D^2} - y)$; this form guarantees that P will decay at large $-y$ where the argument becomes positive. The velocity is then determined by two matching conditions at $y = y_c$ as well as the overall population size N which is given by the integral of P; note that there are three unknowns (v, C, y_c) available to satisfy these three conditions.

Proceeding, the continuity equations take the form

$$C\text{Ai}\left(\frac{v^2}{4D^2} - y_c\right)e^{-vy_c/2D} = P_{cut},$$

$$C\left(\text{Ai}'\left(\frac{v^2}{4D^2} - y_c\right) + \frac{v}{2D}\text{Ai}\left(\frac{v^2}{4D^2} - y_c\right)e^{-vy_c/2D}\right) = \frac{v}{D}P_{cut}, \tag{9.38}$$

and since the contribution from $y > y_c$ is O(1), the population requirement for large N is

$$C\int_{-\infty}^{y_c} \text{Ai}\left(\frac{v^2}{4D^2} - y\right)e^{-vy_c/2D}dy = N. \tag{9.39}$$

Note that this means that C is much greater than 1. For convenience, we will define a new coordinate $z = y - \frac{v^2}{4D^2}$. In terms of z, the probability distribution is $P = \tilde{C}\exp\left(-vz/2D\right)\text{Ai}(-z)$ where the extra factor in the exponential has been absorbed into a redefinition of the coefficient. The matching conditions and population constraint now become

$$\tilde{C}\text{Ai}(-z_c)e^{-vz_c/2D} = P_{cut},$$

$$\tilde{C}\left(\text{Ai}'(-z_c) + \frac{v}{2D}\text{Ai}(-z_c)\right)e^{-vz_c/2D} = \frac{v}{D}P_{cut},$$

$$\tilde{C}\int_{-\infty}^{z_c} e^{-\frac{vz}{2D}}\text{Ai}(-z)dz = N. \tag{9.40}$$

The first equation can only be satisfied if $-z_c$ is close to the first zero $-z_0$ of the Airy function ($z_0 \simeq 2.338$), since the right hand side is O(1) and the left hand side (and in particular $\tilde{C}$ is nominally much larger. It must be the first zero because the probability must remain positive. Denote $\delta = z_0 - z_c$ and then the equations can be approximated as

$$\delta\tilde{C}\text{Ai}'(-z_0)e^{-v(z_0-\delta)/2D} = -P_{cut},$$

$$\tilde{C}\text{Ai}'(-z_0)\left(1 + \frac{v\delta}{2D}\right)e^{-v(z_0-\delta)/2D} = -\frac{v}{D}P_{cut}. \tag{9.41}$$

We can use the first equation to simplify the second and obtain

$$\tilde{C}\text{Ai}'(-z_0)e^{-v(z_0-\delta)/2D} = -\frac{v}{D}P_{cut},\tag{9.42}$$

which when substituted back in the first equation leads to $\delta = 2D/v$. We now know where the matching point z_c is located and since v will turn out to be large, this is indeed close to the zero. We now can obtain directly the coefficient $\tilde{C}$

$$\tilde{C} = P_{cut}\frac{v}{2D\text{Ai}'(-z_0)|}e^{\frac{vz_0}{2D}-1}.\tag{9.43}$$

We now are finally in a position to find the velocity. The normalization condition involves an integral over z and the integral is cut off rapidly at negative z because of the rapid decay of the Airy function $\text{Ai} \simeq e^{-2/3\,x^{3/2}}$ for positive arguments and is cut off at positive z because of the explicit exponential. We can therefore evaluate the integral via a saddle point contribution around the point z_s

$$\frac{v}{2D} = |z_s|^{1/2}.\tag{9.44}$$

where we have used the aforementioned asymptotic form. Evaluating the integrand at this point leads to the factor $e^{v^3/24D}$. When the dust settles, we get that the leading behaviour of the velocity results from matching this factor with the explicit N in the constraint equation, leading to $v \sim (\ln N)^{1/3}$ for large N. As advertised, there is no infinite N limit of the problem, bringing us back to the diverging acceleration found in the chemical kinetics formulation.

The above result was first derived by Tsimring et al. [13] and later re-derived by Rouzine [14] and by Desai and Fisher [15]. Rouzine generalized the cutoff calculation to include the fact that the fitness is actually a discrete variable and showed that this leads to a crossover to $\ln N$ behaviour at very large N. The last of these papers used a more complex scheme of actually trying to solve the stochastic problem at the leading edge of the population instead of just assuming that the leading effect of the "demographic" noise was captured by the cutoff idea. Several other pieces of evidence support the notion that the cutoff gives the correct leading order. First, for the related but simpler problem of a Fisher front [16], Brunet and Derrida [17] showed that imposing a cutoff would led to a correction of the velocity that scaled as an inverse power of $\ln N$; their result has been checked extensively by numerical simulations [18, 19] and later was proven to be precisely correct [20]. Finally, Hallatschek introduced [21] a clever variation of the basic evolutionary model which then could be solved exactly and again recovered the same scaling result.

Returning back to biology, we have learned some extremely important lessons by working out the detailed dynamics of an extremely over-simplified model. The naive expectation that the rate of evolution should scale as N is wildly inaccurate; it is only valid when the product of mutation rate and population size is so low that each mutation gets a chance to become fixed before any other mutant arises. Typical N's are much larger and hence, mutations arise quickly in many genetic backgrounds and only one out of many can take over the population. Speculation that this interference effect would lead to an actual speed limit [22] are also incorrect, but just barely so. Finally, the results point out a possible way out of this conundrum. If the beneficial mutations could somehow be recombined into a single clone, we would recover significantly more rapid fitness advance. This is of course exactly what sexual reproduction accomplishes using chromosomal crossover during meiosis to limit linkage between different genes [23]. Bacteria have their own way of recombining genetic information, for example via the process of competence [24]; even viruses can do some degree of recombination for cases where multiple viruses infect the same cell. Interested readers can refer to several references [25–27] to investigate the extension of the ideas presented here to the more complicated case of including varying degrees of recombination.

In passing, we should mention that the chemical kinetics formulation, much maligned above, actually does work if we are interested in the equilibrium distribution. Let us return for example to the case of the quadratic peak in the fitness versus some quantitative trait. In the absence of clonal interference, we showed above that the fitness distribution becomes infinitely sharp as $N \to \infty$. We leave it as an exercise for the reader to show that in this case the reaction equation reduces to demanding that

$$\mu \frac{\partial^2 P}{\partial x^2} \sim (x - x_{max})^2 P, \tag{9.45}$$

which leads again to a Gaussian distribution but with a width that is now independent of N but rather proportional to $\mu^{1/4}$. Thus as N is increased for fixed very small μ, the width narrows with N as above, but saturates at a finite value due to the mutation rate.

9.5 The Luria-Delbrück Process

We will present one more example of how a detailed analysis of a simple model can offer critical insight into the evolutionary process. For this case, we will return to the case of an exponentially growing population and focus on the accidental emergence of mutants that are resistant to a drug which is lethal to the wild-type. The word "accidental" refers to the fact that until the drug is actually administered there is no fitness advantage to these mutants and in fact in many cases there may be a fitness disadvantage. This problem was first considered by Luria and Delbrück [28] in the context of antibiotic resistance in bacterial colonies. When a drug is applied to such a colony, most cells die but some flourish and continue growing. There are several possible mechanisms that could be underlying this behaviour. One logical possibility is that some cells have adapted quickly enough to the presence of the drug that they avoid death; this can occur and often this requires that cells first transition into a quiescent persister state. However, another possibility is that mutants that happen to be resistant co-exist in the population with the wild-type and that upon drug presentation these mutants take over the entire population. The question posed by Luria and Delbruck was whether there was some characteristic signature of the second scenario.

The model to be investigated consists of wild-type cells growing at rate b, dying at rate d and giving rise to drug-resistant mutants with a probability μ for each birth event. We assume that the population has grown from one individual to size N and we wish to determine the distribution of the number of mutant cells N_m, assuming μ is small. We initially assume that the mutation is neutral (in the absence of drug, of course) and then briefly explain the consequences of allowing the resistance mutation to be slightly deleterious. We will use a simplified approach that just computes the leading asymptotic form of the distribution for large N_m; anyone interested in the full treatment of the problem in the large N limit should see the cited references.

To get started, we treat the wild-type population as deterministic as it grows from a single individual to size N; thus

$$N_W(t) = e^{(b-d)t}, \tag{9.46}$$

and clearly $t_{final} = \ln N/(b - d)$. Let us assume only one mutational event occurs; in any case, the large mutant number tail of the distribution will be determined by the first mutational event that occurs and hence we can ignore all the others (see later). We can then determine the distribution of N_m by recognizing that the size of the mutant population is directly determined by the time t_m at which the mutation occurred. Since

$$N_m(t_{final}) = e^{(b-d)(t_{final}-t_m)} = Ne^{-(b-d)t_m}, \tag{9.47}$$

we find

$$P(N_m) = \tilde{P}(t_m)\frac{dt_m}{dN_m}, \tag{9.48}$$

where $\tilde{P}$ is the distribution of mutation times. Now the probability distribution of mutation times for small μ is just

$$\tilde{P}(t_m) = b\mu N_w(t_m) = b\mu e^{(b-d)t_m}. \tag{9.49}$$

This immediately gives us

$$P(N_m) = \frac{b}{b-d}\frac{\mu e^{(b-d)t_m}}{N_m} = \frac{\mu b}{b-d}\frac{N}{N_m^2}. \tag{9.50}$$

This corresponds to a very long-tailed distribution, inasmuch as it does not even have a mean. Of course, in reality N_m cannot become bigger than N (or equivalently t_m cannot be negative) in this approximate theory, but still this means that the mean is of order $\ln N$ and the variance is even worse. Interestingly enough, one of the first researchers to notice this fact was Mandelbrot [29], who is famous for his work on the power-law tails of distributions in many systems.

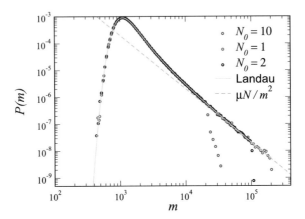

Fig. 9.1 $P(m)$ for the original LD model from a simulation of 10^6 realizations of the stochastic process defined in the text together with the analytic Landau distribution formula derived in [30]. What is shown are graphs for different initial mutant population sizes as well as the $1/N_\mu^2$ tail of the distribution. $\mu = 0.001$, $N = 2 \cdot 10^5$. $N_0 = 1$, 2 and 10

Is this result somehow fixed if we allow μ to be large enough that many mutation events are likely to occur? As already mentioned and without doing any extra calculations, it is clear that this extension should not alter our finding. The mutational events that can generate large N_m happen very early on when the wild type population is small. So, even if μN is large and many events happen towards the end of the allotted time, these do not change the fall-off the distribution. Likewise, there is no reason to expect that a stochastic treatment of the wild-type population should make any difference as far as this point is concerned. And indeed, more explicit calculations [30, 31] as well as numerical simulations (see Fig. 9.1) shows that it does not.

Luria and Delbrück turned calculations based on this model into a celebrated fluctuation test [32] that eventually showed that the experimental data for bacteria was better explained by the hypothesis that the mutants arose prior to antibiotic exposure. Any method based on resistance induction by the drug should lead roughly to a Poisson distribution with variance equal to the mean. Here we have shown that the variance is much bigger than this Poisson value and hence can explain data that is not fit by the induction hypothesis. This argued against any genetic response that somehow was directly adaptive and, at least from experiments with high lethal drug concentrations, against any phenotypic adaptation as well. Partly for this work, they received the Nobel Prize in physiology.

What changes if we assume that the resistance mutation causes a slight decrease in fitness? It is easy to generalize the previous calculation (and also its more sophisticated version) and show that the tail now falls off faster than $1/N_m^2$. The distribution is still heavy-tailed and there will be anomalously large fluctuations, and hence the methodology is robust to the relaxation of the purely neutral assumption.

The Luria-Delbrück model has exhibited a rebirth of importance, due to the advent of targeted therapy for cancer patients [33]. The issues are the same - does the commonly observed emergence of drug resistance arise from selection of already present mutant clones that happen to be resistant or do the cells survive the drug onslaught by finding effective phenotypic countermeasures that may eventually be supplanted by genetic mutations that arise during the treatment. This is an area of immense practical importance and the interested reader is directed to several recent papers [34–36] related to this ongoing research.

Acknowledgements HL was supported by the Center for Theoretical Biological Physics via NSF Grant PHY-1427654. DAK was supported by the BSF, grant number 2015619.

References

1. T. Dobzhansky, *Genetics and the Origin of Species*, vol. 11 (Columbia University Press, 1982)
2. D. Axelrod, M. Kimmel, *Branching Processes in Biology* (Springer, 2015)
3. G. Evans, J. Blackledge, P. Yardley, *Analytic Methods for Partial Differential Equations* (Springer Science & Business Media, 2012)
4. P.A.P. Moran, et al., The statistical processes of evolutionary theory (1962)
5. S. Redner, *A Guide to First-Passage Processes* (Cambridge University Press, 2001)
6. R. Lewin, Science **238**(4823), 24 (1987)
7. J. Haigh, Theor. Popul. Biol. **14**(2), 251 (1978)
8. M. Kimura, *The Neutral Theory of Molecular Evolution* (Cambridge University Press, 1983)
9. M. Kimura, Jpn. J. Genet. **66**(4), 367 (1991)
10. T. Witten Jr., L.M. Sander, Phys. Rev. Lett. **47**(19), 1400 (1981)
11. E. Brener, H. Levine, Y. Tu, Phys. Rev. Lett. **66**(15), 1978 (1991)
12. T.B. Kepler, A.S. Perelson, Proc. Natl. Acad. Sci. **92**(18), 8219 (1995)
13. L.S. Tsimring, H. Levine, D.A. Kessler, Phys. Rev. Lett. **76**(23), 4440 (1996)
14. I.M. Rouzine, J. Wakeley, J.M. Coffin, Proc. Natl. Acad. Sci. **100**(2), 587 (2003)
15. M.M. Desai, D.S. Fisher, A.W. Murray, Curr. Biol. **17**(5), 385 (2007)
16. W. Van Saarloos, Phys. Rep. **386**(2–6), 29 (2003)
17. E. Brunet, B. Derrida, Phys. Rev. E **56**(3), 2597 (1997)
18. D.A. Kessler, Z. Ner, L.M. Sander, Phys. Rev. E **58**(1), 107 (1998)
19. L. Pechenik, H. Levine, Phys. Rev. E **59**(4), 3893 (1999)
20. C. Mueller, L. Mytnik, J. Quastel, Invent. Math. **184**(2), 405 (2011)
21. O. Hallatschek, Proc. Natl. Acad. Sci. **108**(5), 1783 (2011)
22. P.J. Gerrish, R.E. Lenski, Genetica **102**, 127 (1998)
23. M.W. Feldman, S.P. Otto, F.B. Christiansen, Ann. Rev. Genet. **30**(1), 261 (1996)
24. I. Chen, D. Dubnau, Nat. Rev. Microbiol. **2**(3), 241 (2004)
25. E. Cohen, D.A. Kessler, H. Levine, Phys. Rev. Lett. **94**(9), 098102 (2005)
26. R.A. Neher, B.I. Shraiman, D.S. Fisher, Genetics **184**(2), 467 (2010)
27. I.M. Rouzine, J.M. Coffin, Genetics **170**(1), 7 (2005)
28. S.E. Luria, M. Delbrück, Genetics **28**(6), 491 (1943)
29. B. Mandelbrot, J. Appl. Probab. **11**(3), 437 (1974)
30. D.A. Kessler, H. Levine, Proc. Natl. Acad. Sci. **110**(29), 11682 (2013)
31. D.A. Kessler, H. Levine, J. Stat. Phys. **158**(4), 783 (2015)
32. M. Jones, S. Thomas, A. Rogers, Genetics **136**(3), 1209 (1994)
33. I. Bozic, J.G. Reiter, B. Allen, T. Antal, K. Chatterjee, P. Shah, Y.S. Moon, A. Yaqubie, N. Kelly, D.T. Le, et al., elife **2**, e00747 (2013)
34. Y. Iwasa, M.A. Nowak, F. Michor, Genetics **172**(4), 2557 (2006)
35. N.L. Komarova, D. Wodarz, Proc. Natl. Acad. Sci. **102**(27), 9714 (2005)
36. D.A. Kessler, R.H. Austin, H. Levine, Can. Res. **74**(17), 4663 (2014)

Printed in the United States
by Baker & Taylor Publisher Services